Aboriginal Plant Collectors

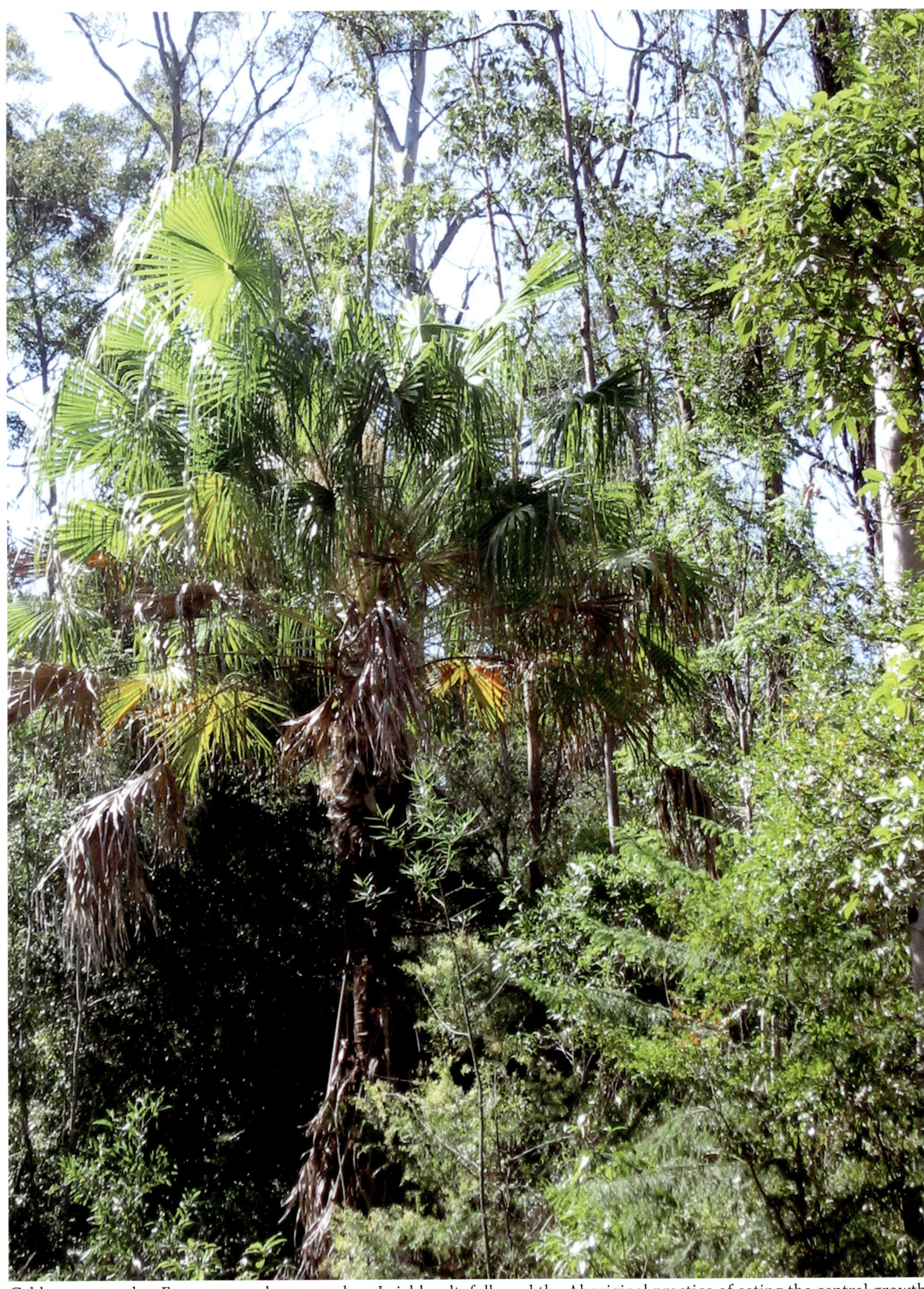

Cabbage tree palm. European explorers, such as Leichhardt, followed the Aboriginal practice of eating the central growth of the palm as greens. Removing the 'cabbage' killed the tree. Philip A. Clarke, Great Dividing Range, southeast Queensland, 2004.

ABORIGINAL PLANT COLLECTORS

Botanists and Australian Aboriginal People in the Nineteenth Century

Philip A. Clarke

ROSENBERG

Dedication

For Judith

Acknowledgments

The following people provided helpful comments to various drafts of this book: Steve Donnellan, Skye Krichauff, John McEntee, Paul Monaghan, Peter Sutton, David Symon and Julie Tan. Lynn Strefford provided research assistance, Kyle Clarke took the author photograph, Ray Marchant produced the map and Lea Gardam assisted with sourcing archival images. As background for the historical research material covered in this volume, Aboriginal people have enthusiastically discussed their use of plants with the author. The South Australian Museum has provided many exciting opportunities to explore Aboriginal culture during over twenty five years of employment. Expeditions organised by the Waterhouse Club of the South Australian Museum have also enabled the author to carry out ethnobotanical research in remote parts of Australia.

First published in Australia in 2008
by Rosenberg Publishing Pty Ltd
PO Box 6125, Dural Delivery Centre NSW 2158
Phone: 61 2 9654 1502 Fax: 61 2 9654 1338
Email: rosenbergpub@smartchat.net.au
Web: www.rosenbergpub.com.au

National Library of Australia Cataloguing-in-Publication data:
Clarke, Philip A.
Aboriginal plant collectors : botanists and Australian
Aboriginal people / author, Philip A. Clarke.

1st ed.
Kenthurst, NSW: Rosenberg Publishing, 2008.

ISBN 9781877058684 (hbk.)
Includes index.
Bibliography.

Human-plant relationships--Australia.
Aboriginal Australians--Social life and customs.
Aboriginal Australians--Food.
Aboriginal Australians--Medicine.
Nature--Effect of human beings on--Australia--History.

333.9530899915

Set in 12 on 14 point Warnock Pro
Printed in China by Everbest Printing Co Limited

Jacket: Front: *Waterfall in Australia*, by Augustus Earle. The artist and his party visited Wentworth Falls in the Blue Mountains in 1826. Aboriginal people provided settlers with guidance and protection in the Australian bush. Augustus Earle, oil on canvas, Blue Mountains, New South Wales, about 1830. Rex Nan Kivell Collection, nla.pic-an2273848, National Library of Australia, Canberra.
Back, top: *Wa-ra-ta* (waratah). This spectacular flowering plant amazed the British colonists arriving in 1788. It is thought that the name is a borrowing from *warrada*, in the Dharug language west of Sydney. George Raper, watercolour, Sydney region, New South Wales, about 1789. The Ducie Collection of First Fleet Art, nla.pic-vn3579250, National Library of Australia, Canberra.
Back: bottom: Heath-leaved banksia. The *Banksia* genus was named in honour of its discoverer, Joseph Banks, during the voyage round the world with Captain Cook in 1770. Aboriginal foragers sucked nectar from the flower spikes and used the smouldering dried seed cones to carry fire (Stewart & Percival, 1997, p. 13). S. Edwards, watercolour, early nineteenth century. *Curtis's Botanical Magazine*, London. Vol. XIX, 1804.

Contents

Preface

This book explores the impact of indigenous people upon the European discovery of Australian plants, most particularly during the nineteenth century. Explorers and plant hunters were amazed at the unique plants and animals they encountered and collected in the 'Great South Land'. Scholars back in Europe were fascinated by these exotic specimens, which challenged the ways they, as scientists, ordered the natural world. One of the imperatives of the British colonists was to discover and secure natural resources of economic benefit to the Empire, in the hope that they might find plants that would come to rival crops from the Americas, like maize, potato and tobacco. Observations of Australian Aboriginal hunting and gathering practices provided the colonists with important clues concerning the productivity of the land, and they tried out many of the plants that Aboriginal hunter-gatherers had utilised for thousands of years. But nothing to rival the economic importance of the American crop plants came to their attention, and adoption of Aboriginal plant uses generally occurred only in times of hardship.

British colonists who came in 1788 to establish themselves in the 'new' country found the indigenous land 'owners' to be both a physical threat and an important source of information about the environment. As the frontiers of settlement expanded, the European plant hunters and botanists were compelled to engage with a people who were being progressively dispossessed of their land. Their opportunities for obtaining information directly from the people they encountered were limited, however, by the severity of the impact of the European intrusion on the indigenous population, and their records illustrate the process of change within Aboriginal Australia.

Throughout the nineteenth century, professional plant hunters sent out from Europe, many of whom were explorers in their own right, were on the frontline of the scientific and horticultural discovery of the world's flora. Plant hunters were a hardy breed of men primarily employed to make collections of dried and living plants in the fledgling colonies and to send them back to Europe. They led exciting but dangerous lives on the fringes of the Empire, more than a few of them dying in the field, sometimes, on their expeditions into the vast unknown, from their encounters with indigenous peoples who had little or no direct prior experience with Europeans. Rather than being university-based scholars, most were trained in the more practical field of horticulture. In addition to the professional plant hunters, a broad spectrum of settlers also became involved with the collecting of Australian plants. Most collectors, whether professional or amateur, worked under the direction of botanists and plant merchants back in Europe, amongst them Britain's Royal Society and James Lee, the co-owner of the Vineyard Nursery in Hammersmith, London.

From the Aboriginal guides who accompanied plant collectors into the field came many of the plants that were pressed and dried as herbarium specimens. These hunter-gatherers' bush skills were derived from their extensive environmental knowledge and first-hand experience of the Australian flora.

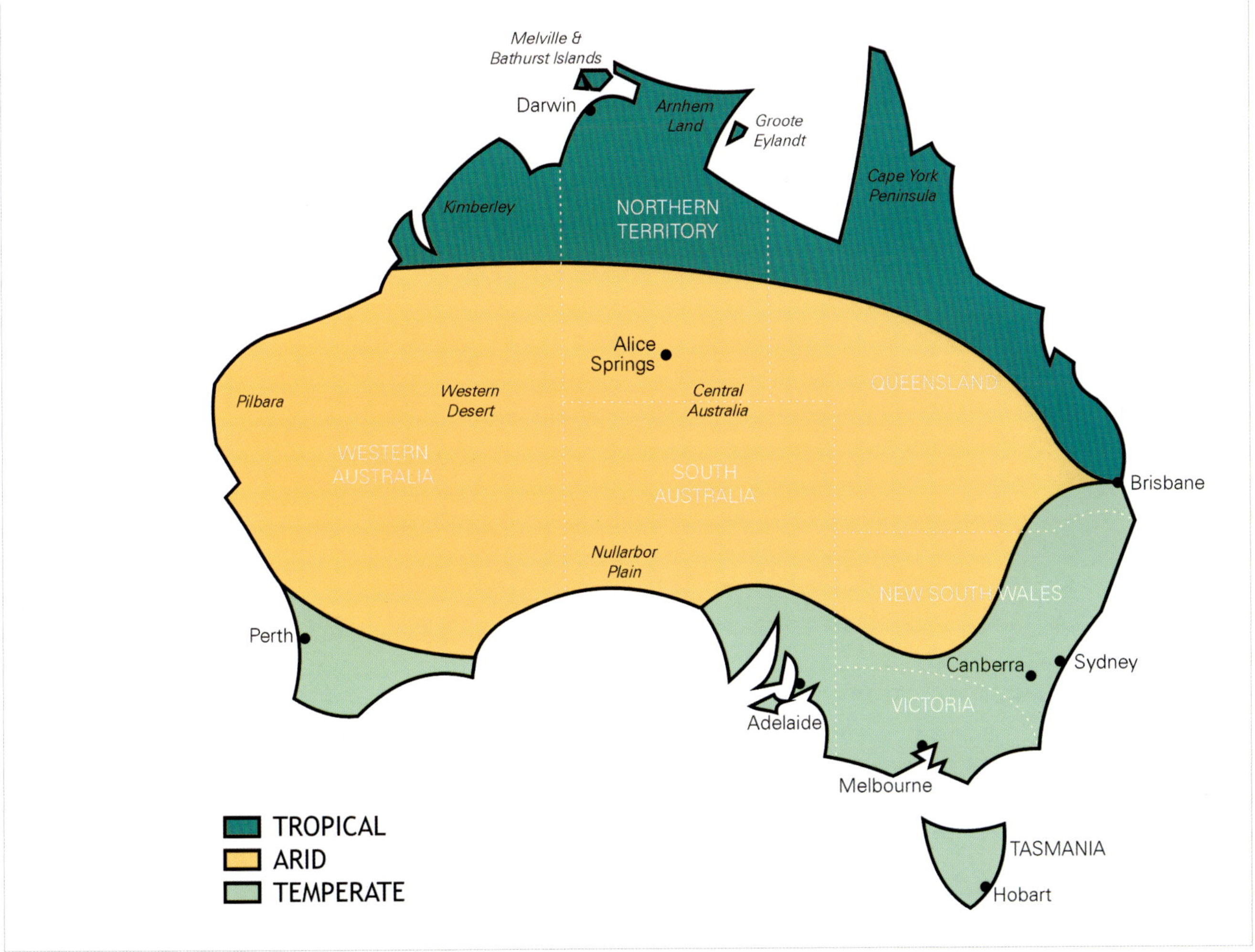

Australia is a geographically diverse continent, with three main climate zones. Map drawn from climate data obtained from the Bureau of Meteorology, Australian Government, 2005.

In published accounts written during the colonial period, indigenous people working with European explorers and plant collectors are often portrayed as silent partners. The present work redresses this imbalance by investigating the role of particular Aboriginal groups and individuals in the botanical discovery of Australia.

Chronologically organised, the book commences with an historical perspective on Aboriginal involvement with Europeans. This is followed by a description of the uses of wild plants that British colonists and their descendants adopted from Australian hunter-gatherers, and an investigation of the long-term influences of Aboriginal culture upon European-Australian perceptions of the flora. The bulk of the book details the interaction between particular plant collectors and Aboriginal people through the nineteenth century, recognising that although they were focused on the flora, the Europeans who actively collected plant specimens in remote regions of Australia were also able to gain rare insights into local Aboriginal cultures. Finally, consideration is given to parallel developments within the sciences and humanities during the nineteenth century from which Australian ethnobotany emerged in the early twentieth century. The further advancement of this field during the twentieth century is outside the scope of the present work.

References in this book to 'European' explorers, colonists and settlers from the seventeenth to the nineteenth centuries relate to people who would, back then, generally have identified themselves as belonging to particular Northern Hemisphere cultural groups, seeing themselves as British, Dutch, French, Prussian, Portuguese, Russian or American. In the case of the British, they would have further differentiated themselves into regional groups such as the English, Irish, Welsh, Cornish and Scottish. In the Australia of the twentieth and twenty-first centuries, European Australians chiefly refers to people of predominantly European extraction who do not derive any of their identity from either indigenous or Asian Australians.

This book spans a period of immense political and social change in Australia. Throughout the nineteenth century, the various colonies were run from Great Britain as separate entities. Federation, the formation of the Commonwealth of Australia in 1901, brought the British colonists and indigenous inhabitants of the mainland and Tasmania under a single home government, giving rise to a nation.

The historical sources quoted in this volume often include phrases and descriptions of indigenous people that would cause offence if used out of context in contemporary situations. As a general rule, Aboriginal people who were sources of ethnobotanic data are acknowledged and standard indigenous word spellings employed when they exist. In addition to terrestrial species, the term 'plants' is used to cover the seaweeds. Fungi are also included under this term, although biologically speaking fungi are not part of the plant kingdom. The common forms of plant names are used within the text, their scientific names being given in the Appendix. The scientific names used here are those recognised by the Australian National Herbarium when the book was written.[1] 'Wild plants' are defined as those that have not been deliberately grown by horticulturalists, and include not only indigenous Australian plants, but also feral species inadvertently or deliberately introduced since Europeans arrived here.

Warning

Many of the plants discussed in this volume are toxic. Therefore, in the absence of adequate supervision it is highly recommended that untrained people do not experiment with the use of any of the plants mentioned in this book.

1 Early Explorers and Aboriginal Guides

Plants, as potential raw materials to boost the economies of the Old World, were key natural resources chronicled by European explorers in remote corners of the New World. The static nature of plants, unlike animals, made them relatively easy to observe and collect, while the full investigation of mineral resources beneath the surface would have to wait until after settlement. As tangible proof of the character of foreign lands, the collection of botanical specimens was an important element of the colonial reconnaissance. Explorers studied indigenous hunter-gatherers for clues to the possible uses of newly discovered plant species.

Discovering the exotic

The first European visitors to Australia found a geographically isolated continent with a flora and fauna largely unlike anything they had encountered elsewhere.[1] The botanical resources of this relatively dry land did not favourably impress the first wave of explorers. In 1688 the widely travelled English buccaneer William Dampier sailed as a crew member on the *Cygnet* along the arid Kimberley coast and landed in the vicinity of King Sound, later claiming that the people living there were without 'Fruits of the Earth'. [2] While maritime explorers became occasional visitors to Australian shores, it would be another hundred years before European settlers were sent to colonise. The first known botanical collection from Australia to reach Europe was made in 1697 when Dutchman Willem de Vlamingh, in command of the ships *Geelvink, Nyptangh* and *Weeseltje*, visited Swan River in the southwest of Western Australia.[3] He is believed to have been the collector of at least two plant species belonging to the genera *Acacia* and *Synaphea*.[4] Vlamingh and his crew would, like Dampier, have left the continent with a poor view of its wild food, having suffered greatly from eating untreated zamia palm nuts.[5] Their interaction with Aboriginal foragers being extremely limited, they were not to know that the fresh nuts could only be rendered edible by a complex process of mashing, leaching and baking.[6]

In 1699 Dampier returned to Australia on the *Roebuck*. He landed at Shark Bay on the central coast of Western Australia, where he collected plants, several years later publishing drawings of them.[7] His botanical collections included a wide range of species from genera such as *Acacia, Dampiera, Solanum, Brachycome* and *Spinifex*.[8] Seventeenth and eighteenth century botanists regarded the botanical specimens that came from such remote places as every bit as important and awe-inspiring as the 'moon rocks' are regarded by geologists today in this era of extraterrestrial exploration. In describing this later visit, Dampier was more charitable in his summing up of the flora of 'New Holland', stating that 'There were also besides some Plants, Herbs, and tall Flowers, some very small Flowers, growing on the Ground, that were sweet and beautiful, and for the most part unlike any I had seen elsewhere'.[9] He discovered a spectacular flowering 'bean' that was later to become known as the Sturt desert pea.[10] Following Dampier, many of the maritime exploration expeditions into the southern oceans made botanical collections along the coasts of Australia.[11]

Pineapple zamia, a cycad also known as 'fool's pineapple'. Many early explorers discovered that cycad nuts are poisonous unless properly treated. Philip A. Clarke, Mount Coot-tha, southeast Queensland, 2004.

The first European visitors found the arid Western Australian coastline to be somewhat desolate, but British explorer Captain James Cook on the *Endeavour* had the good fortune in 1770 to be travelling along the more humid eastern coast.[12] His naturalist on this trip was the independently wealthy and highly influential Joseph Banks, who made a large botanical collection during the voyage. For the next fifty years, Banks was a key figure in Australian natural history, coordinating the collecting of specimens and dispersing them to other naturalists across Europe. As a pupil at Eton College, Banks had developed a keen interest in biology[13], his involvement with plant collectors also beginning early, with the young Banks paying local women who gathered plants for apothecaries to teach him their names and uses. Banks formed a special interest in Australia, and upon his return to England became a strong advocate for the organised settlement of New South Wales.[14] For this reason he has been referred to as the 'Father of Australia'.

During the 1770 voyage of the *Endeavour*, Banks' description of the botanical riches of New South Wales so impressed Cook that he renamed the newly discovered 'Stingray Bay' as 'Botany Bay'.[15] Later, Swedish naturalist Carl Linnaeus suggested that the new southern continent be called 'Banksia', on the basis of the scientific importance of the plant collections made by Banks.[16] Despite the immense diversity of plants the expedition discovered, Cook was not at all enthusiastic about the prospect of finding indigenous plant foods of value to the Empire. He claimed that in Australia:

> The Land naturly [sic] produces hardly any thing fit for man to eat and the Natives know nothing of Cultivation. There are indeed found growing wild in the woods a few sorts of fruits (the most of them unknown to us) which when ripe do not eat a miss, one sort especially which we call'd Apples, being about the size of a Crab-Apple, it is black and pulpy when ripe and tastes like a Damson [bullace], it hath a large hard stone or kernel and grows on Trees or Shrubs.[17]

The editor of his published journal identified the 'Crab-Apple' as the Australian black apple, although a similar account from Banks suggests that it was probably the Burdekin plum.[18] Like the Dutch sailors over seventy years before them, Cook's crew also ate zamia palm nuts after having observed the empty shells around Aboriginal campfires, which resulted in their becoming violently ill.[19]

The main function of naturalists on expeditions of exploration was to discover and collect a broad range of scientific specimens, such as plants, minerals, fossils, invertebrates, vertebrates and indigenous artefacts. Some objects could be retained as dry specimens, while others needed to be kept in spirits. Banks was able to trial various methods of preserving the viability of the seeds he had collected in Australia.[20] This was valuable experience that he would later draw upon when training professional plant hunters to be sent out from the Royal Gardens at Kew. Having naturalists as part of an expedition had the added advantage that they could identify indigenous foods to collect and eat during a voyage to combat scurvy.[21]

Explorers returning to Europe brought back data that became incorporated into the growing body of knowledge concerning the world's geography. From the late eighteenth century, the ships of explorers coming to Australia were equipped not only with maps, but also with libraries of published natural history and geographical works as essential research tools. During the planning for the 1770 trip to Australia, Banks fitted out the *Endeavour* for scientific work at his own expense.[22] He brought on board the best available collecting equipment and an extensive range of books, and as assistants employed the Swedish botanist Daniel Solander and two artists, Alexander Buchan and Sydney Parkinson. Solander had plant collecting and preservation duties, while the role of the artists was to accurately record living plant material before it lost its colour and shape through being pressed and dried for the trip home. By the time naturalist François Péron visited Australia in 1802 as part of the French scientific and exploratory expedition led by Nicolas Baudin on *Le Géographe* and *Le Naturaliste*, he was able to bring with him a library containing the journals of several explorers and colonists such as Dampier, Cook, Phillip and Bligh.[23]

Sturt desert pea. This species was discovered in 1699 when buccaneer William Dampier arrived on the *Roebuck* at Shark Bay on the central coast of Western Australia. Aboriginal people in the Western Desert refer to this plant as *malukuru*, meaning 'red kangaroo eyes' (Clarke, 2007, pp. 15–6). Philip A. Clarke, Alice Springs, Central Australia, 2007.

Claiming Terra Nullius

For the European colonial powers, the acquisition of knowledge concerning the newly discovered lands was an important stage in their preparation for colonisation and Empire expansion. Explorers paid particular attention to botanical resources. Cook, in his trips across the world, gained extensive experience with tropical foods, such as taro and banana, which were key crops for horticultural societies in the Pacific region. When Cook reached Australia, the lack of agriculture or any other obvious signs of the physical alteration of land by humans confirmed for him its wild and undeveloped status, which he took as strong evidence that the indigenous inhabitants were not legal owners of the land. In 1770 Cook declared Australia '*terra nullius*', in other words an unoccupied land, and claimed it as a British possession.[24] Banks, with his knowledge of Australia and his standing with the British Government, advocated the establishment of a British penal colony in New South Wales, which he argued was blessed with indigenous inhabitants who were less war-like than the Maori of New Zealand.[25]

When the first British settlers came to Australia in 1788 the industrial revolution in Western Europe was in full swing.[26] The colonial imperative was to establish Australia as a British outpost in the Pacific before other powers, in particular the Netherlands and France, seized parts of the continent. Apart from the strategic political importance of settlement, the 'new' continent promised to eventually provide the Empire with raw materials and food from introduced crops. As described by historian David Day, colonists believed that this southern land was 'a country in the course of being rescued from a state of nature'.[27] During the eighteenth and nineteenth centuries, Europeans wrongly considered Australia a wilderness, meaning that they thought it was a landscape totally unmodified by people.

The early waves of settlers arrived on the continent expecting to find it populated by 'savages' who had not even acquired the ability of proper speech.[28] (The humanitarian movements which would eventually lead to the British abolition of slavery and the legal recognition of limited rights to indigenous inhabitants of the colonies would not gain traction until the 1830s.[29]) In spite of good intentions by governments in Great Britain, desperate conditions in the settlements on the other side of the world meant harsh treatment was meted out to displaced Aboriginal communities. Historian William J. Lines summed up the attitude of settlers on the frontiers of the southwest of Western Australia in the 1820s and 1830s:

> As Aborigines represented the most fundamental state of human existence, they also represented the most inferior. Therefore, each encounter with the Nyungar [local Aboriginal people] provided settlers with an occasion to reflect on British superiority.[30]

During the settlement phase of its colonies, Australia was not the place to find Europeans appreciating the depth of Aboriginal language and culture or the complexity of Aboriginal links to the land and environment.

Every group of people which sets out to colonise a country will inevitably end up attempting to recreate the cultural landscape of their homeland, layering it over all previous forms.[31] In Australia, the Aboriginal inhabitants and European colonists had vastly different visions of the land they both occupied. For the Aboriginal people, the land was humanised by the deeds of their spirit Ancestors during the Dreaming (Creation), with evidence of their actions still apparent in the landforms, distributions of plants and animals, and in the customs of the different peoples.[32] The newcomers from Europe saw little of this. The British 'pioneer' was not a settler who became 'Aboriginalised' through acquiring hunting and gathering expertise, but a person who transformed the Australian 'wilderness' into a model of Western Europe. As historian David Day puts it, 'Australians were not "white Aboriginals" but transplanted Europeans still groping to accommodate themselves to their surroundings'.[33] The settlers, and those who followed them, accomplished the radical transformation of the landscape by actions such as altering the fire regime, wholesale clearing of vegetation, making roads, ploughing, irrigating, damming, erecting fences, building houses and by introducing a plethora of foreign organisms that quickly overran indigenous plants and animals.[34]

Europeans did not settle Australia evenly, but gradually spread out from the initial settlements located in the southern temperate regions with climates that the British found so enticing.[35] In the Great Dividing Range, stretched out along the east coast of Australia, were forests full of commercially useful timber.[36] Soon after a foothold was established at Sydney, seal colonies were discovered along the coastlines of Bass Strait and the Southern Ocean, which offered the opportunity to develop a skin trade with the Chinese market.[37] From the early 1790s whaling in the Pacific region also looked promising.[38] The temperate parts of Australia also had the climates most suited for the crops and livestock the colonists had brought with them.

Australia had a wealth of space for the early waves of settlers. In 1788 Sydney became the first permanent European settlement in Australia and the capital of New South Wales. In Van Diemen's

Land (Tasmania), the British established Risdon and Hobart Town in 1803 and Launceston in 1806. To the north of Sydney they founded Newcastle in 1804 and Port Macquarie in 1821. In 1814 Bathurst became an inland town on the other side of the Blue Mountains from Sydney. The Moreton Bay settlement was established in 1824 at the site of what developed into Brisbane, the capital of Queensland when it separated from New South Wales in 1859. Western Australia's first settlements were established at Albany in 1826 and Swan River (Perth) in 1829. The early towns of Portland, established in 1834, and Port Phillip (Melbourne), in 1835, were initially within New South Wales, but became part of Victoria when it was proclaimed in 1851. The Colony of South Australia began in 1836, with Adelaide as its capital. Most of these major Australian towns were sited on coasts or along major rivers, which provided harbours for future trade and transport. European settlement progressed quickly; between 1788 and 1845 the settler population of New South Wales alone went from 1024 to 181 541.[39]

While settlers were spreading the European frontier out along the southern edges of the continent, they found the north far more of a struggle. In the region that is now called the Northern Territory, there were short-lived settlements at Fort Dundas on Bathurst Island (1824–28), Raffles Bay, northeast of Darwin (1827–31), and Port Essington (1838–49). The arid and tropical regions were unattractive to the first British settlers, which was a concern for the authorities trying to secure sovereignty over the whole continent.[40] Darwin, the present-day capital of the Northern Territory, was not established until 1869.

In the late nineteenth century, the need to establish telegraph communication systems between Darwin and the cities in the southern and eastern states, as well as the lure of mining, eventually brought large numbers of Europeans into the Red Centre of the continent. In the early twentieth century, the fledgling Australian Government achieved greater control of northern Australian waters, bringing an end to seasonal visits by Asian seafarers, such as the Macassarese and Timorese.[41] The harsh Western Desert region was left largely alone, at least until the mid twentieth century.

As colonists absorbed each 'new' area it became subject to distinct phases of landscape transformation, with the original nature of the physical environment and the distance from parent populations being limiting factors.[42] Following the explorers, who mapped the land, were generally the pastoralists. The setting up of sheep and cattle stations 'opened up' the country, which, in the most favourable areas, permitted later phases of more intensive farming, along with the growth of towns and railway systems to support increasing human population density. Mining also had a major impact in the flow of Europeans into regions such as the Western Australian Goldfields. As an invading culture, the European settlers absorbed key Aboriginal landscape features, such as tracks, camping places and water sources, and then made them part of a rural environment.[43] Urban development, in favourable coastal and riverine areas, became the final phase of settlement. Europeans commenced the physical transformation of Australia from the moment they arrived, and their descendants continue the transfiguring process today.

British colonisation of Australia had devastating consequences for the indigenous inhabitants, although its full impact can only be guessed at. To begin with, we do not know, and can never know, the size of the entire Aboriginal population as it was in 1788. In addition to a large number of unknown people based in remote areas beyond the frontier, most early official Australian Census reports did not list indigenous people, as they did not gain full citizenship in all states of the country until the federal government assumed control over indigenous affairs in 1967.[44] Anthropologists have conservatively estimated the total Aboriginal population of Australia in 1788 as being about 300 000, or one person for every 19 square kilometres across the continent.[45] This takes into account higher rainfall areas, where foragers required less space (1.25 to 8 square kilometres per person in Arnhem Land) and arid regions, where desert dwellers required much more space (31 to 88 square kilometres per person in Central Australia).

As the frontier of European contact with Aboriginal people moved across the continent, indigenous population levels plummeted due to land alienation, introduced diseases and settler conflict.[46] This pattern also occurred in areas remote from the capital cities. For example, it was estimated that between the 1870s and the 1940s, the Aboriginal population of the eastern Lake Eyre Basin region in Central Australia dropped to less than 10 per cent of its pre-European level as a result of the establishment of cattle stations.[47] Apart from population decline, the indigenous survivors of colonisation in many regions experienced cultural change in terms of language loss, cessation of religious practices, decline in artefact and artistic traditions and movement away from key places in the landscape.[48] In many parts of Australia, particularly in the temperate region, the recent recovery of indigenous population levels in the Census reports has been due to growth in the number of people who claim both European and Aboriginal descent.[49]

Aboriginal guides

Nineteenth century plant collectors in Australia were amongst the first Europeans to realise the extent to which Aboriginal hunter-gatherers utilised a diverse flora. Professional plant hunters considered Aboriginal guides, who also acted as trackers and collectors, to be essential members of their expeditions into the unmapped regions. Convicts were sometimes members of these parties as well, but they generally offered little other than their labour. Historian Anne Moyal claimed 'the Aborigines were a constant presence in the scientific reconnaissance of the 19th-century science. They were, after all, the original inhabitants of the territory into which the naturalists plunged'.[50] Indigenous uses of plants, established over tens of thousands of years, provided valuable evidence for determining the properties and potential uses of plants that Europeans were encountering for the first time. Notwithstanding, colonial naturalists were highly selective in the knowledge they appropriated. Historian Colin Finney remarks: 'While a formidable reservoir of knowledge about the plants and animals of the land existed in the minds and lore of the Aborigines, with the exception of a few individuals who sought to tap this knowledge the development of Australian natural history for the Europeans and their descendants began anew'.[51]

From the point of view of the British authorities, Australia was a 'new' land to explore, claim, settle and develop. The survival of European explorers hinged upon how the indigenous inhabitants received them. Aboriginal guides were relied upon for their knowledge and experience of the different environments and indigenous peoples.[52] In their service to the explorers they found pathways, established protocols with Aboriginal land owners to allow expedition parties safe passage, and were even bodyguards. Aboriginal interpreters were essential, as social conditions in the newly established towns did not favour the colonists learning to speak a variety of indigenous dialects.[53] In times of shortage, explorers called upon their Aboriginal guides to use the land as a means of

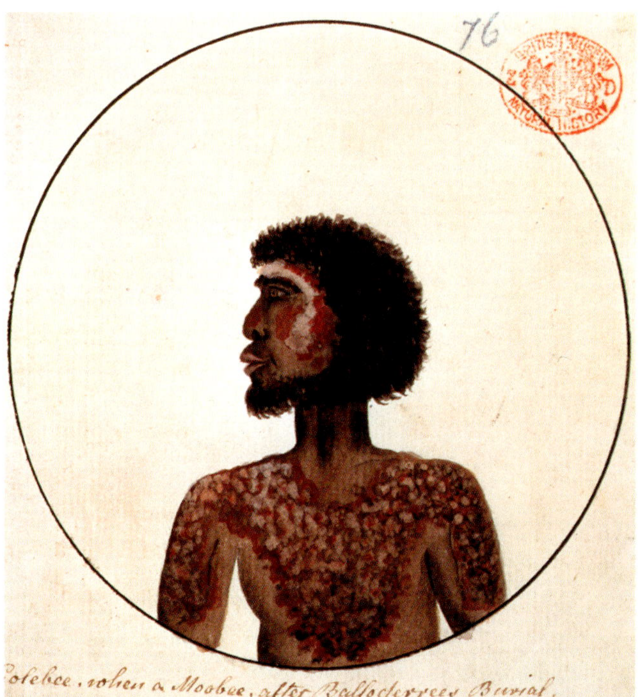

Colebee [Colbee], when a Moobee after Balloderee's Burial. Colbee was a guide for Governor Arthur Phillip. Their intimate knowledge of the country and its indigenous cultures meant that Aboriginal men were often taken on expeditions of exploration and plant collecting. Thomas Watling, watercolour, Sydney region, New South Wales, about 1792–94. V12067/R, British Museum of Natural History, London.

Balloderee [Boladeree]. This man acted as a guide for Governor Arthur Phillip in April 1791, when Phillip was exploring the river systems inland from Sydney. Thomas Watling, watercolour, Sydney region, New South Wales, about 1792–94. V12058/R, British Museum of Natural History, London.

providing them with food and water, as well as for making shelter and water craft. When present in large numbers, local Aboriginal people were considered both a physical threat and a good sign that the land was fertile. Explorers looked for signs of foraging in the country they traversed.

In employing indigenous guides, explorers could either recruit Aboriginal men from settled areas and take them on as part of their expedition, or opportunistically fall in with those they met during their travels and acquire local knowledge directly from them. Some explorers used both strategies simultaneously. By employing Aboriginal guides at the outset, expedition leaders gained companions who had time to develop familiarity with their needs. A disadvantage with this strategy was that the expertise of these guides was most concentrated upon the culture, language and environmental resources of their own territory, which the expedition would sooner or later leave behind. On the positive side, Aboriginal people who as individuals had lived as hunter-gatherers before Europeans arrived, possessed bush skills and a general knowledge of landscapes that could be broadly applied across the continent.[54]

The pattern of British settlers taking advantage of local Aboriginal landscape knowledge was established early. In April 1791, Governor Arthur Phillip and other colonists travelled inland from Sydney with the aim of discovering whether the Hawkesbury and Nepean rivers were one and the same.[55] The party included Sydney Aboriginal men Colbee and Boladeree who, although they were from the coastal region, nonetheless proved useful in making contact with inland groups.[56] On 11 April 1791, Phillip's party was near Rose Hill when they came across an Aboriginal band whose members were strangers to their guides.[57] A man, about thirty years old and holding only a firestick, approached the colonists. Colbee walked out to him and introduced himself, 'I am Colbee, of the tribe of Càdigal'. The lead man replied that he was Bèreewan of the Boorooberongal people. Colbee then took Bèreewan by the hand and led him to the colonists, where further introductions took place.

Aboriginal people in the Sydney region who were adults when the British first arrived remained deeply embedded in their own culture after colonisation, despite having established close associations with the settlers. While Colbee and Boladeree understood English well enough to interpret for the colonists when meeting with other Aboriginal people, there were topics they would not suffer any questions about. Both men refused to discuss anything related to the ritual of extracting a front tooth, commonly performed by coastal people and related to male initiation.[58] This silence is still the response in parts of Aboriginal Australia when outsiders ask inappropriate questions about the secret realm of sacred knowledge and practice.[59]

In the early years of British settlement there was a scarcity of horses, donkeys and bullocks for use as pack animals. Thus many of the early inland excursions had to be undertaken on foot, severely limiting the quantity of supplies that could be taken, particularly over the rough country typical of the Blue Mountains.[60] Explorers in the arid zone did not use camels as a form of transport until as late as 1846.[61] Paths, where they already existed due to Aboriginal land use, were an advantage. As European settlement spread across the landscape, Aboriginal tracks were incorporated into stock routes and eventually as parts of a modern European-type road system.[62] The existence of most former Aboriginal tracks is poorly recorded, as colonists of 'new' lands generally do not fully acknowledge their debt to the landowners they displaced.

Few of the first colonists were capable of surviving alone for more than a couple of weeks in the Australian bush, although the explorers eventually gained a few Aboriginal survival skills. For instance, in 1831 Surveyor-General Major Thomas L. Mitchell was exploring inland New South Wales when his Aboriginal guides showed him a method of purifying foul water from waterholes to render it suitable for drinking.[63] This was achieved by digging a hole in sand near the main water body and letting the water seep into it. This effectively filtered it, while also making it cooler. The water was drunk through a layer of grass, which made it sweeter. To be successful on long inland expeditions, explorers needed to learn all they could of living off the land to maintain the health and strength of their men and pack animals.

In the wake of the inland explorers came the settlers. Aboriginal guides were still necessary, making sure that the settlers did not get lost in the bush, which to many newcomers seemed featureless and hard to interpret. Apart from providing directions and finding stray livestock, indigenous hunter-gatherers could read the country in other ways to the settlers' benefit. Aboriginal people possessed detailed knowledge of the pattern of seasonal changes in the landscape, which was very different to what the British were familiar with in northern Europe.[64] Settlers often consulted certain Aboriginal people as weather forecasters. According to Aboriginal Protector George A. Robinson in 1830, the Tasmanians 'had attained to such celebrity that my people, i.e. white men, would consult them on this subject [the weather], and always appeared satisfied at what the natives told them.'[65] In these ways, although European settlement was overall detrimental to indigenous societies, particular individuals survived in the new regime by offering their skills and land-based knowledge in service to the usurpers of their country.

Boongaree, a Guringai man from Broken Bay, wearing his official breastplate. Boongaree had unique links with the European exploration of Australia, having accompanied both Matthew Flinders and Phillip Parker King on voyages around the continent. Thaddeus Bellingshausen, Sydney region, New South Wales, about 1820. Reproduced from *The Voyage of Captain Bellingshausen to the Antarctic Seas 1819–21,* 1945.

The experience of Englishman Daniel Bunce provides an example of the depth of interaction between European plant collectors and Aboriginal guides. Bunce came out to Australia in 1835, first landing at Hobart Town in Van Diemen's Land (Tasmania).[66] Relocating to the mainland in 1839, he travelled in what is now Victoria from Port Phillip to Western Port, with a local Aboriginal band to lead him. Bunce spent a night trying to understand the language of his guides, and then:

> We were afoot again by early dawn. At my request, my companions tarried a short time, while I collected

> specimens of the flora of the place; and, as soon as they perceived the reason of our delay, rendered willingly what assistance they could, by bringing to me various leaves, herbs, &c. They had names for many of them, which I carefully noted, for future reference.[67]

Bunce's botanical interests eventually led him to establish a plant nursery at St Kilda in Melbourne. Such trips provided Aboriginal people and Europeans alike with opportunities to become more familiar with each other's culture.

Boongaree

The life of Boongaree typifies the role that Aboriginal members had on European expeditions of exploration and scientific investigation. Boongaree was a Guringai man who was born before British settlement, probably about 1765, and during his tender years lived in the Broken Bay area to the north of Sydney. As a hunter-gatherer, he had excellent all-round survival skills. In 1799 Boongaree accompanied the navigator, Matthew Flinders, on the *Norfolk* to Hervey Bay, to the north of present-day Brisbane in Queensland.[68] On this trip he proved to be of great value, particularly in dealing with Aboriginal groups which were unfamiliar with Europeans. Flinders was chiefly attracted to Boongaree because of his 'good disposition and manly conduct'.[69]

From 1801 to 1803 Boongaree sailed with Flinders on the *Investigator* during the first recorded circumnavigation of the Australian continent.[70] Amongst the party were former surgeon's mate cum botanist Robert Brown and the gardener Peter Good. The expedition produced a large collection of preserved plants, although Brown explained that with so many species new to botany, he often had to assign specimens 'nick-names' until they could be checked with other specimens already at the Kew Herbarium.[71] On this trip Boongaree would often go naked and unarmed to parley with hostile Aboriginal groups armed with spears.[72] On one occasion, on 31 July 1802 at Sandy Cape at the northern end of Fraser Island in southeast Queensland, Flinders noted that several Aboriginal people approached them, before falling back. In response:

> Bongaree stripped off his clothes and laid aside his spear, as inducements for them to wait for him; but finding they did not understand his language, the poor fellow, in the simplicity of his heart, addressed them in broken English, hoping to succeed better.[73]

Boongaree was in the habit of taking his own traditional weapons along with him during his expeditions.

While Boongaree had some difficulty in communicating with the Aboriginal groups they encountered during their trip, the European expedition members saw much physical uniformity amongst the indigenous peoples. Good noted in his journal that the Aboriginal men at Sandy Cape, who were probably Badtjala people, had:

> no weapons either offensive or defensive that we saw – Bungery [Boongaree] (a Native of Broken Bay which we had with us) amused them by throughing [sic.] his Spear, which seemed to surprise them and it appears they are unacquainted with the use of the Wumora [woomera, i.e. spearthrower] they did not differ much in their persons or appearance from the Native of Port Jackson but have a different Language for Bungery could not understand them ...[74]

The people at Sandy Cape did not use spearthrowers, but rather than not having any weapons at all, it is likely that they had kept them hidden and that they were simply unfamiliar with the type of spear that Boongaree brought with him.[75] Through Boongaree's exper-iences, the Flinders expedition gained insights into the cultural diversity of Aboriginal Australia.

After completing the circumnavigation of the continent with Flinders, Boongaree returned to Sydney, where he subsequently had better luck than the explorer. During the return voyage to England, Flinders was imprisoned on Mauritius in December 1803, arriving home eventually in October 1810 in poor health, and dying in July 1814.[76] Boongaree remained in the districts around Sydney until volunteering to be part of Lieutenant Phillip Parker King's voyage on the *Mermaid* to the western coasts of Australia, commencing in 1817.[77] He was an obvious recruitment choice, since King's mission was to fill blanks that Flinders had left on the coastal charts. The plant hunter, Allan Cunningham, also became part of the expedition team. Of their guide and companion Boongaree,

King stated: 'This man is well known in the colony as the chief of the Broken Bay tribe; he was about forty-five years of age, of a sharp, intelligent, and unassuming disposition, and promised to be of much service to us in our intercourse with the native ...'[78] Boongaree's exploration experience was unique amongst Aboriginal people.

Boongaree's deep interest in exploring was later demonstrated when he and his family took their boat to greet the *Vostok*, under the command of the Russian explorer, Captain Thaddeus Bellingshausen, who visited Sydney in March 1820 during his expedition to Antartica.[79] Boongaree was described by the Russian as wearing 'wore-out trousers of a British sailor, and on his forehead there was a plaited band, decorated with red clay and mud, on his neck he wore a copper plate, in the shape of a crescent moon, with the inscription: BOONGAREE Chief of the Broken-Bay-Tribe 1815.'[80] The practice of issuing breastplates, many of them brass, to reward Aboriginal people who were in service to colonial administrations, commenced in New South Wales during the early nineteenth century.[81] The plates often labelled them 'Chiefs', 'Kings' and 'Queens' of 'tribes' of certain places. Such titles were mainly based upon the perception of colonists, rather than reflecting any indigenous hierarchy, and contained a strong element of ridicule.[82] Boongaree spoke Pidgin English and was able to explain to Bellingshausen that the north shore of the harbour was his traditional land and that he had accompanied both Flinders and King around Australia.

The Aboriginal adventurer and his family inspected the Russian ship, and while on board asked for 'tobacco, old clothes and ropes, and whatever they happened to notice.'[83] Bellingshausen responded by giving them grog, sugar and tobacco. He also offered to employ Boongaree's family as natural history collectors, promising to give them 'clothes and ropes if they brought fish, live birds, a kangaroo and other animals' as specimens for his expedition.[84] During his time in the colony Bellingshausen was able to make a scientific collection of the plants of New South Wales, crossing the Blue Mountains with Cunningham.[85]

Bellingshausen recorded that he and his crew often visited the camp of 'King Boongaree' and his wife Matora ('Queen Gooseberry').[86] The Russian was able to observe various Aboriginal hunting and gathering techniques, as well as record cultural practices like removing part of the little finger on the female left hand.[87] Bellingshausen believed that this was done to prevent interference with the winding of fishing lines. Although Aboriginal people may have given Bellingshausen this explanation, it would also have been done for social and cultural reasons. From Reverend Samuel Leigh's account from the same period in Sydney, it is known that the operation was only performed upon female children.[88] The two joints of the little finger on the left hand were removed by tying hair string tightly around the joint to stop circulation; eventually the

Matora, Boongaree's wife, was known by the colonists as 'Queen Gooseberry'. In 1820, visiting Russian explorer Captain Thaddeus Bellingshausen offered to pay Matora and her family to collect natural history specimens. Thaddeus Bellingshausen, Sydney region, New South Wales, about 1820. Reproduced from *The Voyage of Captain Bellingshausen to the Antarctic Seas 1819–21*, 1945.

Aboriginal camp, showing fishing implements. When a severe food shortage threatened the survival of the colonists, they gave Boongaree (second from the left) and Matora (third from the left) a boat on the condition that they provided the colonists with fish. Thaddeus Bellingshausen, Sydney region, New South Wales, about 1820. Reproduced from *The Voyage of Captain Bellingshausen to the Antarctic Seas 1819–21*, 1945.

dead flesh and bone fell away. According to Leigh, '[women] who do not suffer this loss are treated with contempt'.

The settlers considered Boongaree, who was often seen garbed in a discarded soldier's uniform and wearing a cocked hat, a somewhat ridiculous figure.[89] In spite of this, Bellingshausen had an extremely high opinion of his character:

> He is about 55 years of age; he has always been noted for his kindness of heart, gentleness and other excellent qualities and has been of great service to the colony … He has often endangered his life in his efforts to keep the peace within his tribe. A few years ago an escaped convict fell into the hands of another tribe. They robbed him, took his axe away from him, and were about to kill him. Boongaree appeared on the scene, took the man under his protection, secured his freedom, and then for three days carried him on his back to Port Jackson, taking him across rivers and feeding him roots. He asked for no reward, save the fugitive's pardon. The Government of the colony gave Boongaree a long boat as a present. He is a generous man, generally beloved for similar kind actions.[90]

Such was their close relationship that Boongaree presented Bellingshausen with a set of weapons, including a shield, spear and a 'fork' for catching fish.[91]

In terms of their preference for a nomadic lifestyle, Bellingshausen equated the indigenous inhabitants of New South Wales to the Tzigani (Gypsies) of Eastern Europe.[92] The Russian explorer gained an impression of the colonial management of Aboriginal affairs, remarking that:

> Governor Macquarie, anxious to break the natives of their nomadic life and to accustom them to a fixed dwelling-place, presented Boongaree with a house and garden in

> Broken Bay specially built for him, and gave him the title of 'chief' of the place ... But the magic charms of drink and tobacco, the greatest of all temptations to these natives, are stronger than all the joys of a fixed, plentiful and quiet life and still attract them to the town of Sydney.[93]

The government had initially provided Boongaree and his family with a small boat, as they had other Aboriginal fishermen, on the condition that they give up part of their daily catch.[94] Fish they netted were also exchanged for alcohol and tobacco. Boongaree died at Sydney on 24 November 1830 and was buried at Rose Bay.[95]

Nanbaree and Abaroo

In 1802 another Aboriginal man, Nanbaree, already well known in New South Wales, joined the Flinders expedition around Australia.[96] Nanbaree's relationship with the colonists had begun in 1789 when he, then very young, and an Aboriginal girl named Abaroo (Booroong) were found suffering from a life-threatening disease, possibly smallpox.[97] The children were nursed back to health with the help of Arabanoo, the Sydney man who was captured and detained by the British in 1788 for the purposes of translating.[98] The disease had spread rapidly amongst the Aboriginal population, starting at Port Jackson in 1789. In Sydney, Captain Watkin Tench was alarmed at the existence of the epidemic:

> An extraordinary calamity was now observed among the natives. Repeated accounts, brought by our boats, of finding bodies of the Indians [Aborigines] in all the coves and inlets of the harbour, caused the gentlemen of our hospital to procure some of them for the purposes of examination and anatomy ... Pustules, similar to those occasioned by the smallpox, were thickly spread on the bodies, but how a disease to which our former observations had led us to suppose them strangers could at once have introduced itself, and have spread so widely, seemed inexplicable.[99]

The severity of their sickness probably forced many of the survivors to depend on the British colonists for shelter and food.[100] At the time of the epidemic, Nanbaree was estimated to be about six or seven years of age, Abaroo about ten.

When Arabanoo was himself stricken and died from the disease, the two children moved in with colonists, and according to Captain John Hunter

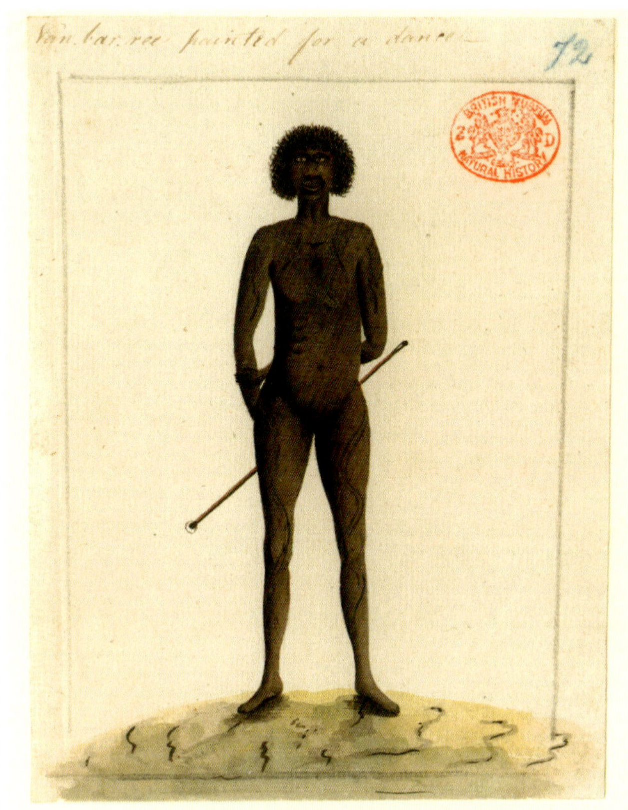

Nan-ba-ree Painted for a Dance. Nanbaree, like Boongaree, accompanied Matthew Flinders on his circumnavigation of Australia. Nanbaree also acted as an interpreter for the colonists and the Aboriginal people around Sydney. Thomas Watling, watercolour, Sydney region, New South Wales, about 1792–94. V12063/R, British Museum of Natural History, London.

and Dr Daniel Southwell appeared to have picked up speaking English relatively quickly.[101] The British were keen to establish friendly relations with local Aboriginal clans, and Nanbaree lived with surgeon John White's family and was christened Andrew Sneap Hammond Douglass White.[102] He appears to have maintained strong links to the local Aboriginal community, as naval officer David Collins was able to describe his initiation at a ceremony in January 1795.[103] This included a ritual that involved the removal of Nanbaree's front tooth, which Colbee's wife Daringha asked Collins to present to White. Colbee treated the initiate's gums, which suffered from blows during the extraction, by applying broiled fish to them. On one occasion in 1797 Aboriginal relatives of a dead man planned to kill Nanbaree in retribution, but they were thwarted by the appearance of a soldier.[104] As the result of other killings by his kin, Nanbaree was speared in 1798,

Abbarroo [Abaroo] a Moobee after Ballodaree's Funeral. Abaroo's face and body are decorated for mourning rituals. This woman helped William Dawes when he was compiling his Aboriginal word lists. From people like Abaroo, the colonists learnt the names and uses of indigenous Australian plants and animals. Thomas Watling, watercolour, Sydney region, New South Wales, about 1792–94. V12045/R, British Museum of Natural History, London.

but survived his wounds.[105]

Nanbaree often acted as an interpreter when colonists met Aboriginal people who had little or no understanding of the English language. Of the two orphans, Nanbaree appears to have adapted better to colonial life. In September 1790 he went along with a party of colonists by boat to Manly Cove, where they came across a group of 200 Aboriginal people feasting upon the carcass of a whale. Here, Nanbaree acted to 'interpret on both sides, [but] showed little desire to return to their [Aboriginal] society, and stuck very close to his new friends [the British].'[106] It was on this occasion that they met Bennelong and an altercation with another Aboriginal man led to Governor Arthur Phillip being speared in the shoulder.[107]

Abaroo appears to have been less interested in the British and returned to her clansmen to find a husband when she was about thirteen.[108] The prospect of her leaving bothered her guardian, Reverend Richard Johnson, since he did not believe that she had yet learnt enough English to 'explain their [the colonists'] intentions to the natives.' On one occasion after visiting the lower part of Sydney Harbour, Abaroo returned to Johnson's house extremely ill, suffering from an alleged sorcery attack from the Botany Bay people, who despised her. According to Collins, she claimed 'that the women of Cam-mer-ray had made water in a path which they knew she was to cross, and it had made her ill.'[109] Surgeon White attended Abaroo and bled her arm, but this did not appear to be effective and led to Aboriginal people treating her using a customary healing ritual, to which she responded more favourably.

When Abaroo finally left Johnson's care, she joined a band that occasionally came into Sydney and played an active role in local Aboriginal affairs. Collins noted that Abaroo was a midwife at the confinement of Bennelong's sister.[110] On this occasion, she 'was employed in pouring cold water from time to time on the abdomen.' When the tracker Boladeree died, Abaroo was involved in the mourning process and out of respect allowed her head to be cut by his mother.[111] She was also asked by the men not to eat any fish or meat on the day of Boladeree's burial.

The marine, William Dawes, used both Nanbaree and Abaroo as informants when compiling his Aboriginal word lists.[112] Governor Hunter remarked, 'Through the means of these children, if they should retain their native language, a more intimate and friendly intercourse with the people of this country may in time be brought about.'[113] From Aboriginal people like Nanbaree and Abaroo, the colonists would have learnt some of the indigenous names and uses of Australian plants and animals. The initial willingness by early colonists to learn about their new surroundings is reflected in the relatively high number of indigenous words in contemporary Australian English that appear to have come from Aboriginal languages based around Sydney.[114] Nanbaree died at the home of a European friend on 12 August 1821, and was later buried in the same grave as fellow Aboriginal people, Bennelong and his wife, at Kissing Point (Ryde).[115]

Blundell

In 1821 Boongaree had intended to accompany King on a second voyage, to the northern Australian coasts on the *Bathurst*, but on the morning they were to sail another Aboriginal man, Blundell, replaced him.[116] Blundell was about 40 when he went on this expedition.[117] When still a young boy, his father had been killed in battle and a shark had taken his mother. As an orphan, in 1791 Blundell accompanied Captain William Hill to Norfolk Island on board the *Supply*. Collins stated that here 'he seemed to have gained some smattering of our language [English], certain words of which he occasionally blended with his own.'[118] Blundell became a seasoned traveller, sailing as a crew member with Captain William Rook on the *Rosetta* in 1816.[119] This ship went from Sydney to Kangaroo Island in present-day South Australia to collect animal skins and salt, a voyage that went well beyond the existing frontier of official British settlement, occurring twenty years before the establishment of the Colony of South Australia.

Explorers found that Aboriginal people in remote districts generally reacted better to seeing Europeans for the first time when they could see that they were travelling with other indigenous people, even though they spoke a different language and had strange artefacts. King noted that during the 1821 expedition an Aboriginal group approached them on a shore near Lizard Island in the Barrier Reef of northern Queensland. Although Aboriginal people often shunned Europeans when first seeing them, this time they did not, as 'the appearance of Blundell, who on these occasions always took his clothes off, perhaps gave them greater confidence.'[120]

Indigenous cultural reactions

For Aboriginal people with no direct prior contact with Europeans, the sight of white-skinned explorers wearing clothes, carrying strange objects and walking alongside or riding exotic beasts, must have been a shocking experience. The explorers were aware of their initial impact when meeting indigenous people. Plant collector Charles Fraser recorded in his journal an encounter with Aboriginal people in the Brisbane area, during a trip to southeast Queensland in 1824:

> It is remarkable how much better the condition of the aboriginal inhabitants appears upon the coast than it is in the interior. While at Moreton Bay, I fell in with natives who had never seen an European. One old chief put his hand all over my arm and shoulder to feel if my cloths were part of myself, when the ecstacy [sic] of some was beyond my powers of description. They had no weapons but long spears, and perhaps, if left to themselves, would not arrive at the possession of bows and arrows for some centuries. They had never seen iron or steel, and when I presented them with tomahawks, knives, and scissors, it produced the most extraordinary surprise, one of the natives throwing himself down on the sand, rolling over and over, roaring and making a hideous noise, but all through pure delight.[121]

There were many such first encounters between Europeans and indigenous people.[122] Groups of Aboriginal men frequently demanded of the first Europeans they met that they remove their clothes to help in their identification.[123] Aboriginal people first took Europeans to be the spirits of their own dead relatives returned from the spirit world with new technologies, rather than total strangers. Knowing its impact, explorers often used the strategy of directing Aboriginal members of their party, and occasionally subordinate European men, to remove their garments to demonstrate that they were all humans in physical form.

While larger groups of Aboriginal men in hunting parties would sometimes approach European explorers, smaller bands and those with members of mixed age and gender generally kept well away from strangers. The first contacts British explorers had with indigenous Australian communities were generally mediated through senior men who kept women and young people away from view.[124] In early New South Wales, Collins remarked of the Aboriginal people: 'In our early intercourse with them (and indeed at a much later period, on our meeting with families to whom we were unknown) we were always accosted by the person who appeared to be the eldest of the party, while the women, youths, and children, were kept at a distance. The word which in their language signifies father was applied to their old men.'[125] Although Aboriginal society did not have hereditary chiefs, older men, by virtue of their life experience and ceremonial status, had considerable authority amongst their kin.[126]

As the frontier of British settlement swept across the Australia, Aboriginal people meeting Europeans for the first time had similar reactions to those described in New South Wales during the 1780s. On the Trans-Australian Exploration Expedition, William Wills recorded in his journal an incident that occurred in February 1861 as his party made their way back to Cooper Creek from the Gulf of Carpentaria. The explorers were in the Billy Creek area of western Queensland when:

> About half a mile further, we came close on a black fellow, who was coiling up by a camp fire, whilst his gin [wife] and piccaninny [young child] were yabbering alongside. We stopped for a short time to take out some of the pistols that were on the horse, and that they might see us before we were so near as to frighten them. Just after we stopped, the black got up to stretch his limbs, and after a few seconds looked in our direction. It was very amusing to see the way in which he stared, standing for some time as if he thought he must be dreaming, and then, having signalled to the others, they dropped on their haunches, and shuffled off in the quietest manner possible.[127]

Sometimes, the sudden appearance of Europeans meant that the exposure of Aboriginal women and children was unavoidable.

An insight into the perceptions indigenous people had when meeting Europeans for the first time can be obtained through looking at post-contact forms of language. The association of Europeans with deceased kin is illustrated in the Lower Murray region of South Australia by the Ngarrindjeri term *grinkari*, which was taken to mean both a 'human corpse' with its skin peeled off and a 'European.'[128] Similarly, in southern Western Australia the Nyungar people used their word for the spirits of dead people, *djanga*, for all Europeans.[129] Beyond the settlement frontier of the late eighteenth and early nineteenth centuries in New South Wales, Aboriginal people sometimes took escaped British convicts they met to be the 'ghosts' of their own returned relatives.[130] Elsewhere in the world, other isolated cultures meeting Europeans for the first time have used similar logic to explain their arrival.[131]

Prior to European settlement, Aboriginal people living at or near the coast of northern Australia had seasonal exposure to Asian visitors who had come to harvest resources in their country and for this reason were probably better equipped both psychologically and culturally to deal with the British arrival. From the 1700s, possibly as an effect of European colonisation in the broader global region, there was a growth in trade centred on mainland China, which spread out across the Pacific Rim.[132] As a result, Indonesian seafarers (Macassans) were seasonally visiting northern Australia as part of their trading cycle, to collect animal-based items such as trepang (also known as sea slug, *bêche-de-mer* or sea cucumber), turtle shell and trochus shell.[133] The returning praus were also loaded with materials derived from plants, such as northern ironwood and northern cypress pine timber for building construction and boat repairs, and northern sandalwood for the making of incense sticks to burn in temples.[134] The Asians also collected root bark from the cheesefruit tree, to make red dye, and possibly Australian nutmeg as a spice.[135] During the last few hundred years, Torres Strait Islanders also became involved in this regional trade. Torres Strait Islanders are predominantly of Melanesian origin, but have maintained cultural contacts with both Australian Aboriginal people and Papuans.[136]

In northern Australia, the legacy of these pre-European relationships with exotic cultures is reflected in Aboriginal mythology, ceremony and the material culture.[137] Aboriginal people met these visitors on their own terms, which did not involve loss of land, and were selective in what they absorbed from foreign cultures. The Macassans and Torres Strait Islanders were incorporated into Aboriginal ceremonial life. It is likely the same would have happened with European visitors had they not eventually come in relatively large numbers to settle permanently. Aboriginal people in the northern Cape York Peninsula region were in direct contact with the horticulturists of the Torres Strait, and yet they did not take up their practice of growing crops, although they did take on other Melanesian cultural elements.

We can only speculate upon what indigenous people, particularly in the south, would have thought about European plant hunters arriving in their country to collect plants. They may have thought that pieces of plant were being gathered for the same reason they collected them, that is, to make artefacts or for use in rituals such as for healing, honouring ancestors or sorcery.[138] If Aboriginal observers noted that plant hunters were taking poisonous or non-economic species, this may have contributed to their rationalisation that Europeans were spirits, for it is a widely held Aboriginal belief that spirits use plants in their dealings with humans. For instance, at Elliot in the Northern Territory it was Aboriginal tradition that evil spirits fed the toxic caustic-vine to human babies to kill them.[139] In the Cobourg Peninsula area of the Northern Territory, Aboriginal people do not eat the 'cabbage' (central growth shoot) of the Wendland palm, although they believe devils and ghosts do so.[140] Whatever the reasoning, Aboriginal people would have initially viewed Europeans within the confines of their own worldview.

The opportunities that European explorers and early plant collectors had to observe Australia's indigenous people produced fragmentary glimpses of Aboriginal life as it would have been in the decades prior to British settlement. In remote regions the language difficulties and cultural differences between the observer and the observed were extreme. Many of the brief accounts of Aboriginal life written by colonists contain data, albeit slender, that is useful to modern researchers of Aboriginal plant use. All too often today, due to major cultural and ecological changes, it is knowledge that can no longer be obtained directly from Aboriginal people, and what is available for analysis still needs to be considered in light of the period in which it was produced.

Throughout nineteenth century Australian history Aboriginal guides maintained a crucial role in guiding Europeans, even when the increased availability of pack animals made travel much easier. Aboriginal travellers such as Boongaree, Nanbaree and Blundell were part of a group that became known as 'Sydney blacks', who moved with Europeans pushing forward the colonial frontier.[141] They were survivors of the first wave of British settlement and were present at key moments in early Australian exploration history. The course of European expansion across the continent would have been much different without their involvement. Explorers routinely took on Aboriginal guides, forming partnerships such as George Grey and Kaiber, John Mitchell and Barney, Edward Eyre and Wylie, Edmund Kennedy and Jackey Jackey, Ludwig Leichhardt and Charley Fisher, Peter Warburton and Charley, Ernest Giles and Jimmy, John Forrest and Windiitj.[142] As discussed in later chapters, there were equally close relationships between professional plant hunters and their guides. It is apparent that for the Europeans in the Australian bush their safety and the likelihood of success was often placed in the hands of their Aboriginal partners.

2 Settlers and Australian Plants

The First Fleet took thirty-two weeks to travel from Portsmouth in England to Botany Bay on the eastern coast of Australia.[1] From Cook's published account and advice from Banks, the British knew before setting out that the land here would offer them little by way of food initially. The settlers would therefore have to rely heavily upon supplies brought out with them from England and the additional goods picked up along the way when they called in at Rio de Janeiro and the Cape of Good Hope. It was planned that on their arrival at Botany Bay, vegetable gardens and farms would be established quickly.

To Europeans, Australia seemed particularly strange and exotic. The night skies of the Southern Hemisphere had many unfamiliar constellations, and here the seasons were reversed. The First Fleet arrived in January 1788, which for temperate Australia is the middle of summer. In spite of the aridity of the season, Governor Arthur Phillip was forced to sow crops as soon as he arrived, instead of waiting for the following spring.[2] The early results were poor and the immediate future of the colonists looked grim. Through sheer necessity the colonists turned their attention to wild resources in the surrounding bush. They did not initially expect to gain much from the indigenous land users, who they knew to be hunter-gatherers. Judge-Advocate David Collins remarked: 'We found the natives about Botany Bay, Port Jackson, and Broken Bay, living in that state of nature which must have been common to all men previous to their uniting in society.'[3]

Finding bush tucker

In Australian English, 'bush tucker' is defined as 'wild, not domesticated food'.[4] While it can refer to any food taken from the bush, it has also been used to describe the 'simple fare', such as billy tea and damper, which the colonists lived on when they were establishing themselves in the bush. In Aboriginal English, the term generally refers to food that was traditionally eaten by indigenous hunter-gatherers prior to European colonisation, as distinct from the 'station' or 'mission' foods, like flour, salted pork and sugar that the colonists provided for them.[5] In nineteenth and early twentieth century Pidgin English, a related term, 'tuckout', was applied to all food.[6] For example, in the 1840s, an Aboriginal person in Adelaide was recorded to exclaim 'Piccaniny [little] tuckout – only bread'.[7] In 1901 at Ooraminna Range, south of Alice Springs in Central Australia, an Aboriginal person was quoted as defining 'bushy tuckout' as 'seeds and edible bulbs of various kinds'.[8] In Australian English today, 'bush tucker' more often refers specifically to wild types of food and drink consumed by Aboriginal people living in remote parts of Australia. To emphasise this indigenous connection, tucker is sometimes written as 'tukka' to make it appear non-English.[9] Bush tucker is also sometimes referred to as 'bush food'.[10]

Extreme hardship was the driving force behind the colonists using bush tucker in Australia. Replenishment of supplies from Great Britain during the first decades of settlement was infrequent, while relief from sibling colonies in South Africa and China was irregular. By mid-

1790 Phillip needed to reduce the workload on the convicts as they became progressively weaker from hunger.[11] Colonist Richard Atkins remarked that lack of food supplies in Sydney put them in a 'very ticklish situation'.[12] From Phillip's account, the daily meals for the colonists were extremely basic: 'They did not include tea; there was nothing to drink but water and the bad Portuguese rum taken on board at Rio for the soldiers and their wives'.[13] These shortages, most acutely felt by the poorest settlers and convicts, were incentives to augment their meagre supplies with material from the bush.

The colonists looked towards Australian plants to help battle scurvy, which many of them suffered after the long voyage from Great Britain. Parties of marines and convicts risked attack from the Aboriginal owners of the land they trespassed upon as they pushed into the bush surrounding Sydney in search of greens they could eat. It was believed that Aboriginal people had ambushed and killed many Europeans who went missing in the first few years of settlement while searching for edible plants and collecting rushes for thatching.[14] But there were also more positive cultural contacts. First Fleet surgeon John White recorded in his journal for May 1788 a chance meeting between colonists and an Aboriginal group, north of Botany Bay:

> As they [the Aboriginal people] conducted us to water, a toadstool was picked up by one of our company, which some of the natives perceiving, they made signs for us to throw it away, as not being good to eat. Soon after I gathered some wood-sorrel, which grew in our way, but none of them endeavoured to prevent me from eating it; on the contrary, if a conclusion may be drawn from the signs which they made relative to the toadstool, they shewed, by their looks, that there was nothing hurtful in it.[15]

White mentioned that two months later a party of convicts collected a 'plant resembling balm', which they found to be a 'good and pleasant vegetable'.[16] On another of his trips into the bush around Sydney, an Aboriginal man pointed out edible figs.[17] Aboriginal inhabitants on occasion showed their compassion towards the newcomers.

Settlers found some of the bush tucker appealing to their taste, with a few wild substitutes for their traditional foods considered worthy of consuming in their own right. On a trip from Sydney to Botany Bay in October 1788, colonist William Bradley recorded that his party 'never met with the smallest appearance of any kind of Cultivated ground', although they found wild plants that resembled spinach, samphire, parsley, sorrel, celery, cabbage, broccoli and currants.[18] Once identified, the main problem the colonists had was one of dispersal, with most edible plants being widely scattered through the bush rather than concentrated in areas like cultivated fields. Wild animals, already accustomed to Aboriginal predation, required considerable effort in hunting. In spite of the difficulties in locating and collecting it, bush tucker undoubtedly saved many of the early colonists in New South Wales from starvation and scurvy.

In Van Diemen's Land (Tasmania), where settlement commenced in 1803, early settlers also suffered from severe food and beverage shortages. Botanist and plant nurseryman Daniel Bunce noted in his historical summary of the colony for 1807:

> During this and the previous year there was a great dearth of provisions in this colony. Kangaroo flesh was sold for one shilling and sixpence a pound, and that species of sea weed, or salsolaceous plant [salt bushes] ... called Botany Bay Greens, being the chief support of the inhabitants.[19]

British settlers in other southern Australian colonies also experienced shortages during their establishment phase, although probably not as severely as New South Wales and Van Diemen's Land. By the time settlements were being established in South Australia, Swan River (present-day Western Australia) and Port Phillip (present-day Victoria), the colonists had ready access to supplies from already established towns such as Sydney and Hobart Town.[20]

Many of the wild plants used by Europeans would have been largely unobtainable without the assistance of the indigenous people. Their intimate knowledge of the local environment meant that they knew where to locate particular food sources, the seasons they were available, and the best techniques for obtaining them. And sometimes they kept the knowledge to themselves. For example, in 1836 an Aboriginal woman at Glenelg, near the city of Adelaide in South Australia, regularly supplied the newly arrived colonists with a type of 'watercress' for food.[21] The colonists reportedly had no knowledge of where it was obtained. The settlers' adoption of Aboriginal plant uses was mainly restricted to consumables like foods, beverages and medicines, although in the colder southern regions an exception was the settlers who took on the Aboriginal practice of carrying fire inside dried punk fungus.[22] The colonists came from an industrialised country and would generally have preferred to rely upon their own tools and weapons, rather than use the artefacts of supposed 'primitives' whose land they had come to take away.

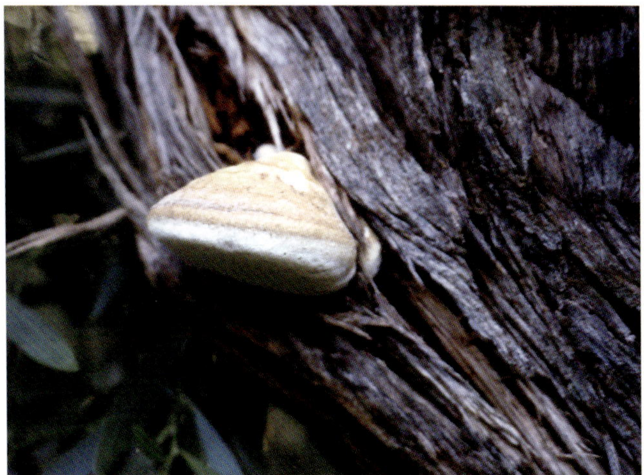

Bracket fungus, called punk by the settlers, on the trunk of a brown stringybark tree. In Aboriginal fashion, the settlers used bracket fungi to carry fire. Punk is an English word meaning a material that will smoulder rather than bursting into flame. Philip A. Clarke, Aldgate, Mount Lofty Ranges, South Australia, 1985.

Roots, tubers and stem starch

The settlers would have tried many of the Aboriginal sources of roots and tubers. Along the east coast they made a starchy and fibrous meal, in Aboriginal fashion, from the swollen underground stems of the cunjevoi, after much soaking, pounding and roasting.[23] This lengthy preparation was necessary, as the plant is poisonous if eaten raw, causing the mouth and throat to swell. Even when indigenous food plants were not poisonous, they often required processing to remove tannins or soften tough fibres. During the early days of settlement in southern Australia, for example, the settlers cooked and ate the young shoots of bracken, boiling them in water to remove the bitter taste and render them palatable.[24] Plants could also provide drinking water to travellers in need. The adventurer Louis de Rougemont claimed to have obtained water by cutting into the trunk of the Queensland bottle-tree when he was stranded in northern Australia in the late nineteenth century.[25]

Of the fleshy roots of the yam-tree (black kurrajong), botanist Joseph H. Maiden reported: 'A correspondent tells me they have been got 8 to 10 lb. [3.6 to 4.5 kg] in weight, and are not despised by Europeans ... The inside is beautifully white, a little sweetish in taste ...'[26] Nineteenth century gardener and plant collector Anthelme Thozet claimed that Aboriginal people and Europeans gained emergency food from the Queensland bottle-tree. He said 'that the soft juicy tissue of the stem can be eaten, and that many a wanderer in the bush has staved off hunger by its means'.[27] In general, because of the difficulties in collecting and processing wild sources of roots and tubers, settler use of them was probably less than for other bush food categories, particularly greens and fruits, which required minimal preparation before consumption.

Early settlers in the southwest of Western Australia used a complex process to extract starch from the trunks of by-yu nut trees, a type of cycad.[28] The stem pith was separated out, baked in an oven, shredded, then soaked in water for six hours. After being shaken up and filtered, the resulting sediment was washed several times, dried gradually, and then finely powdered, ready for use. For thousands of years, Aboriginal foragers had harvested the seeds of the by-yu nut[29], but it is claimed that they universally ignored the stem starch. However, the process the settlers used was very similar to that used by the indigenous foragers to make cycad seeds edible.[30] It is therefore possible that the settlers adapted their techniques from Aboriginal ones. In the 1920s and 1930s in New South Wales, starch was commercially produced from the central pith of the burrawang, another cycad species.[31] In some districts of New South Wales it was reported that colonists collected the fine soft wool that densely covers the upper part of the burrawang's trunk for stuffing beds.[32]

During the early phases of establishing the 'settled' districts, trade developed around the use of particular wild roots. In central New South Wales, settlers made arrowroot (a starch used in biscuits and puddings) from the bulbs of the Darling lily (sandover lily). At Wilcannia, one settler reputedly 'earned a handsome sum by making this substance when flour was all but unobtainable'.[33] Although Aboriginal people in some parts of Australia reputedly ate the Darling lily bulbs after roasting and pounding, in Central Australia it was regarded poorly as a food and only used medicinally, for skin complaints.[34] Aboriginal foragers were often employed as food collectors. Settlers paid Ngarrindjeri people of the Lower Murray River area to collect bulrush roots, which were also processed into arrowroot.[35] Police Trooper George Mason, who was also Aboriginal Protector in the Lower Murray district during the 1850s, would lend his boat to Point McLeay Aboriginal Mission residents so that they could gather these highly fibrous roots for a storekeeper at Wellington. South Australian colonists suffering hard times gathered Aboriginal foods, such as bulrush roots and yam daisy tubers, to help them manage shortages on the frontier of settlement.[36]

Fruits and nuts

In the temperate zone, many of the Aboriginal food plants with edible fruit and berries grow along the coastal dune systems. Due to their proximity to early British settlements, they were amongst the first indigenous foods that came to the settlers'

The Darling lily's dark green leaves shine in the sunlight. The early settlers in outback New South Wales extracted arrowroot from the bulbs of this lily. Aboriginal foragers regarded the plant as 'hard time' food, meaning that it was eaten only when most other foods were unavailable. Philip A. Clarke, south of Birdsville, Central Australia, 1986.

Yam daisy. South Australian settlers suffering hard times gathered and ate yam daisy tubers, but for Aboriginal foragers in southeastern Australia it was a major food source (Clarke, 2007, pp. 16, 20, 38, 40, 73). Philip A. Clarke, Bridgewater, South Australia, 1986.

Apple-berry. This woody vine was known in the Ngarrindjeri language of the Lower Murray region of South Australia as *kundowi*. Homesick settlers named many indigenous food plants after species from their homeland, often for only superficial similarities. With the apple-berry it was colour and taste, rather than form, that reminded them of the true apple. Philip A. Clarke, Kingston, southeastern South Australia, 1987.

White apple. This is a type of bush apple that bears its fruit along the trunk. Settlers utilised it as bush tucker in tropical rainforest areas. Norman B. Tindale, Danbulla rainforest, near Cairns, northern Queensland, 1972. N.B. Tindale Collection, AA338/6/35/79, South Australian Museum Archives, Adelaide.

attention. In the Sydney area, colonists found edible fruit on various plant species, such as those they called sour currant-bush, native raspberry, wild fig, native cherry and the apple-berry.[37] In 1836 Quaker missionary and botanist/horticulturalist James Backhouse was on a trip to what is now southeast Queensland, where he discovered the Moreton Island sand hills thickly covered with midyim bushes. As a trained plant nurseryman, he was interested in their purple-spotted white fruits, which were 'the most agreeable, native fruit, I have tasted in Australia; they are produced so abundantly, as to afford an important article of food, to the Aborigines.'[38]

The settlers were particularly attracted to wild fruits that did not require processing to render them edible. Maiden claimed that at Cape Byron in northern New South Wales, settler children ate the fruit of the mushyberry, which 'has a peculiar aroma, and could be preserved in the same way as oranges, citrons, and lemons.'[39] Maiden claimed that in southeastern Australia, the low-growing native cranberry (ground-berry) bushes had 'apple-flavoured' fruits that were 'much appreciated by school-boys and aboriginals.'[40] Along the east coast of Australia, settler children ate the small, red, slightly sweet fruit of the smooth rambutan (wild quince), as well as gummy fruits of the geebung.[41] Here, as well as in coastal Northern Territory, settlers harvested the fruit from lilly pilly and bush apple trees for eating.[42] In the Kimberley of northwestern Australia, Aboriginal people and Europeans ate the slightly acidic pulp extracted from nuts of the boab tree.[43] The emu apple was important in the outback of New South Wales. Maiden stated that the 'sub-acid fruit of this tree relieves thirst, enabling travellers to endure the inconvenience of want of water for many hours. It is eaten both by colonists and aborigines, is of the size of a small nectarine, and of a crimson colour.'[44]

In early New South Wales, British colonists made jams, tarts and jellies from wild fruits, many of which came into season around November, making the best of fruit too low in sugar and too high in acid to be eaten raw as table fruit.[45] Maiden noted that in the inland region of New South Wales and southeast Queensland the desert kumquat 'produces an agreeable beverage from its acid fruit', and 'a fair preserve may be made of the fruit.'[46] Similar use was made of the Macquarie Harbour grape (coastal sarsaparilla or coastal lignum) in Tasmania.[47] Along the east coast, the native tamarind tree produced clusters of slightly acidic fruit, which settlers used for their preserves.[48]

Many of the wild fruits have unpleasant properties for eating. Tasmanian colonist Ronald C. Gunn discussed the use of the prickly currant-bush in his colony and claimed that 'Some years ago, when our British fruits were scarce, it was made into puddings by some of the settlers; but the size and number of the seeds were objectionable.'[49] On the western coast of South Australia in the nineteenth century, some settlers ate the fruit of oondoroo solanum, which is a small arid zone bush.[50] The fruit causes a hot burning taste in the mouth and can, if eaten in large quantities, lead to sickness. The Burdekin plum is dominated by a large stone, but the flesh is pleasant tasting if the fruit is buried in the ground for 48 hours for ripening.[51] Colonists used some hot-tasting plants as condiments, with the black berries of the Tasmanian pepper tree proving good substitutes for pepper.[52]

The European lack of knowledge concerning preparation was initially a problem for the use of certain Aboriginal food plants. Colonist William Bradley, who was exploring around Sydney in March 1788, reported finding burrawang (cycad) nuts:

> [In a] ... Cove we met with a kernel which they [Aboriginal people] prepare & give their Children, I have seen them eat it themselves, they are a kind of nut growing in bunches somewhat like a pine top & are poisonous without being properly prepared the method of doing which we did not learn from them.[53]

Bradley claimed that one of the convicts had been poisoned through eating the nuts.[54] Apart from the early maritime explorers mentioned in the previous chapter, the list of victims who made the mistake of eating raw cycad nuts obtained from the burrawang and zamia palms includes nineteenth century travellers such as Matthew Flinders in 1802, George Grey in 1839 and John McDouall Stuart in 1864.[55]

Near Braidwood in southeast New South Wales, the settlers learned how to make arrowroot from burrawang nuts.[56] Charles Moore noted that in some districts of New South Wales:

> a good starch has been obtained from the seeds [of the burrawang], which also, when washed, or sliced and steeped for some days in running water or roasted, were largely used by the aborigines for food. Without some precaution of this kind they are in a fresh state dangerously acrid.[57]

East coast settlers referred to zamia palm nuts as 'blackfellow's potatoes', a name presumably based on their light appearance and soft texture after cooking.[58] In the Brisbane River area of Queensland during the 1820s, convicts ate the nuts of the black bean, probably after baking them to remove toxins.[59]

Through the nineteenth century, wild sources of fruit supported a number of cottage industries. Settlers made large quantities of jam from the fruit of the Bamaga satinash in the northeast of Cape York Peninsula in Queensland.[60] In 1853 the coming into season of 'native currant' fruit on the mainland of South Australia was of enough interest for it to be reported in an Adelaide newspaper.[61] Although the scientific name is not given, from the description it appears to refer to the wiry ground berry. The report states that the fruit was made into jellies and jam to extend the period of its use. Aboriginal collectors also provided food for European consumption. Point McLeay missionary George Taplin recorded in his journal for 15 August 1864 that he had bought one hundredweight (about 51 kilograms) of 'native currant' fruit from Aboriginal women for the price of '2 pence per lb' (454 grams).[62]

Taplin's comparison with a currant would normally suggest that the fruit came from the coastal bearded heath, more commonly known today as 'native currant' in the Lower Murray region. But since the fruiting season of the bearded heath is much later in the year, it is more likely that he was referring to either the leafless currant-bush or the wiry ground berry.[63] In the case of the wiry ground berry, botanist Frederick M. Manson Bailey recorded that in Adelaide during the 1840s they were a favourite food for settler boys and were commonly found in the Mount Lofty Ranges.[64] Establishing plant identities from historical sources alone is often difficult, as common names varied greatly across the continent while also coming in and out of use.

New Zealand spinach or warrigal cabbage, eaten as a green in eastern Australia by both Aboriginal foragers and European settlers. It was introduced into Europe as a food plant during the nineteenth century. Philip A. Clarke, Adelaide Plains, South Australia, 2004.

Greens

British settlers were eager to locate wild substitutes for greens they had grown in Europe. They often checked with local Aboriginal people whether the plants they found were poisonous or not, and many different plant species were tried across Australia.[65] In 1788 an anonymous female convict in Sydney described how 'there is a kind of chickweed so much in taste like our spinach that no difference can be discerned'.[66] Maiden stated that the coast saltbush was 'once used as [a] pot-herb in New South Wales'.[67] This is probably the same edible plant that medical man John White had described as 'growing on the seashore, greatly resembling a sage'.[68] The botanist at Kew, William J. Hooker, claimed that earlier settlers along the Nepean River near Sydney used the 'native cabbage' (besser marsh watercress) and that this and other wild greens would 'afford excellent pot-herbs when luxuriant and flaccid'.[69]

Another Sydney plant that the colonists ate is a species now commonly called New Zealand spinach.[70] The pre-European distribution of this plant was southern Australia and New Zealand, but it has since become naturalised in many parts of the world. Settlers included New Zealand spinach within a category they variously referred to as 'Botany Bay greens', 'Sydney greens' or 'warrigal greens'. The European use of New Zealand spinach commenced prior to Australian settlement, with Captain Cook using it during his Pacific voyages in the form of 'sour kroutt' to treat scurvy.[71] From seeds that Joseph Banks took to the Royal Gardens at Kew, the species became the only vegetable food that occurred in Australia to be taken in the eighteenth century prior to British settlement and cultivated internationally.[72]

The indigenous origin of one particular use of a bush tucker plant has been debated. The cabbage-tree was a source of greens for the settlers. Watkin Tench, who was Captain of Marines during the early years of settlement in Sydney, remarked:

> That species of palm tree which produces the mountain cabbage is also found in most of the freshwater swamps within six or seven miles of the coast. But it is rarely seen farther inland. Even the banks of the Hawkesbury are unprovided with it. The inner part of the trunk of this tree was greedily eaten by our hogs, and formed their principal support.[73]

While on a trip in June 1788 from Broken Bay to Manly Cove, White recorded that his party had camped near a cabbage-tree swamp, where they ate 'cabbage' along with their salt rations.[74] Although initially common in the Sydney Basin, cabbage-trees became scarce through European over-exploitation.[75] Colonist Clement Hodgkinson claimed that along the east coast of Australia in the 1840s, cabbage palm was eaten by both Aboriginal foragers and Europeans, and that it had a 'sweet taste of Spanish chestnut'.[76] Unfortunately, the removal of the edible terminal bud kills the plant.

An anonymous author gave another account of colonists using cabbage-tree as a food plant. He claimed that on one occasion during his travels in the 1830s through New South Wales, local Aboriginal guides took him to a remote mountain hut. Here, 'The two blacks left me about a couple of hours before sundown, giving me one of their tomahawks ... also a good bundle of cabbage-tree for my supper, and an opossum's skin cloak'.[77] Botanist Ferdinand von Mueller disputed an indigenous origin of this food source when he claimed that Aboriginal people did not know of its culinary value prior to European settlement.[78] The widespread Aboriginal use of palm cabbage as food in the tropics suggests that he might have been wrong.[79]

Colonists also tried eating the growth centres, or 'cabbage', of ferns. In 1832, botanist and horticulturalist James Backhouse travelled through Tasmania in the company of Old Boatswain, the Aboriginal 'wife' of a European sealer. She collected rough tree-fern for Backhouse, who claimed 'It is in substance like a Swedish-turnip, but is too astringent in taste to be agreeable, and it is not much altered by cooking'.[80] Indigenous Tasmanians would split the top end of the rough tree-fern and turn one half back to expose the soft centre for collection as food.[81] In Victoria and New South Wales, Aboriginal foragers ate the 'cabbage' from the soft tree-fern as well.[82] The sea also provided the first wave of colonists with greens. Early settlers in Van Diemen's Land had been compelled to eat seaweed.[83]

In the late nineteenth century, settler Annie Richards claimed that for Europeans living along the arid western coast of South Australia the low growing parakeelya herb was 'considered good eating with bread'.[84] While the indigenous influence upon the European use of this food is not stated, it is known that Aboriginal hunter-gatherers used this annual herb as a source of edible seed, with its foliage serving as a thirst quencher.[85] In this area, Europeans considered at least some wild plants to be very tasty, with a settler commenting that 'during railway construction days, "saltbush soup" was a favorite dish at some of the boarding-houses'.[86] In outback New South Wales, the fragrant saltbush was also found to be a useful vegetable after boiling in water to extract the excessive salt.[87] During drought time, settlers in the inland regions of the eastern states would use portulaca as a table vegetable.[88]

Colonial food

All people are naturally conservative about changes in their diet, with substitution of similar items occurring more readily then major switches in food types. Anthropologist Mary Douglas has described food as a system of communication, in terms of what and how we eat being heavily influenced by our cultural perceptions of food in general.[89] In the case of the European settlers in Australia, hunter-gatherer foods, or 'bush tucker', would not have been considered appropriate for an agriculturally based society. Anthropologists Peter Farb and George Armelagos claim 'that by knowing how people eat, anthropologists can know much about them and their society'.[90] After the initial colonisation of New South Wales, the settlers' use of bush tucker gradually lessened with the establishment of crops from the Northern Hemisphere which were more suited to horticulture than any of the wild plants.

Despite the European preference for consuming more familiar types of food, wild sources continued to have a crucial role in times of hardship. Europeans lost in the bush tried Aboriginal foods. In 1792 botanist Claude Riche, who was part of the French expedition led by Antoine Bruni D'Entrecasteaux, was stranded on the South Australian coast for three days.[91] He reportedly survived by eating the small white fruits of the carrot-wood (coastal bearded heath). In 1853 botanist John C. Bidwill was lost in the southeast Queensland bush for eight days and managed to survive by extracting water from the native grape vine.[92] When Europeans settlers in remote areas suffered hardship, their survival sometimes depended upon their ability to collect wild food and medicine and to process it correctly. Aboriginal assistance was an advantage in such situations.

In the early nineteenth century, European sealers were active along the southern coasts.[93] They lived lonely lives in bases on previously uninhabited offshore islands, where they were protected from Aboriginal attack. In the main, they were former seamen from a diverse range of national and cultural backgrounds, often in command of a workforce of Aboriginal women abducted from Tasmania and mainland Australia. Sealers living on Kangaroo Island in South Australia during the 1820s and 1830s collected wild fruits for making puddings and jam.[94] Depending on the season, these fruits would have included those from the monterry, coastal bearded heath, wiry ground berry and blueberry tree. While the records of sealers' lives are scanty, it is known that Aboriginal women hunted and gathered for them.[95] Sealers, perhaps to an extent more than any other group of settlers in Australia, relied heavily upon bush tucker and Aboriginal knowledge of how to collect and process it.[96]

A few former convicts based in the woods near Sydney managed to live in an Aboriginal fashion, subsisting on wild foods. In the mid-1790s, James Wilson was such a man, who 'preferred a vagrant life with the natives'; he was known by Aboriginal people as Bun-bo-e.[97] Wilson submitted to his Aboriginal companions ritually scarring his shoulders and breast, and in the woods he went about wearing nothing else but a kangaroo skin.[98] The availability of bush tucker enabled outlaws from the colony who had fled across the frontier to evade the authorities. In 1815 it was reported that a group of 'ignorant and deluded' convicts, who had tried escaping from Sydney across the Blue Mountains to New Guinea, had gained food from Aboriginal people before their eventual recapture.[99] In 1820 the Russian explorer Thaddeus Bellingshausen visited Sydney and recorded in his journal: 'It is said that many white people live amongst the natives on the other bank of the Nepean River with their wives and children.'[100] They apparently lived on escaped cattle that had gone wild and bred. Along the Hawkesbury River, escaped convicts helped Aboriginal people organise attacks upon settlers.[101]

In the Lake Macquarie district of New South Wales during the early nineteenth century, colonial authorities feared that bushrangers and Aboriginal inhabitants would form a coalition against law-abiding settlers.[102] With British settlement rapidly expanding, it would be difficult for the government to maintain control over a large frontier if several outbreaks of conflict occurred simultaneously. In the same period, gangs of Tasmanian bushrangers, whose ranks comprised convict 'bolters' who had fled to the woods, plundered the estates of inland settlers. These gangs relied heavily upon wild foods and medicines. Some of the bushrangers had Aboriginal women travelling with them, and when on the run they were greatly assisted by local Aboriginal bands.[103]

From about 1826 to 1831, escaped convict and bushranger George Clarke organised Aboriginal people along the Namoi River in the Gunnedah-Narrabri region of New South Wales into parties for the purpose of cattle duffing.[104] Settlers referred to Clarke as the 'white Aboriginal', due to his relationships with indigenous people and the fact that he lived in the bush beyond the reach of the authorities. Surveyor-General Mitchell considered him to be a 'degenerate white man' and gave an account of his exploits:

> A runaway convict named George Clarke, alias The Barber, had, for a length of time escaped the vigilance of the police by disguising himself as an aboriginal native. He had even accustomed himself to the wretched life of that unfortunate race of men; he was deeply scarified like them and naked and painted black, he went about with a tribe, being usually attended by two aboriginal females, and having acquired some knowledge of their language and customs.[105]

The Gamilaraay people initiated Clarke into their culture, leaving him with permanent decorative scars on his back and breast.[106] The Aboriginal community would have considered this, and other rituals, to be necessary before the bushranger could be allowed to have wives. Mounted police troopers eventually captured Clarke, with the help of Aboriginal trackers. Later, Mitchell went into the field to investigate the bushranger's account of travelling along a large river that the Aboriginal people called 'Kindur' (probably the Gwydir River), as far as the sea.[107]

Across Australia, Aboriginal people were present on both sides of the law. In 1844 bushrangers on Kangaroo Island in South Australia evaded the police with the aid of Tasmanian Aboriginal women who had formerly lived with the sealers.[108] Such was the perceived threat of these outlaws that troopers from Adelaide were sent to the island to apprehend them. The South Australian police eventually caught the criminals by employing as a tracker another Aboriginal woman living there, known as 'Old Wauber' or 'Waub'.[109] Colonial authorities often conscripted the aid of Aboriginal trackers when searching for villains. Generally, convicts treated the indigenous community poorly, which helps explain why Aboriginal people on some occasions actively helped in the apprehension of convict runaways, who they called 'croppies'.[110] The New South Wales authorities employed Aboriginal trackers as 'special constables', who were each signified a 'chief' by a metal breastplate attached by chain around their neck.[111] On the fringes of settlement, trackers were also useful in recovering stock lost in the bush.[112]

There are a few cases of Europeans becoming stranded in remote parts of Australia and being adopted into local Aboriginal bands, when bush foods provided their sole sustenance. A prominent example is William Buckley, a runaway convict who lived with the Watourong people in the Port Phillip Bay region of Victoria from 1803 to 1835.[113] His lengthy disappearance reputedly led to the creation of the Australian English idiom of having 'Buckley's chance', meaning none.[114] Similarly, from 1829 to 1842 convict James Davis (Duramboi) lived as a hunter-gatherer amongst Aboriginal groups in southeast Queensland, after running away from the Moreton Bay penal settlement.[115] Another case was the sailor James Morrill, who in 1846 was cast upon the Queensland coast at Cleveland Bay (Townsville) after his ship, which was travelling from Sydney to China, was wrecked on a shoal in the Pacific Ocean.[116] He lived with Aboriginal people for seventeen years, and in later writing his reminiscences identified bush foods, particularly roots and fruits, by their Aboriginal names.[117]

After the initial hard times, as the land was being developed for agriculture, European Australians became more dismissive of bush tucker, regarding it as barely palatable in spite of its earlier importance on the frontier. Tom Bellchambers, a South Australian, was forced to eat Aboriginal foods when travelling through the Murray River country. He later said that during his wanderings in the 1880s and 1890s there were 'times when each day sees you taking up the slack in your belt. Under such conditions you become less fastidious as to your tucker. Three days on grass tree cores and sheoak apples will, I guarantee, cure the greatest dallier at the table'.[118] For settlers, bush tucker was seen only as emergency food.

Beyond the frontier period, European Australians continued in a limited way to eat wild fruits and nuts, particularly those that do not require elaborate processing. In the southwest of Western Australia, they collected the fruit of the broom ballart to make jam.[119] In the Adelaide area, colonial botanist Frederick M. Bailey recorded that during the 1840s a related species, the native cherry (ballart), was 'eaten in quantities, both by the blacks and by the whites, when really nothing else was to be met with in the wild country'.[120] Across Australia, the fruit of the quandong was made into jams, jellies and pies, and also dried and stored for future use.[121] In northern Queensland, settlers developed a taste for the nuts of the Kuranda quandong (Johnstone River almond), which had been much prized by Aboriginal foragers.[122]

Europeans in southeast Queensland followed the Aboriginal practice of eating the nuts of the bunya pine, from the first days of settlement and as recently as the Great Depression of the 1930s.[123] At Borroloola on the southern shore of the Gulf of Carpentaria, it is reported that during the 1940s, when stores ran out, Europeans and Aboriginal people alike resorted to bush tucker, obtained from cycads and water lilies.[124] When settlers lived alongside an Aboriginal population that maintained knowledge of the local flora, the likelihood of their using wild plants when their own normal sources ran out increased. In times of shortage, Aboriginal gatherers had knowledge of where to look for bush tucker and how to prepare it.

Tea for everyone

Eighteenth and nineteenth century British immigrants coming to Australia brought with them their custom of drinking tea. In 1788, with supplies of plantation-grown tea extremely low,

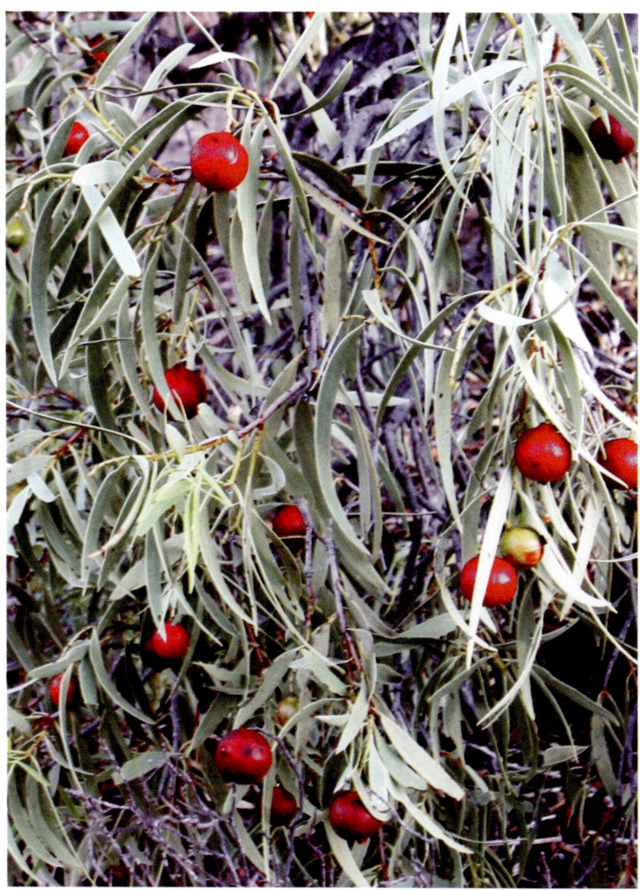

Quandong or wild peach. The fruit of this arid zone tree is a favourite among Aboriginal people. European settlers made it into jams and jellies, and used it in pies. Its Australian English name is thought to derive from *guwandhaang* in the Wiradhuri language of southwest New South Wales. Philip A. Clarke, Great Victoria Desert, Western Australia, 2006.

Coastal sarsaparilla. Aboriginal people have used the leaves to make a tea that is drunk to treat digestive disorders. Prior to British settlement in southern South Australia they probably chewed the leaves. Its common name reflects a similar use by British colonists. Philip A. Clarke, Kingston, southeastern South Australia, 1987.

a female convict in Sydney was reported to have stated that 'something like ground ivy is used for tea', and hoped that proper tea would soon come from China.[125] The account probably refers to the 'sweet tea' or 'sweet sarsaparilla' which colonists around Sydney used for tea.[126] In 1789 a sailor lost in the Careening Cove area on the outskirts of Sydney lived entirely on 'sweet tea' leaves for three days before being found.[127] Bundles of 'sweet tea' were seasonally sold in the old George Street Markets in Sydney up until the 1930s.[128] Colonists similarly used native lilac (false sarsaparilla) as a tea-based medicine, although its effectiveness is in doubt. Botanist Frederick M. Bailey remarked that the 'roots of this beautiful purple flowered twiner are used by "bushmen" [Europeans working in the bush] as a substitute for the true sarsaparilla, which is obtained from a widely different plant. I cannot vouch for any medicinal properties'.[129] Since British settlement in southern South Australia, Aboriginal people used coastal sarsaparilla (coastal lignum) leaves to make a tea which was drunk to treat digestive disorders.[130] The common name indicates that settlers also similarly used this plant.

Apart from tea, Europeans coming to settle in Australia yearned for alcoholic beverages.[131] Given the expense of importing them, early settlers probably tried fermenting many of the wild fruits, particularly those that are heavy cropping, to make their alcohol. In Queensland, Maiden recorded that with the crows apple (sour plum) a 'beverage is produced by boiling the fruit, which after going through certain processes, is denominated wine, and forms an agreeable beverage'.[132]

In Tasmania, indigenous people and European stock-keepers alike utilised the resin of the cider gum to make an intoxicating drink.[133] The sweetness of the liquid would have greatly added to its attraction. Presumably the settlers adopted this use from the Aboriginal people. The amount procured from each tree was substantial. In 1846 Lieutenant William H. Breton noted its use in the Western Range, which lies west of Launceston:

> The shepherds and stock-keepers ... are in the habit of making deep incisions wherever an exudation of the sap is perceived upon the bark. The holes are made in such a manner as to retain the sap that flows into them, and large enough to hold a pint [0.57 litres]. Each tree yields from half a pint to a pint daily during December and January; but the quantity lessens in February, and soon after ceases. The cider, or sap of the tree, has an agreeable sub-acid taste, and sometimes is of considerable consistency. It is said to have an aperient [mild laxative] effect on those who drink much of it.[134]

In 1859 botanist and plant nurseryman Daniel Bunce confirmed the seasonality of cider gum procurement and described Aboriginal extraction practices. He claimed:

> [the tree] at certain seasons, yields a quantity of slightly saccharine liquor, resembling treacle, which the stockkeepers were in the habit of extracting, and using as a kind of drink. The natives had also a method, at the proper season, of grinding holes in the tree, from which the sweet juice flowed plentifully, and was collected in a hole at the bottom, near the root of the tree. These holes were kept covered over with a flat stone, apparently for the purpose of preventing birds and animals coming to drink it. When allowed to remain any length of time, it ferments and settles into a coarse sort of wine or cider, rather intoxicating if drank to excess.[135]

Cider gum beverage was sent overseas. In 1844 colonist Ronald C. Gunn offered to send Sir William J. Hooker at Kew Gardens some 'more Cider of the Cider Tree if you desire it.'[136]

The colonists made fragrant oils and drinks out of Australian mint species such as the native pennyroyal.[137] Early British settlers in Van Diemen's Land and European sealers on Kangaroo Island used leaves from teatrees as a tea supplement.[138] This practice was largely discontinued after the passing of the colonial frontier, probably because, as Backhouse stated, 'the flavour is too highly aromatic to please the European taste.'[139] According to Gunn, a species of burr was also 'said to be an excellent substitute for tea.'[140] Backhouse recorded that the leaves of 'Cape Barren tea' (white correa), 'which is common all along the sea coast, forming a shrub from two to four feet [61 to 122 cm] high, have been used by the sealers on the islands in Bass's Straits, as a substitute for tea.'[141] East coast colonists also made tea from the leaves of the teatree, as well as from running postman.[142]

In contemporary Australia, 'teatree' refers to many different species of the plant family Myrtaceae, particularly those classified in the genera *Leptospermum* and *Melaleuca*.[143] Given that European colonists used wild plants as tea substitutes, the use of such colloquial names is not in itself surprising. The use of 'ti-tree' as an alternative spelling requires some explanation. It appears that this came about through the confusion

Heath teatree. Plants such as this owe their common Australian English names to the practice of early British settlers using their leaves as a substitute for tea. Philip A. Clarke, Aldgate, Mount Lofty Ranges, South Australia, 1985.

that Sydney colonists over the relationship between teatrees and the Pacific Island plant known in many Polynesian languages as *ti*.[144] Today, gardeners know the *ti* as ti-palm, cordyline or the good luck plant. Details of when and how the word entered Australian English are not evident, although it is known that the Hawaiians made an intoxicating beverage by fermenting the juice of ti roots. The first settlers of Pitcairn Island in 1790, who were *Bounty* mutineers and Tahitians, had also taken up the practice.[145]

During the early stages of European settlement, many wild Australian plants would have been experimented with as sources of drink. Bailey reported that the lemon aspen tree 'produces a pleasant acid fruit, which may be utilised in forming cooling drinks. A form of the species is met with on the Logan [River in northern Queensland], and the usefulness of its fruit was brought under notice some years ago by the late Rev. B. Scortechini.'[146] There were similar uses of Australian species of *Citrus*, such as the Queensland lime, wild lime and native lime.[147] Bushmen in northern Queensland and the Northern Territory made an acidic drink by boiling the fruit and foliage of the brown Indian hemp (kenaf).[148]

Bush medicine

Coming from a nation convinced of the innate superiority of the white man and his civilisation, the first British colonists were not ready to concede that any hunter-gatherer society could have well-developed systems for managing their health and wellbeing, and would also have had their own folk remedies for most common ailments.[149] There were significant cultural blocks against European colonists widely adopting Aboriginal healing practices, beyond using a few indigenous medicinal plant sources that they saw as in some way equivalent to their own herbal remedies. Each culture possesses its own set of notions about health, in particular about what causes sickness and therefore how best to treat it, and the use of medicinal plants can only be fully understood in terms of that culture's worldview.[150]

In Aboriginal cultures, sickness was often attributed to such things as sorcery, breaches of religious sanctions and rules of behaviour, intrusions of spirits and disease-objects, or loss of a person's soul. Cases of swift and inexplicable onset of deadly illness were attributed to supernatural causes. Prior to European colonisation there appear to have been far fewer fatal diseases in isolated parts of the world such as Australia and the Americas.[151] A large number of indigenous medicines have been found by pharmacologists in more recent times to have a valid chemical basis.[152]

Western European notions of the causes of poor health shaped the first British settlers' response to the Australian environment. Historian Jennifer Hagger described how they persisted in using flannel underwear, in spite of the hot summers, because it was believed to prevent colds and rheumatism.[153] Because of poor diet and hygiene, the general standard of health and cleanliness amongst the first settlers arriving from England was probably far inferior to that of the hunter-gatherers living around Sydney at the time. The introduction of exotic diseases, along with frontier conflict, quickly reversed this situation.

Generally, Europeans in the eighteenth and nineteenth centuries considered that Aboriginal attempts at healing belonged in the area of trickery and sorcery. In 1788, Sydney colonist Captain Watkin Tench described a 'superstitious ceremony' whereby a *caradyee*, or 'doctor of renown', treated a sick man by acting as if he had sucked out a river pebble from his breast.[154] This view continued into the nineteenth century, with Mounted Constable Samuel T. Gason in Central Australia claiming in 1879 that Diyari healers were 'impostors' and that their treatments amounted to 'trickery'.[155] Similarly, Taplin in the Lower Murray of South Australia stated that the 'doctor's methods' lay:

> more in incantations than anything else. There are certain men amongst them sometimes called 'Kuldukkis', sometimes 'Wiwirrarmaldar', and sometimes 'Puttherar' – but all mean doctors, and they profess to cure the sick. They blow and chant and mutter over the sick person, all the while squeezing the part affected by the disease, and after many efforts will produce a bit of wood, or bone, or stone, which they declare has been extracted from the place, and is the cause of the ailment.[156]

Taplin also was not convinced of the value of

Aboriginal healers, believing that they were 'great impostors' who operated by 'sleight-of hand'.[157]

Given the early lack of interest in recognising indigenous healing practices, there does appear to have been at least a concession by some observers that they could be effective in limited ways. In 1820 Bellingshausen reported in New South Wales that an elderly Aboriginal man had saved a soldier bitten in the leg by a deadly snake, by sucking and scraping the poison out while using a tourniquet.[158] The cured man was said to have been so grateful that he gave his entire wealth, amounting to five pounds, to his healer. In 1837 Danish adventurer Jorgen Jorgenson stated that he was in awe of the capacity of Tasmanian Aboriginal people to recover from serious physical injuries, particularly spear wounds.[159] Similarly, western Victorian colonist James Dawson claimed in 1881 that 'Every tribe has its doctor, in whose skill great confidence is reposed; and not without reason, for he generally prescribes sensible remedies. When these fail, he has recourse to supernatural means and artifices of various kinds'.[160] Despite these sympathetic accounts, however, many colonists presumed that Aboriginal healing practices were generally ineffective.

In spite of evidence that Europeans came to use some of the same medicinal plants as Aboriginal people did, we do not always know whether each recorded use by colonists was due to their having acquired Aboriginal knowledge or due solely to their own independent investigations. Colonists often experimented with particular plants as medicines, regardless of whether or not they knew about Aboriginal uses of them.[161] Wild aromatic plants were targeted as replacements for herbal teas and for various healing herbs. A bitter taste was a strong indication for potential uses as a tonic or the treatment of indigestion. For instance, in 1788 surgeon Dennis Considen wrote from New South Wales to Banks, saying: 'this country produces five or six species of wild myrtle [species of *Melaleuca*, *Kunzea* and *Leptospermum*], some of which I have sent you dried. An infusion of the leaves of one sort is a mild and safe astringent for the dysentery'.[162] In Queensland, both Aboriginal people and European bushmen suffering from headaches and colds would inhale the crushed leaves of headache vine, which are highly aromatic.[163]

In early New South Wales, a type of 'acid berry', which has been identified as the sour currant-bush, was used as an antiscorbutic medicine to treat convicts who had arrived from England suffering scurvy.[164] In southeastern Australia, Aboriginal people and colonists applied a poultice made from the old man's beard creeper onto aching joints of the legs and arms.[165] In the Kangaroo Island sealing communities of the 1820s, 'bush ti-tree' was used as a medicine to 'purify' the blood.[166] Here, Aboriginal women living with the sealers were in habit of employing their own plant-based remedies, as well as using trinkets as medicinal charms, to treat themselves and their European partners.[167] Across the arid inland regions of Australia, Aboriginal people and earlier settlers alike used the fleshy leaves and stems of the munyeroo, which is a creeper, both as a cooling diuretic medicine and as a food.[168]

The historical records are incomplete on the origin of most colonial healing practices, for few settlers would acknowledge Aboriginal knowledge as the source of their early remedies. Poet and Utopian idealist Dame Mary Gilmore (1865–1962), who grew up in rural New South Wales, said in her reminiscences:

> the white forgets the uncounted ways in which he [was] ... unintelligent (and still would be unintelligent) but for what the blacks taught. As parallels to the treatment of snake-bite by sucking, take the use of eucalyptus, the application of weak wattle tan-water for burns and blisters, of clean mud as poultices, of native gums in dysentery, the eucalyptus beds and steam pits for colds and rheumatism, and ask was it a black or a white intelligence that was first to find and apply these.[169]

Gilmore also stated that a whole industry in making medicines owed its existence to Aboriginal practices:

> It is true there is a eucalyptus extract industry now; but the knowledge that led to that was originally derived from the natives, who used eucalyptus leaves in steaming, and for wounds. For rheumatism steam pits were made, heated by fires, raked out, lined with leaves and then possum-rugs laid over the top. Another use of the leaves was as a strapping for wounds that needed closing in order to heal. These uses came to the pioneers from the blacks.[170]

Old man's beard. Aboriginal people and colonists in southeastern Australia used runners from this creeper, which has hair-like seed awns, to make a poultice for aching joints. It will cause severe blistering if left on for too long. Philip A. Clarke, Lakes Entrance, Victoria, 1988.

Part of the reason for this lack of acknowledgment probably stems from differences between Aboriginal and European methods in preparing their medicines. Before British settlement, Aboriginal people did not possess the technology to boil water for making European-style teas, although they employed many aromatic plants in rubs and steam-bath aromatherapy.[171] Aboriginal healers altered some of their practices when billycans became available to them, which meant that water could be maintained at boiling temperature and more concentrated medicines could be prepared.[172]

It is claimed that one of the first medicinal plants to be used by the British immigrants of 1788 was the Sydney peppermint gum.[173] Medical man John White, who was a collector of botanical specimens during his seven years in Australia, published an account of the Sydney peppermint gum, claiming that oil removed from the leaves could be used for colicky complaints.[174] He also suggested the potential for medicinal use of other eucalypts occurring around Sydney, the gums from which were later exported as 'Botany Bay Kino'.[175] The colonists obtained the bitter gum known by botanists as kino from a number of eucalypt species and employed it in the treatment of diarrhoea and scurvy.[176] The manna exuding from the manna gum was said to work as a mild laxative. Bushmen also used the bark and gum of many acacia species, which are rich in tannins, to stop diarrhoea.[177]

Given the harsh conditions the immigrants experienced on the voyage out, and to a slightly lesser extent on the land once they had arrived, it is not surprising that scurvy was a common ailment. In March 1789 Collins recorded:

> A convict belonging to the brick-maker's gang had strayed into the woods for the purpose of collecting sweet tea; an herb so called by the convicts, and which was in great estimation among them. The leaves of it being boiled, they obtained a beverage not unlike liquorice in taste,

> and which was recommended by some of the medical gentlemen here, as a powerful tonic. It was discovered soon after our arrival, and was then found close to the settlement; but the great consumption had now rendered it scarce. It was supposed, that the convict in his search after this article had fallen in with a party of natives, who had killed him.[178]

In 1790 White described a 'sweet tea' tonic used to treat scurvy that was made from a wild 'creeping kind of vine' (sweet sarsaparilla) and which had a sweet taste, 'exactly like the liquorice root of the shops'.[179] Seamen arriving in Australia in particular required relief from scurvy. French naturalist Jacques-Julien La Billardière was a member of the Bruni D'Entrecasteaux expedition on *La Récherche* and *L'Espérance*, which reached Tasmania in 1792; there La Billardière collected many plants and described a new species of 'parsley', which was used by the sailors to improve their health and to vary their diet.[180]

Long after initial British settlement, circumstances sometimes forced colonists to try wild sources of medicine. In Tasmania in the 1830s, Bunce recorded Europeans using the sassafras tree to prepare a medicine:

> We observed that, in many cases, large flakes of bark had been stripped from the trunks of the largest trees. On inquiry, we discovered that the [European] inhabitants were in the habit of thus collecting and boiling the bark, and using the decoction as an antiscorbutic, as a substitute for the sarsaparilla.[181]

Gunn also described sassafras tea as having the effect of a mild laxative.[182] Linguist Luise Hercus noted that during her 1960s fieldwork in Victoria amongst the Aboriginal people, the 'leaves of this plant [sassafras] were still used by the people from Healesville for flavouring tea'.[183] Bunce described how in Tasmania the leaves of stinkwood could be used 'to relieve a nervous headache'.[184] Through the 1850s, hard-pressed Victorian goldminers ate the very sour leaves of the Australian wood sorrel to prevent scurvy.[185]

Many of the species used in New South Wales by the colonists to make medicinal teas were restricted to the temperate zone, which meant that settlers arriving in the deserts had to rely on the use of other plant species. In 1898 pastoralist Max Koch described the use in the Mount Lyndhurst region of northern South Australia of a plant generally known today as the fruit salad bush. He lists it as:

> Local name, 'Horehound.' Aboriginal name, *Yunga-yunga*. The decoction of the leaves of this perennial plant is used by bushmen for colds. Others flavour their tea by putting a leaf or two in it.[186]

In spite of Koch's comparison to 'horehound', a commonly used herb of the Northern Hemisphere, it is likely that settlers derived their use of this desert plant from Aboriginal people. In Central Australia, the fruit salad bush is a highly favoured medicinal plant used by Aboriginal healers in a variety of ways to treat colds, such as inserting the leaves through a hole bored through the nasal septum, using the leaves wrapped as a pillow, or crushed and mixed with animal oil to make a massage ointment.[187] Another Aboriginal use of a medicinal plant adopted by bushmen in Central Australia was the application of the sap, or 'milk', from the milk-bush to heal sores.[188]

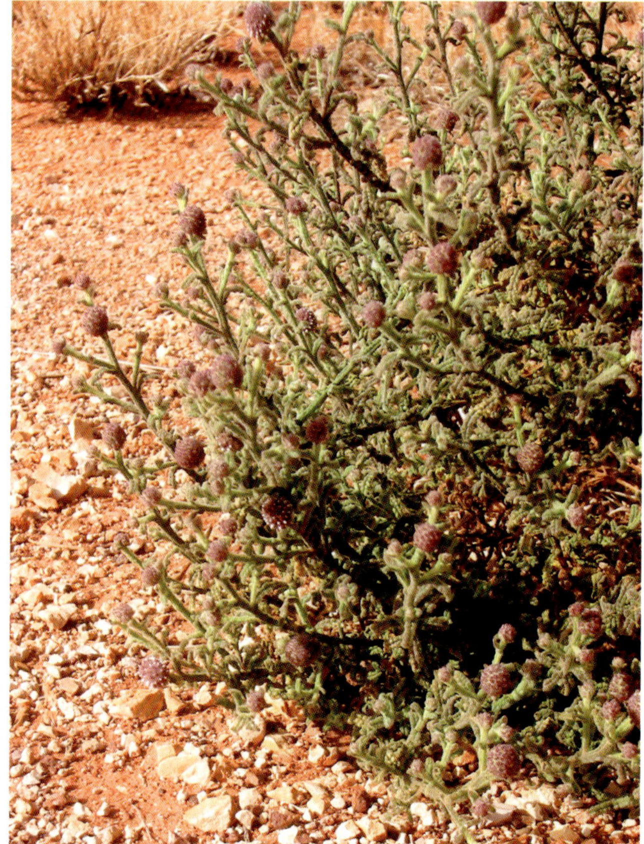

Fruit salad bush. Aboriginal people in Central Australia used strongly smelling plants, such as this, in aromatherapy. A 'tea' made from the leaves of this perennial plant was used by European bushmen for colds. Philip A. Clarke, Marla, northwest South Australia, 2007.

In the tropics, botanist Leonard J. Webb claimed that some Aboriginal 'remedies entered the "medicine chest" of bushmen, drovers, and timber cutters, whilst others became popular with Chinese herbalists'.[189] Early Queensland settlers dried the leaves of sacred basil, which is also found in South-East Asia, to make a bush tea, possibly following the Aboriginal practice of drinking a brew made from the plant to treat colds.[190] Colonists in northern Queensland drank oil extracted from the leaves of the northern stringybark to treat malarial fever, and used citronella from the lemon-scented gum as a deterrent for mosquitoes.[191] Another medicinal plant adopted by Europeans from Aboriginal healers in northern Queensland was the sandpaper fig, the milky juice from the young shoots being applied externally for healing wounds.[192] Northern Australian bushmen used infusions of milkweed to treat serious dysentery and fever.[193]

Generally, the level of hardship that the Europeans experienced determined whether and when they consumed bush tucker (wild foods and beverages) and utilised Aboriginal medicines. Their greatest usage occurred during the frontier and station periods of British settlement, falling away as the landscape was transformed by agriculture. In a few developing rural regions, Aboriginal people maintained close relationships with the environment through participating in European-controlled activities, such as working on cattle stations, in the whaling and fishing industries, and as collectors of gum and bark.[194] In more remote areas, hunting and gathering activities continued, albeit in a restricted way. In both situations, the resources of the bush remained accessible to Europeans via contemporary indigenous knowledge. This was less likely in areas where closer settlement forms, such as dairy farms, replaced earlier, more extensive forms of agriculture, resulting in the cessation of Aboriginal hunting and gathering practices and the widespread loss of indigenous vegetation.[195]

The sources of the wild foods, beverages and medicines that the colonists preferred to use were those relatively easy to collect and required little or no processing. Settlers were more likely to use an Australian plant species if it had some physical, albeit superficial, resemblance to a Northern Hemisphere species. The perceived similarity was not restricted to physical appearance alone, but included properties such as taste and smell. Colonists would have been attracted to edible fruits that had heavy crops. They were undoubtedly less interested in indigenous foods which had a short season and grew in a scattered fashion, or were located in areas away from the best farming land. Aromatic wild plants were attractive as potential medicines to people primed by their previous use of highly scented plants in the Northern Hemisphere. The lifestyle of a farmer settler was sedentary, making movement to collect dispersed foods more difficult than it was for mobile hunter-gatherers, so that the availability of Aboriginal labour probably encouraged European use of some wild foods.

The Australian bush was important to the colonists as a supplementary larder and medicine chest during the earliest phases of settlement. British settlers were forced to rely upon bush tucker in times of hardship, but generally found the indigenous flora to be impoverished in terms of staple foods of a type suitable for agricultural development. Unlike the situation in the Americas, when Europeans first arrived in Australia there were no fields of cultivated plants like maize or potato which had been developed by indigenous horticulturalists over thousands of years.[196] For this reason, by the late nineteenth century the regular use of wild plants by the descendants of the first European settlers had largely foundered.

3 Making Plant Names

The colonists who arrived in New South Wales in 1788 found a vastly different country from their homeland, with a largely unknown flora and fauna. At the outset, they filled the void with names derived from English equivalents, with a smattering of terms from local indigenous languages. The expanded use of English common names for the Australian flora and fauna imparted a hint of familiarity to the 'new' landscape although, by using them, the colonists were more often than not making superficial comparisons between biologically unrelated organisms. The adoption of indigenous words for Australian species offered less confusion. In the case of plants, proper collecting and the results of botanists' taxonomic studies would eventually provide links between English common names, Aboriginal terminology and scientific classification.

English common names for Australian plants

The records of the first wave of colonists to reach New South Wales reflect the lack of any coherent system for identifying Australian plants. The First Fleet did not have with them a trained botanist. Captain John Hunter remarked:

> The vast variety of beautiful plants and flowers, which are to be found in this country, may hereafter afford much entertainment to the curious in the science of botany; but I am wholly unqualified to describe the different sorts with which we find the woods to abound.[1]

Colonists attempting to describe Australian plants in terms of analogous examples from their homeland created many English-derived common names, while the early involvement of local inhabitants in the colonisation process led to the adoption of a few Aboriginal terms. More such names may have been incorporated into Australian English had the settlers possessed a better understanding of local indigenous languages. One of the earliest attempts at recording an Aboriginal language spoken in the Sydney region was made by Lieutenant William Dawes around 1790. He developed a close personal relationship with a young Aboriginal woman called Patyegarang, producing a large word list that contains indigenous terms for plants, with simple translations such as 'the name of a fruit' or 'the fruit of the potatoe plant'.[2] Like so many colonists, Dawes was ill-equipped to botanically identify the species of plants that Aboriginal people described to him.

Danish adventurer Jorgen Jorgenson provided an example of the frustration involved in describing plants experienced by colonists who were generally without any botanical training. Reminiscing in 1837, he wrote how Tasmanian gatherers 'used to dig out of the lagoons *wattalapee* or *pomalle*, here called the native potatoe, but it bears no resemblance to the English root'.[3] Due to the dearth of written records for Tasmanian languages, the identity of the 'native potatoe' remains problematic today. If not for the habitat, this record could have referred to potato orchids, which a later recorder, plant collector Ronald C. Gunn, also termed 'native potatoes'.[4] A more likely possibility is the small-flowered geranium, which Gunn stated was known around Launceston as 'native carrot'.[5] Aboriginal foragers were said to have dug up its large fleshy roots and roasted them. But Jorgenson's record may have referred to an entirely different plant. Today, the full identification of many plants mentioned in early historical records is not possible, precisely because a high proportion of available accounts lack properly determined and consistent plant names. The best Aboriginal plant use records are those linked to a physical specimen collected at the time of the recording, with an indigenous informant.

The historical records of Australian plants are full of terms based upon similarity to European species.[6] Examples include Australian sarsaparilla, bush plum, native cherry, desert raisin, wild currant, native guava, wild pear, bush tomato, native raspberry, wild gooseberry, native walnut, tropical almond, desert banana, wild grape, native potato, pencil yam, bush carrot and native truffle.[7] The index of any regional Australian flora handbook will furnish many more examples. Names of this type have two elements, the first part making it apparent that it is the indigenous version, and the second indicating its resemblance, in either appearance or use, to a particular European species. In a smaller

Wild currant, also known as carrot-wood or coastal bearded heath. Many unrelated species of Australian plants were called 'currants' by the European settlers, which subsequently has made their precise identification from the literature difficult. Philip A. Clarke, Lakes Entrance, eastern Victoria, 1988.

Nasturtium tree, a small tree whose common name is based on the similarity of its leaves to the true nasturtium. In some Queensland Aboriginal languages the nasturtium tree was known as *tumkullum*. Aboriginal toolmakers made their spears from its trunk, a fact noted by Europeans searching for commercial timbers. Philip A. Clarke, Great Dividing Range, southeast Queensland, 2004.

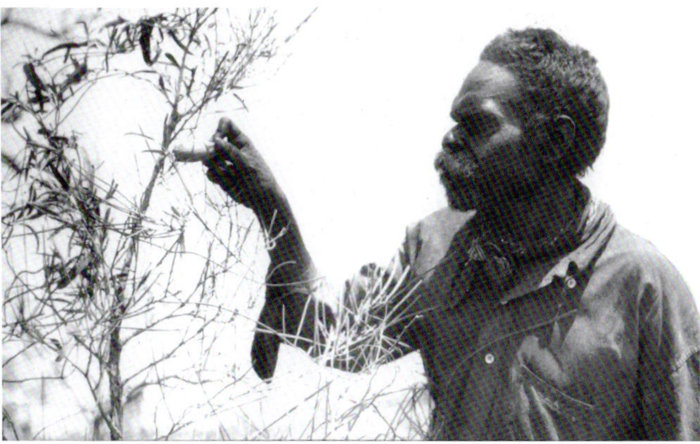

Pik Pik, a Luritja man, collecting *ipala* (desert bananas), which in the past have also been called 'wild cucumbers' and 'desert pears'. They are variously described as tasting like green almonds or apple cucumbers when ripe. William D. Walker, Central Australia, 1928. W.D. Walker Collection, AA357/2/65/16E, South Australian Museum Archives, Adelaide.

Desert banana. Western Desert people know this plant as *kalgula*, from which the Western Australian Goldfields town, Kalgoorlie, possibly derives its name. Many Aboriginal placenames relate to plants (Clarke, 2007, p. 17). John B. Cleland, Central Australia, probably 1950s. A48981, South Australian Museum Ethnobotany Collection, Adelaide.

Bush tomato. This plant has a common name with two parts – the second referring to a similar European plant, the first drawing attention to the fact that it is a wild equivalent. For desert foragers, the outer rind of the fruit is a food source that can be kept dry on wooden skewers (Clarke, 2007, pp. 42, 68, 83). Philip A. Clarke, Musgrave Ranges, Central Australia, 1994.

number of cases the resemblance is stretched, with names such as 'nasturtium tree' and 'peanut tree'.[8] In the case of the 'native apricot', the name is particularly misleading, as the apricot-looking fruit is too bitter to consume.[9]

The adoption of a terminology based upon English folk systems has led to difficulties with identifying many Australian plants, with individual names often referring to a number of botanically unrelated species. Taking the case of 'plums', colonists coined names such as 'billy goat plum' (or 'Kakadu plum'), 'black plum', 'Burdekin plum', 'damson plum', 'green plum', 'gulf plum', 'nonda plum', 'plum pine', 'sour plum', 'wongay plum' and at least several others.[10] While most of these names refer to just one plant species, the main difficulty occurs when they are spoken of simply as 'wild plums'. In taxonomic terms, none of the ten different species named above are closely related to the true domestic plums originating in Eurasia, and in fact are representative of six different plant families.

There are many instances where a single species has a multitude of common names. Botanists know an arid zone tree which is an Aboriginal food source as *Capparis mitchelli*. This scientific name refers to a caper and honours its discoverer, the explorer Major Thomas Mitchell. The common names for *Capparis mitchelli* listed in published floras include 'bumble tree', 'desert caper', 'native pomegranate' and 'wild orange'.[11] Mitchell stated it 'resembles a small lemon'.[12] He claimed that they were to be found 'beyond the Bàrwan [River] there at the "Morella Ridges," to which the natives were in the habit of resorting at certain seasons, by a path of their own, to gather a fruit of which they were very fond, named by them "Moguile," and which I had previously ascertained to be that formerly discovered by me, and named by Dr. Lindley *Capparis Mitchellii*'.[13] As shown by this example, colloquial names and descriptive terms, such as when an author is making comparisons between species for the reader's benefit, are easily confused. Because language is constantly changing, independently of written records, we often do not know the status of given names and descriptions.

The indigenous origin of plant names is not always apparent. In the above example, linguists and etymologists believe that 'bumble' was derived from *bambul*, the term for the plant in the Kamilaroi language of northeast New South Wales.[14] If so, 'bumble tree' has nothing to do with the English meanings of being clumsy or associated with the European bumblebee. Plant species that occur over a wide range are more likely to gain different names in comparison to those with a more restricted distribution. The co-existence of different common names for the same type of plant reflects variety within Australian English, in terms of geography, social context and the period.

Quite apart from the influence of indigenous languages, plants can gain common names through

Spearwood. A plant so named because it is the source of the wood that desert peoples utilise in making spear shafts. The twisted canes are collected green and straightened over the fire. Philip A. Clarke, Victory Well, Everard Ranges, northwest South Australia, 2007.

European perception of associations with hunter-gatherers. For example, blackboys (grasstrees) and blackgins (giant kingia) are possibly so named due to the blackened and Aboriginal-looking appearance of their stumps after fire.[15] As children growing up in the Mount Lofty Ranges, my friends and I believed that the 'blackboy' name related to the appearance of the dead flower stalks, which looked to us like Aboriginal spears being held vertically. The name blackboys may also refer to their Aboriginal uses as a source of nectar, artefact cement, spear shafts and firemaking material.[16] The spearwood or spearbush of western Central Australia is aptly named as the plant source of the wood that desert people utilise in making their spears.[17] Waddy-wood of eastern Central Australia is so named because its wood was made into Aboriginal clubs (waddies).[18] Similarly, in tropical Australia, the digging-stick tree was a source of wood for making tools.[19] In northern Australia, the fish killer tree has a name that bears testimony to the Aboriginal use of its bark to poison fish in waterholes.[20]

Many of the common names for Australian plants are quite ambiguous. To the casual observer, some sets of alternative common names for a single species appear at first glance inconsistent. For instance, the plant that botanists know as *Acacia melanoxylon* has a species name derived from Ancient Greek *melanos* (black) and *xylon* (wood), referring to the highly figured dark brown timber which is highly prized for cabinet-making and for fashioning into gunstocks.[21] In Tasmania,

Waddy-wood, a species of wattle that occurs in eastern Central Australia. It is so named because of its Aboriginal use for making clubs (waddies). Philip A. Clarke, Birdsville, Central Australia, 1986.

this species is commonly, and confusingly, known as either the blackwood or lightwood.[22] While the former term relates to wood colour, the latter is a reference to the fact that the timber, in comparison to eucalypt, is low in weight.[23]

In 1882 John E. Brown, the Conservator of Forests for the South Australian Government, gave an example showing the extent to which the existence of a variety of common names within particular groups of plants can exacerbate the confusion over species identity. He despaired that the South Australian blue gum was known as just 'blue gum' by the splitters of the Wirrabara Forest in the mid north, while it was called 'white gum' or 'pink gum' around Adelaide and in the Mount Lofty Ranges, and 'bastard red gum' in the southeast of the colony.[24] The colours indicated by these common names could possibly refer to leaves, flowers, bark or heartwood. The reliability of plant naming systems is not just a matter for academic botanists. Brown claimed that in the case of the 'South Australian blue gum', which was his preferred term, the muddling up of European common names had led to serious mistakes with the use of timbers by the construction industry.

Indigenous classifications of the flora

All the world's languages contain encoded information about how speakers interact with their environment. Eighteenth and nineteenth century European scholars considered most indigenous languages to be 'primitive' and were largely ignorant of the complex classification systems possessed by hunter-gatherer societies. The modern field of anthropological linguistics uses the study of language to gain insights into how cultures order the universe.[25] When Europeans first arrived here, there were at least 200 distinct languages in Aboriginal Australia, and within them numerous speech varieties or dialects.[26] The manner in which languages reflect how people see their environment is a topic far too vast for detailed discussion here, although examples of language terms are given to demonstrate that there is a multitude of ways to classify the flora.

The importance a culture places upon specific objects and phenomena within the environment is highly influenced by the manner in which they gain a living from the land. The importance of reading the country leads to proliferation of words in categories that are highly relevant to the speakers. In Aboriginal languages, as with English nomenclature, the different parts and growth stages of useful plants may each have separate names, so that a European recorder of an indigenous language may need to document a string of indigenous terms for a single plant species. An example from the Nunggubuyu language of eastern Arnhem Land is the water lily (*yangguri*), which is an important food source[27], and has over a dozen terms associated with it. For example, the edible root corm is known as *wudan*; when the corm is very immature, it is called *jirigilil*; more advanced but still immature, it is referred to as *yiwujung*; and when aged and bitter tasting it is *nindan*. Similarly, in the Mitchell River area of the Gulf of Carpentaria, Aboriginal foragers called the giant water lily *arnurna*; the roots were known as *thoongon*; the seed-stalk was *urgullathy*; and the round seed-head was *irrpo*.[28]

In some Aboriginal languages, plants have both secular and sacred names.[29] There may also be different terms for forms of plants that botanists would regard as subspecies or varieties, rather than separate species.[30] The vocabularies of Aboriginal languages are crucial repositories of environmental data. The loss of any of the world's languages has serious implications for the continuity of knowledge concerning cultural landscapes.[31] The unique cultural perceptions of the environment can be demonstrated in other ways. Aboriginal trackers who worked with colonists and the police were able to read the landscape to gain insights into what had occurred because they saw things that Europeans normally would not notice.[32]

Aboriginal hunters and gatherers generally recognise two main classes of food: animal and vegetable. In 1861, the seedsman and plant collector J.F.C. (Carl) Wilhelmi provided an account of how Aboriginal people living in the vicinity of Port Lincoln in South Australia classified their food. He claimed that the 'natives divide all their articles of food into two classes – the "paru" and "mai," the former including all animal and the latter all vegetable articles of food.'[33] The same division is found in the Western Desert languages spoken further north across the Great Victoria Desert, where I have done fieldwork, although here the animal and plant foods are categorised as *kuka* and *mai* respectively.[34] Such distinctions are relevant to the gender division of labour, with women and children gathering ground-based foods like plants as well as lizards and witchetty grubs, while men focus their attention upon hunting larger and more mobile animals that are meat sources.[35] Similar food classifications are to be found in many other languages in Aboriginal Australia.[36]

Aboriginal languages contain terms that are applied to generalised categories of plants and animals that are of little or no direct interest to the people. In Aboriginal English such plants are generally referred to as 'rubbish plants,' which translates into Australian English as 'weeds.'[37] In the Nunggubuyu language of eastern Arnhem Land, linguist Jeffrey Heath recorded the term *madinjar* for several species of shrub, including the feather-plant, jacksonia (broombush) and turkey-bush. All of these plants grow about one or two metres high, are without value in terms of Aboriginal use, have nuisance spiky leaves and sharp twigs, and in the dry season bear small but brightly coloured flowers.[38] In some contexts, *madinjar* is extended to mean a whole scrubland association that is chiefly comprised of such plants, such types of vegetation offering very little in terms of food or medicine. When it was appropriate, Aboriginal hunter-gatherers would have actively modified dense stands of scrub dominated by 'rubbish' species by deliberately burning them.

A few plant names are based either upon their resemblance to other things, or as a reference to a custom or tradition. Linguist John McEntee has made a study of plant names in the Adnyamathanha language of the northern Flinders Ranges in South Australia.[39] Here, he noted the example of puffball fungus, called *vurdli-vuthi*, reputedly with the literal meaning of 'star dust.'[40] A young man kicking a puffball to release its yellowish spores into the air was said to be 'pulling down the stars,' a reference to falling in love with a girl in the wrong kinship category. In the same language, the Sturt desert pea is called *ngarapanha*, possibly meaning 'little liar,' while *varpawarta* is the samphire, which is said to translate as 'south wind bush.'[41] Other plant species blown in as seeds by the strong south winds were also known by this term. Indigenous plant use is sometimes referred to in the names. An example is *miya-vuthi*, the woolly cloak fern, said to mean 'sleep dust' in reference to the Adnyamathanha custom of brushing its leaves over a child's eyelids to induce sleep.[42]

From Aboriginal terms to Australian English words

Evidence for the close interaction between early European settlers and indigenous peoples is found in present-day Australian English. While it has been resistant to the absorption or, as linguists say, 'borrowing,' of most indigenous words, there is a major exception in the adoption of Aboriginal placenames.[43] An early example in New South Wales was Parramatta, based upon the local Aboriginal placename, Baramada.[44] Relatively few indigenous terms for plants and animals have found their way into Australian English. In the case of the flora, there would have been potentially many thousands of names available from the vocabularies associated with all the Australian Aboriginal languages spoken in 1788.[45] The relatively small number of plant words imported into Australian English mostly came in during the frontier periods of European colonisation, with a strong bias towards indigenous languages from eastern and southern Australia, where British settlement first commenced and has been most intense. In recent decades, however, there has been a growing movement amongst botanists and zoologists to use indigenous common names for Australian plants and animals.

The adoption of Aboriginal names appears most often amongst those species that do not closely resemble plants already familiar to Europeans. Many of the indigenous terms are associated with species that have edible parts, particularly fruits, or are prominent elements of the landscape, such as large trees or spectacularly flowering bushes. Similarly, the adoption of Aboriginal terms for mammals favours the southern and eastern Australian source languages and is greatest for prominent and, to European eyes, unusual species such as the kangaroo, koala, numbat, quokka, tammar, wallaby and wombat.[46] In 1918 biologist and anthropologist Baldwin Spencer described the process by which Aboriginal language terms pass into wider and more common usage:

> A white man living for a time in one locality hears a native name applied to some special object. When he travels on to another place he carries this name with him, and, as likely as not, applies it to some other thing to which, either in general appearance or in regard to its use, it is apparently similar, and the native, wishing to please the white man, adopts the new name, with the result that confusion arises in consequence of the same name being applied to two different objects. Or, again, a name that is applied originally in one special tribe to one special object may become widely used for the latter, because it carried on from tribe to tribe by white men, or even by natives working for them, as the country is opened up or settlement extends.[47]

Spencer also suggested that some translating mistakes by Europeans taking on Aboriginal plant names came about through asking their informants leading questions.

Words that are borrowed by one language from another are often transformed in the process, as each language has its own set of sounds (phonetic system), with which imported words must conform.[48] When investigating historical recordings of Aboriginal terms, there are often problems in relating the written forms to spoken words. Indigenous words may appear mangled when recorded by Europeans who are not linguistically trained. An approach that Australian linguists now employ when working with certain historic word lists is to find out where in Britain (or Europe) the recorder was raised, in the hope of determining their regional dialect (or accent) and thereby being able to predict how they would have heard sounds in the indigenous language and attempted to write them down.

Once words have been incorporated into an Australian English variety of speech, they may persist there long after the indigenous source language has ceased being spoken, such as has happened in many parts of southeastern Australia. From my own fieldwork during the late twentieth century, I found that European Australian fishermen and duck shooters of the Lower Lakes and Coorong in South Australia still recognised and used some Aboriginal terms. These were predominately for local animals and plants, such as taralgi (estuary perch), punkeri (hardhead duck) and mumuruki (bulrush). At the same time, the local Aboriginal community utilised a rich variety of Aboriginal English, known by the speakers as *Ngarrindjeri yanin*, but I rarely heard in general use any of the indigenous terms commonly used by fishermen and duck shooters.[49] This situation reflected the early involvement of Aboriginal people in the fishing and wild fowl canning industries of southeast South Australia, followed by a period since the 1960s when their participation largely ceased.

As suggested by Spencer above, it is likely that the spread of the use of certain indigenous plant terms within Australian English came about through Aboriginal movements that occurred after

European settlement. Many indigenous loan words were part of Pidgin English spoken on the advancing European frontier.[50] During the colonisation period, the initial mixing of indigenous and European peoples around rapidly spreading settlements produced an environment requiring a common language. In these conditions, a pidgin language generally develops. In 1798 Judge Advocate and Secretary David Collins remarked:

> Language [indigenous], indeed, is out of the question; for at the time of writing this, (September 1796) nothing but a barbarous mixture of English with the Port Jackson dialect is spoken by either party; and it must be added, that even in this the natives have the advantage, comprehending, with much greater aptness than we can pretend to, everything they hear us say.[51]

On the frontier, Aboriginal people appear to have learnt new languages much faster and better than was the case with the newly arrived Europeans.[52] Pidgin English is a simplified language formation, often assumed by colonists to be a bastard form of English.

Widespread plant species have tended to attract the adoption of many indigenous names into Australian English, some of them with more localised use than others. An example is a wattle species known by botanists as *Acacia stenophylla*, which is widespread through Central Australia. In the botanical literature, its common names have been recorded as munumula, balkura, gurley, gooralee, ironwood, dalby wattle, river cooba, river myall, belalei, eumong, native willow, black wattle and dunthy.[53] At least half of these appear to have an indigenous language origin. Which particular term is used and by whom today would be linked to existing variations in Australian English speech style across the region where the plant occurs.

There are many examples of Aboriginal plant terms that have had at least some limited life within Australian English. A multitude of locally used names are given in early botanical publications. In his economic botany book, Maiden listed hundreds of indigenous plant terms, some of which had become colloquial names in Australian English. For example, in his description of the 'blue gum' of New South Wales coastal districts, he claimed that the species:

> also bears the names of 'Bastard Jarrah,' and occasionally 'Woolly Butt.' Sydney workmen often give it the name 'Bangalay,' [pronounced 'bang alley'], by which it was formerly known by the aboriginals of Port Jackson. It is called 'Binnak' by the aboriginals of East Gippsland.[54]

This example illustrates the extension of the meaning of 'jarrah,' which was originally derived from an indigenous language in southwest Western Australia, based upon the perceived properties of the plant.

Indigenous words from the east coast

Aboriginal languages formerly spoken around Sydney were the source of many words still in common use today. English language scholar William Ramson calculated that one-tenth of the Aboriginal vocabulary of about 250 Dharug language words recorded by Hunter in 1790 remains current in Australian English.[55] This level of borrowing is at least five times greater than for any other Australian Aboriginal language. The colonists' bias in adopting words from the first indigenous languages they encountered is the pattern reflected elsewhere in the world. For example, many of the contemporary American English words derived from Amerindians were borrowed from languages spoken in coastal areas of the North Atlantic where Europeans first settled.[56]

The official floral emblem of the State of New South Wales is the waratah, a name thought to be a borrowing from *warrada* in the Dharug language west of Sydney.[57] The species had cultural significance for hunter-gatherers, shown on one occasion when Sydney Aboriginal people placed a waratah flower alongside the body of an Aboriginal man being buried.[58] Aboriginal foragers also obtained nectar from the waratah by sucking the tubular flowers.[59] Europeans also use this term for a botanically related plant, the Tasmanian waratah, which is only found wild in Tasmania[60], in preference to using a Tasmanian Aboriginal term. There was some colonial use of the term 'tulip tree' for both these species, but this was of limited value, considering that the structure of the flower bears only passing resemblance to the tulip familiar to gardeners. Another plant with a common name probably derived from a Dharug word is the geebung, from the original form *jibung*, which in the indigenous language was specifically applied to the fruit.[61]

The term 'burrawang' is used in contemporary Australian English for a group of east coast cycads, also called zamia palms.[62] According to Maiden, the food commonly known by Europeans as 'Burrawang Nut [is] so called because they used to be, and are to some extent now, very common about Burrawang, N.S.W.'.[63] Linguists and etymologists disagree, arguing that the term is a loan word derived from *buruwang* in the Bandjalang language of northern coastal New South Wales, and that it originally belonged to a cycad species.[64] The burrawang has pineapple-shaped fruit, the seeds of which Aboriginal people ate after elaborate preparations.[65] British settlers referred to these and other cycads as 'fool's pineapple', in recognition of the seed's toxicity in its raw state.[66] Today, burrawang is used somewhat haphazardly in the plant nursery trade in reference to Australian cycad species in general, particularly those originating from New South Wales. Linguists and etymologists also assume that the name of the cunjevoi, a lily-like plant that grows in moist forests along the eastern seaboard, has been derived from another Bandjalang word.[67] Cunjevoi roots are very poisonous, but northern Aboriginal people still used them as food after extensive preparation.[68]

Some Aboriginal words that Europeans adopted for plants did not originally refer to an individual species, but to the Aboriginal use to which they were put. Scholars believe that the name of the kurrajong trees was derived from *garrajung*, a Dharug word from the area inland from Sydney that referred to 'fishing line'.[69] The black kurrajong has a fibrous bark that Aboriginal artefact-makers used as raw material to manufacture string for their lines and carry-bags.[70] The plant species that early plant collector Allan Cunningham called 'currajong' was a hibiscus and therefore unrelated, although Aboriginal people were using it as a source of bark fibre to make cord.[71] In Australian English, kurrajong became more broadly applied to a group of botanically related tree species across Australia.[72] In Tasmania, an unrelated species was called the Tasmanian kurrajong, presumably because of its strong bark fibre.[73] In his list of commercially useful fibres, Maiden lists various 'kurrajong' species, covering the genera *Hibiscus, Sterculia* and *Brachychiton*.[74]

The Kamilaroi language of northeast New South Wales might be the source of words such as myall (weeping myall) and brigalow, which both describe species of acacia (wattle) tree. The term 'myall' is possibly derived from *maial*.[75] In the case of brigalow, the original term, *burigal*, may have more specifically related to a cluster of plants in the landscape, because the word *buri* was applied to the brigalow-spearwood, while the plural suffix *-gal* referred to there being many, such as in a scrub.[76] Both the brigalow and the weeping myall are dominant species in parts of southeast Queensland and central New South Wales respectively. A dominant acacia species growing in arid western South Australia has been called western myall. The application of this name must be due to the tree's similarities to the myall of New South Wales, since Aboriginal people in this part of South Australia call this species *kardia*.[77] Bindi-eye, a name applied to several species of nuisance burr daisies, is thought to be derived from *bindayaa* in the Kamilaroi language.[78]

The region stretching from southeast Queensland to southeast New South Wales is the home of the bangalow or piccabeen palm, which is today a major ornamental plant in temperate parts of the world.[79] Linguists and etymologists believe that 'bangalow' was derived from *banggalu*, a word for the palm species in the Tharawal language at Shoalhaven in New South Wales, while 'piccabeen' came from *bigibin*, which is the equivalent term in the Yagara language originally spoken around Brisbane in Queensland.[80] The euphonious, or agreeable-sounding, quality of Aboriginal words like these must have heavily influenced their borrowing. It is thought that the Australian English name bunya pine is derived from another Yagara word, *bunya-bunya*, which is applied to this tree.[81] In southeast Queensland, Aboriginal people consumed the edible bunya pine nuts during major feasts.[82] The midyim, which has an edible fruit, may also have a Yagara-derived name.[83]

Southern Aboriginal plant terms

From the southeast Tasmanian word *bubiala* possibly comes the common plant name 'boobialla'.[84] In Tasmania, plant collector Ronald C. Gunn stated:

> Of the genus *Acacia*, the Aborigines were in the habit of collecting the pods of the species *Sophora* [coastal wattle] or *Boobialla*, (which is a common shrub, growing from 6 to 15 feet [1.8 to 4.6 m] high, on the sand-hills of the coast,) when the seeds were ripening, and, after roasting them in the ashes, they picked out the seeds and ate them.[85]

This name appears to have originally just applied to the coastal wattle, whose 'beans' the Aboriginal people ate.[86] But settlers have also used it as a name for various large species of native juniper (*Myoporum*), most of which, but not all, have edible berries.[87] Moandik man Ron Bonney, with whom I did much ethnobotanical work in the southeast of South Australia during the 1980s, was aware of the shared name in Australian English for both species and suggested that it might be explained by the similarity they had in growth habit and leaf shape.

Boobialla or blue-berry tree. This large bush has an edible stone fruit. Its common Australian English name was probably derived from an indigenous Tasmanian word, *bubiala*. Philip A. Clarke, Middleton, Encounter Bay, South Australia, 1989.

From southern Victoria possibly comes the name 'murnong', also known as 'yam-daisy'. While scholars believe that the plant name was derived from *mirnang* in the Wathawurung and Wuywurung languages of the western side of Port Phillip Bay, this word, or variations of it, was found in many indigenous Victorian languages. For example, Bunce was travelling with local Gippsland Aboriginal people in 1839 when his guides gathered for him 'Some long tuberous roots, of a composite plant [yam-daisy], ... of which we partook. These plants produced a bunch of tubers like the fingers on the hand, from whence they were called *myrnong-myrnongatha*, being the native word for "hand"'.[88] The murnong is the likely identity for many of the temperate plant foods simply described as 'edible roots'. The sweet-tasting tuber is recognised as one of the major food sources for Victorian Aboriginal foragers.[89] The first settlers found numerous earth mounds, known by them as 'mirrn-yong heaps', that had been created on open plains and river flats by the 'accidental gardening' of Aboriginal women regularly digging up murnong tubers on the same patch of earth.[90] In 1878 historian Robert Brough Smyth suggested that the murnong might be suitable for cultivation in temperate countries as a food crop.[91]

The name of the dillon bush (nitre-bush) is thought to have been derived from *dilanj*, a term for this plant in the Wemba Wemba language of western Victoria.[92] This species appears to have an association with humans. In the late nineteenth century, at Swan Hill in northern Victoria, it was noted that large dillon bushes were typically found growing around Aboriginal camps, probably because of their seeds lodging there after being eaten.[93] Aboriginal foragers ate the salty dillon fruits, which look like European grapes, raw, including the stones.[94] The dillon bush prefers disturbed situations, a characteristic that has become more apparent in recent times as this species spreads along stock routes in overgrazed areas in the semi-arid and temperate zones.[95]

Dillon or nitre-bush. The name dillon is thought to have been derived from *dilanj*, a term for this plant in the Wemba Wemba language of western Victoria. Aboriginal people ate the fruit whole, stone included. Philip A. Clarke, Eucla, Western Australia, 2006.

Lerps on river red gum leaves. At first glance the insect origin of this sweet flaky substance is not apparent. Its name is probably derived from the Wemba Wemba word *lerep*. Philip A. Clarke, Adelaide Plains, South Australia, 1986.

The Wemba Wemba language has the term *gambang*, from which 'cumbungi', an alternative common name for bulrush, may have come.[96] For southern Aboriginal foragers, bulrush roots were a major source of food and string fibre.[97] Wemba Wemba is also considered to be the language source for 'mallee', from *mali*, which is used today in Australian English for several species of small, multi-stemmed and relatively low-growing eucalypts that typically occur in the semi-arid region stretching from southern Western Australia, through parts of South Australia, to central Victoria and outback New South Wales.[98] Mallee is also used as a description for the whole plant community of which they form a dominant part.

Some plant terms relate specifically to the insects that live upon them. The lerp mallee (ridge-fruited mallee) of southern temperate Australia is so called because of the lerp manna that is found on its leaves. This sweet white flaky substance occurs on many eucalypt species and is actually the small protective shelters produced by a sapsucker insect. Smyth explained:

> In summer the Aborigines of the Mallee country eat *Larap, Larp*, or *Lerp* – a kind of manna. It somewhat resembles in appearance small shells; it is sweet, and in color white or yellowish white ... It is a nutritious food, and is eaten with various kinds of animal food. 'This saccharine substance,' says Baron von Mueller, C.M.G., in a letter to me, 'is obtained from one, or perhaps from several, species of *Eucalyptus* of the Murray and Darling districts. It is not a real manna, but is known as lerp, a name given to it by the Aborigines.'[99]

Its Australian English name was therefore probably derived from a western Victorian Aboriginal language, perhaps based on the Wemba Wemba word *lerep*.[100] Lerps can be scraped off simply by running the leaf through the teeth. It was observed

Ballart, known also as wild cherry. Aboriginal people highly valued the fruits of this bush as food during the summer months when they camped near the coast. In Aboriginal English the fruits are known as 'dolls eyes'. Philip A. Clarke, Kingston, southeastern South Australia, 1987.

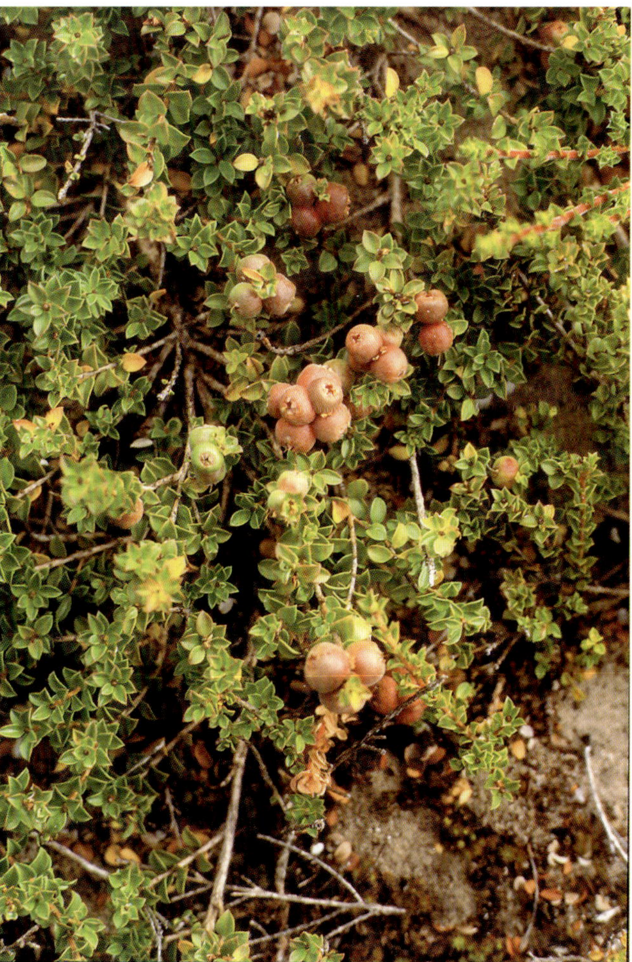

Monterry, a low-growing creeper which bears a large late summer crop of edible fruit once used by European settlers for jams and tarts. It is thought that the name derives from the word *mandharri* in the Ngarrindjeri language of the Lower Murray. Philip A. Clarke, Kingston, southeastern South Australia, 1987.

that when in season Aboriginal people could collect from 18 to 23 kilograms of lerp in a single day.[101]

A name that scholars believe to have come from an eastern Victorian language is 'ballart', referring to a plant commonly known as 'cherry ballart' or 'native cherry'.[102] The original word may well have been *balad*. The exact origin is unclear, since in the Gippsland district of eastern Victoria the Aboriginal term for the plant was recorded as 'ballat', and in the Lake Condah area of western Victoria as 'pallert'.[103] In Australian English, 'ballart' is also extensively used for other taxonomically closely related species, such as the coastal ballart and broom ballart.[104] From the Ganay language of the Gippsland in eastern Victoria we probably have *gunyang*, a word applied today to various species of kangaroo apple found in southeastern Australia and Tasmania.[105]

An example of a word borrowing by Australian English from a southern South Australian indigenous language is 'yacca' for grasstrees, possibly derived from one of the languages around the Adelaide region, where *yaku* or *yakko* was recorded to mean resin.[106] In many parts of temperate Australia, grasstrees were a major source of resin for adhesive and cement used in Aboriginal artefact manufacture.[107] Another candidate as an imported word into Australian English is 'monterry', probably related to *mandharri* from the Ngarrindjeri language of the Lower Murray.[108] In this part of South Australia today, both Aboriginal people and Europeans commonly use this word for a plant which grows as a dense groundcover in exposed situations, such as the tops of coastal sand dunes.[109] Europeans have called the fruit 'native apple' and also described it as 'a sort of raspberry'.[110] Another possibility for a plant term having originated from a Lower Murray Aboriginal language is 'wirilda', from *wurrulde*.[111] Australian English may have borrowed the name for this plant, which is a wattle tree, sometime after British settlement via the Adnyamathanha language of the Flinders Ranges, where it has also been recorded. This could have come about through the documented late nineteenth century movement of Aboriginal workers in the pastoral industry between stations in the Lower Murray and the Mid North.[112]

A commonly used name for a cycad species growing in southwest Western Australia is by-yu nut, the term '*by-yu*' considered to have come from the Nyungar language term *bayu*.[113] As with other cycads, the fruit of this plant is edible only after treatment to remove harmful toxins.[114] It was one of the few seeds in the region obtained from trees, and is in season from February to March. Individual cycads are long-lived, with the dating of by-yu nut trees suggesting ages in excess of 600 years for some specimens.[115]

The Nyungar term *warran* is very likely the source of the name for the warran yam, which was an important Aboriginal food source.[116] Plant collector James Drummond had a very high opinion of this plant, claiming that 'the native yam [warran yam], of the class *Dioscorea*, is stated to be the finest esculent vegetable the colony produces'.[117] Inland from the central Western Australian coast there are areas where it was intensely harvested. In 1838 explorer George Grey at Hutt River came across extensive diggings by Aboriginal foragers looking for warran yams. He reported:

> for three and a half consecutive miles [5.6 kilometres] we traversed a fertile piece of land, literally perforated with the holes the natives had made to dig this root; indeed we could with difficulty walk across it on that account, whilst this tract extended east and west as far as we could see ... more had here been done to secure a provision from the ground by hard manual labour than I could have believed it in the power of uncivilised man to accomplish.[118]

Grey's observations suggest that here Aboriginal people were acting as horticulturalists, although it is not clear from his report whether or not they were aware of the improvement of the soil through digging and thus performed this act consciously and systematically.

One of the most commercially valuable timber trees in southwest Western Australia is the jarrah. The origin of its common name is thought to be the Nyungar word *jarily*, referring to a eucalypt species.[119] In the commercial timber trade, jarrah is prized for use in construction, notably for railway sleepers and floorboards.[120] Other eucalypts with names believed to have originated from Nyungar dialects are 'karri' from *karri*, 'tuart' from *duward*, and 'wandoo' from *wandu*.[121] In the case of tuart, there is a suggestion that the Nyungar people used the same term for the york gum as well.[122] Rather than believe that the term originally applied to quite different tree species, Maiden offered another solution, which was 'perhaps they [Aboriginal people] gave the name originally to the York Gum, and afterwards the white man fixed it on the modern "Tuart"'.[123] If so, this represents an imperfect borrowing of a plant term.

Words from arid regions

The remote and arid outback region of New South Wales has produced several plant terms based on indigenous words. One widely used contemporary name is 'quandong', which linguists and etymologists think is derived from *guwandhaang* in the Wiradhuri language of semi-arid central New South Wales.[124] This is a small, parasitic, arid-zone tree, sometimes called 'native peach', which bears edible fruit up to the size of an Australian 20-cent piece. Often used by settlers, the quandong has more recently been developed for commercial production.[125] The term is also commonly used for a closely related species known as the bitter quandong (ming), and for other botanically unrelated plants, such as the blue quandong (Brisbane quandong), grey quandong and the Kuranda quandong, all of which bear edible fruits or nuts.

The wilga is a small tree of the dry inland regions, its name thought to be derived from *wilgarr* in the Yuwaalaraay and Wiradhuri languages of central New South Wales.[126] Another imported word candidate from this area is 'gidgee', an offensive-smelling species of acacia. This name possibly comes from *gijirr*, a term that refers to this plant in local languages.[127] Europeans use the common name gidgee for several other inland species of acacia. Further examples of plant names that may have been derived from Aboriginal languages in the arid zone of New South Wales are 'mulga' from *malgu*, and 'coolibah' from *gulabaa*.[128] These trees are dominant features of the Australian arid-zone vegetation. The mulga is an acacia that occurs on open plains, sometimes in dense stands, while the coolibah is a large eucalypt, made famous through featuring in A.B. 'Banjo' Paterson's bush ballad, 'Waltzing Matilda', written in 1895.

A plant with a name derived from a southwest New South Wales Aboriginal language is ming, also known as the bitter quandong. The term was recorded in its original form in the Muthi-Muthi language of the region as 'mingun'.[129] Although of dubious value as a source of edible fruit, the tree was probably highly sought after in the pre-European economy as a narcotic for trade. In the semi-arid regions of the Murray–Darling Basin, it is recorded that Aboriginal people gathered the root and bark of the ming to make a 'stupefying' (narcotic) drink.[130]

The arid zone of Central Australia was settled much later than the southern and eastern parts of the continent, at a time when Australian English appears to have been less permeable to borrowed indigenous words. One of the few Central Australian indigenous words imported into Australian English is 'pituri', the name of a large desert bush which the people of eastern Central Australia formerly used as the source of a narcotic and treated as a highly prized trade item.[131] It is believed that this common name was derived from an indigenous language in southwest Queensland, at the centre of the trade, *pijiri* being a word for the plant in the Pitta Pitta language.[132] Among Aboriginal men, pituri had recreational uses, but it was also said 'to excite their courage in warfare' and was given to male initiates before ceremonies to heighten their sense of revelation.[133] After European settlement in Central Australia, trade in pituri collapsed. The term itself came to mean in Australian English all desert species of chewable wild tobacco (of the genus *Nicotiana*), which are generally prepared in a similar way to true pituri (*Duboisia hopwoodii*).[134] Although several species are involved, those most valued by desert people today are the tobacco species now commonly known by European Australians as sandhill pituri and rock pituri.[135]

Coolibah, a widespread eucalypt species found across inland Australia. The tree's common name was possibly derived, in the original form of *gulabaa*, from an Aboriginal language in northern central New South Wales. Philip A. Clarke, Clayton River, Central Australia, 1987.

There is also the case of a desert Aboriginal word being adopted as a common name into Australian English because its meaning had been misunderstood.[136] The name 'munyeroo' (common pigweed) was apparently derived from *manyurra*, a term originally used by Diyari people of northeast South Australia for a somewhat similar plant generally known today as 'parakeelya'.[137] Anthropologist Baldwin Spencer claimed in 1918 that the use of words like munyeroo (for parakeelya) was being widely spread among Aboriginal languages in Central Australia by European settlers. He said that at Alice Springs, the 'the real native name is "ingwitchika" [*ingkwityeke*], but the name "munyeroo" has been adopted from the white man, and is, or was, almost universally used'.[138] Parakeelya is also thought to be of Aboriginal origin, possibly derived from *baragilya* in the Guyani language spoken west of the Flinders Ranges in South Australia.[139] Both munyeroo and parakeelya have edible seeds and foliage.[140] The witchetty bush, however, derives its Australian English name not from an indigenous plant name,

Parakeelya, a low-growing plant with fleshy foliage that desert dwellers used to quench their thirst. Colonists ate the leaves with bread. Its Australian English name is possibly derived from *baragilya*, a word from the Guyani language spoken west of the northern Flinders Ranges in South Australia. Philip A. Clarke, western Macdonnell Ranges, Central Australia, 2004.

Witchetty bush. Its common name refers to the edible grubs that Aboriginal foragers gather from the roots. Philip A. Clarke, western Macdonnell Ranges, Central Australia, 2004.

but from the edible grubs that Aboriginal foragers gather from its roots. The term 'witchetty' is probably derived from *wityu* (grub hook) and *varti* (grub), from the Adnyamathanha language of the Flinders Ranges.[141]

Adopting tropical indigenous plant terms

Tropical examples in the published floras of Australian English common plant names derived from Aboriginal languages appear to be generally fewer, in comparison to those from desert and temperate regions. One example is 'conkerberry', which may have come from the term *ganggabarri* of the Mayi-Yapi and Mayi-Kulan languages spoken in the Cloncurry River district, south of the Gulf of Carpentaria[142], where the small reddish fruits were gathered from the prickly shrubs in large quantities towards the end of the wet season, around February, and eaten raw.[143] The similarity of *bari* to 'berry' may have enhanced the chances of Australian English adopting this indigenous word.

The wongay plum is believed to have had its name taken from *wongay*, which is a word from the western Torres Straits language known as Kala Lagaw Ya, the speakers being Islanders rather than Aboriginal people.[144] 'Jitta' is a name for the saffron heart tree which scholars believe comes from *jidu* of the Dyirbal and Warrgamay languages of the Herbert and Tully rivers district of northern Queensland.[145] Here, the jitta is also commonly called kerosene wood, because as a torch the wood will burn for a long time. Aboriginal toolmakers used jitta timber for making musical instruments, spears, fishhooks and daggers. Another northern example of the adoption of tropical indigenous terms is 'dundathu', an alternative common name for the Queensland kauri pine.[146]

Botanical classification

One of the key functions of herbaria and natural history museums has been taxonomic research; that is, classifying the past and present flora and fauna of the world.[147] Scientific taxonomy establishes a standard nomenclature and reference system for the large amount of information recorded about specific plants and animals which is independent from the many folk classification systems that have existed in the world. As scientists, botanists arrange the Plant Kingdom into a 'tree' which is chiefly branched into 'families', then further subdivided into 'genera', and finally to individual 'species'. For example, in botanical texts the messmate stringybark is referred to by the binomial *Eucalyptus obliqua*, which indicates that it is a species within the *Eucalyptus* genus that is in turn grouped with other genera, such as *Leptospermum* and *Melaleuca*, into the Myrtaceae family. When referring to a particular plant, the convention is to use a genus name, followed by its species epithet (name). Reproductive features of individual species, such as the structure of flowers, fruits and seed shoots, are major characteristics used in determining the taxonomic place of each one. The scientific taxonomic system as it is practised today incorporates the notion of evolutionary relationships between species.

For biologists to properly assign a newly discovered organism a scientific name, a full description of its physical attributes must be published in a recognised science journal or book.[148] A specimen of the new species, from which the description was based, must also be permanently lodged in a collecting institution, such as a herbarium or museum, where it remains readily accessible to other scholars. While the scientific names are generally derived from Ancient Greek or Latin words with some relevance to a description of the organism, they may also honour a person or place. The need for taxonomy became apparent when natural scientists began ordering and classifying the natural world. During the expansion of the European empires it became an imperative to organise the records to logically place the vast numbers of unfamiliar organisms being discovered.[149] Taxonomic research continues today, aided by the technology of extracting and analysing DNA to help describe new species.

Taxonomy has had a positive impact upon the European descriptions of Aboriginal relationships with the flora. In situations when skilled botanists were involved in the recording of Aboriginal plant uses and language terms, the identity of the species involved is more certain. Botanist Robert Brown was able to record a plant's common name and associated Aboriginal terms, and then scientifically classify it. In 1803 Brown went on a collecting trip south of Sydney, where he recorded a number of Aboriginal plant words; for example, the white mangrove was listed as *ranganee* and the woody pear was *mogoko*.[150] In view of the unreliability of English colloquial names as applied in Australia, Brown's botanical specimens and original identifications are important today for maintaining the identity of such plants, both in terms of science and Aboriginal linguistics. Because all languages change through time, it is only the scientific binomial system that promises to preserve the identity of plants through the ages.

The incorporation, or borrowing, of indigenous plant terms into Australian English has not been perfect, with changes occurring in pronunciation and, due to the expansion of the use of loan words, often covering additional species sometimes growing in other regions. The use of indigenous terms in Australian English has the advantage, however, that it avoids the further overuse or stretching of European words. The tendency of British colonists in the early years of settlement to recycle English terms for naming unrelated plants has caused much confusion to present-day scholars reading historical sources. Although scientific names are often employed as common names, for instance 'eucalypt' from *Eucalyptus*, some vagueness remains as they are generally taken from the generic, rather than specific, level. The additions from indigenous languages have enriched common speech and imparted distinctiveness to Australian national identity. Even though European settlement displaced Aboriginal use of the land over much of the continent, the past existence of Aboriginal cultural landscapes is still apparent in the use of certain indigenous terms for plant names in Australian English.

4 George Caley in New South Wales

One of the outcomes of the exploration of the world that began in the seventeenth century was a lucrative plant trade. In northern Europe, gardens were becoming more aesthetically orientated, rather than being just a source of culinary and healing herbs, and different types of plants from exotic places with climates similar to those in Europe were sought to enrich the estates of wealthy landowners. For the Dutch during the 1630s this desire was expressed in the famous craze for tulips, bulbous plants which originated in the cool mountainous areas of the Middle East.[1] A little later, with the technological advances of the Industrial Revolution, it became possible to keep even tropical plants in colder European countries. Pineapples and palms were grown in glazed 'stove houses', heated by stoves burning wood or coal to allow the plants to survive the harsh winters.[2] By the early nineteenth century, the heating of these glasshouses was achieved chiefly by piped steam[3], which from the 1830s allowed wealthy European plant enthusiasts to indulge a passion for orchids.[4] Large-scale transportation of plants also took place outside Europe as colonies established in warmer climates acquired from other parts of the world more suitable crops, like sugar cane and tea, to support their local economies.

Botanists were keen to acquire specimens from the distant colonies in their quest to taxonomically describe the whole plant kingdom. From the early eighteenth century, developments in this scholarly discipline were spurred on by the almost non-stop discoveries in remote regions of hundreds of exotic plants hitherto undescribed. The finding of so many new species, chiefly due to exploration, forced botanists to devise more complex systems to catalogue them. While Renaissance scholars recognised some five or six hundred different plants, by 1700 the French botanist Joseph Tournefort had ascertained that there were over 10 000 known types.[5] In 1735 Carl Linnaeus (the 'Father of Taxonomy') introduced the two-part (binomial) system for classifying every member of the Plant and Animal Kingdoms.[6] This is essentially the same set of rules that scientists use today to identify species. At the same time, the study of botany was gaining its independence from medicine, although maintaining strong links.[7] Gardens became places to study living examples of plants. Historian Richard Aitken notes: 'as botany moved from its medicinal origins and increasingly embraced Renaissance scientific thought, so botanic gardens outgrew their early role as *horti medici*'.[8]

Plant collecting and the colonial economy

British settlers arrived in Australia at a time when the Industrial Revolution in Europe was well advanced and the Northern Hemisphere powers were competing with each other to discover, and then to plunder, the entirety of natural and human resources in the world. There was an almost insatiable desire for land and empire: to drive industries, relocate populations, generate wealth, and extend political and religious influences across the globe. Amateur and professional plant hunters actively searched the newly discovered lands for species that would be useful to develop as cash crops intended for use as foods, spices, medicines, fibres, dyes, oils, tanning agents, perfumes, building materials, forage plants and garden ornamentals.[9] The growth of interest in botanical science also led many naturalists to seek out new plant species to study. It was, as put by historian Tyler Whittle, 'a scramble for green treasure'.[10]

In the first decades of British settlement in New South Wales, many settlers became collectors for the European nursery trade, a hotbed of rivalry which saw numerous businesses actively seeking new varieties to expand their sale catalogues.[11] While potted plants had a low success rate during long sea voyages, carefully packaged seeds generally remained viable. William Curtis in London published articles on new plants for the horticultural trade in *The Botanical Magazine*.[12] A major player in the early movement of indigenous plants from New South Wales to Europe was James Lee, the co-owner of the Vineyard Nursery in Hammersmith, London.[13] Maybe to better position himself for the expected trade from Australia, Lee wrote a pamphlet called *Rules for Collecting and Preserving Seeds from Botany Bay*, which he published around 1787.[14] He was particularly interested in obtaining new plant species from temperate parts of the globe to propagate and sell. Through his business, Lee became a driving force in the popularising of Australian flora during the late eighteenth and early nineteenth centuries. Among his list of wealthy and influential customers was the Empress Josephine, who cultivated many exotic plants at Malmaison near Paris in France.[15]

In addition to the plant nursery trade, settlers in New South Wales became involved in the large-scale movement of natural history specimens to scholarly European collectors, who paid top price for minerals, seeds, pressed plants and preserved animals from Australia.[16] Aboriginal artefacts, many of them stolen from their makers, were also considered items worthy of sending back to Europe.[17] Visitors passing through New South Wales made collections too. For example, Dr Joseph Arnold, a surgeon aboard the *Hindostan*, which made landfall at Port Jackson, Sydney in 1810, claimed it was his intention to 'bring many curiosities home with me, could they be had'.[18] In bartering for objects he paid the settlers and convicts in rum, rather than cash. In the case of the local Aboriginal people, he stated: 'We only can get things cheaply of the Indians [Aboriginal people], who bring coral, shells, &c., and who are glad to take old clothes, biscuits or wine for them'.[19] Arnold must have formed his private business plans before arriving in New South Wales, because when his ship called in at the Cape of Good Hope in South Africa he had purchased a hogshead (about 236 litres) of rum.[20]

The organised scientific plant-collecting enterprises of the late eighteenth and early nineteenth centuries were under the direction of influential botanists, notably Sir Joseph Banks and, later, Sir William J. Hooker. Both men were associated with the Royal Gardens at Kew, and members of the highly prestigious Royal Society. The first plant hunter Banks sent out to the far reaches of the British Empire was Francis Masson, a trained gardener, who arrived at the Cape of Good Hope in 1772.[21] Among the 1700 new species he discovered in South Africa, a number were destined to become familiar garden plants, including the

Sir Joseph Banks, the 'Father of Australia'. From the Royal Gardens at Kew, Banks directed the collecting of plant specimens, both living and dried, around the world. Thomas Woolnoth, engraving, early twentieth century. Rex Nan Kivell Collection, nla.pic-an9283225, National Library of Australia, Canberra.

drooping agapanthus, pelargonium, belladonna lily, bird of paradise flower, king protea and red-hot poker. By the late eighteenth century, the focus of many European-based botanists had shifted to Australia. Their interests were not just academic, as the search for useful plants was considered economically important for the newly established colonies. As Banks well knew, it was an imperative for the Colony of New South Wales to establish valuable exports to help reduce its burden upon the British treasury[22], for it was intended that this colonial possession would ultimately become an asset to the Empire.

Most of the early plant hunters sent out to the far corners of the world were not scholars, but gardeners.[23] This was because in addition to making collections of dried herbarium specimens, these men were under instruction to send living material back to their employers. The second half of the eighteenth century was an era which saw the establishment of numerous botanical gardens across Europe which were intended to match the success and scientific importance of Kew.[24] There was intense competition for new plants to showcase in these new gardens.

Many of the plant hunters became explorers in their own right, with successful expeditions making both botanical and geographical discoveries. Historian Colin Finney remarked: 'In future years of Australian exploration this was to become a common theme – naturalists yearning for the greater fame accorded to explorers – while the contrary was also evident, explorers occupying much of their time in the collection of natural curiosities'.[25] The plant hunters were part of the first wave of Europeans at the frontier to encounter indigenous cultures before they had been subjected to the full and detrimental impact of colonisation.

Banks considered that plant hunters had to be of a certain temperament and personality in order to be successful in their endeavours. The character traits he demanded they espouse included diligence and humility, with bachelors perhaps making the best plant collectors of all.[26] Historian Charles Lyte summed up the plant hunters as having 'a singleness of purpose that seemed to inure them from considerations of danger and discomfort'.[27] Similarly, Richard Aitken claimed that they required the characteristics widely recognised in Scottish gardeners, in particular determination, reliability and a keen pair of eyes.[28] Mostly men, they surrendered life's comforts in order to lead an exciting life, albeit one which experienced danger and climate extremes, and usually provided only a poor livelihood.[29]

Wa-ra-ta (waratah). This spectacular flowering plant amazed the British colonists arriving in 1788. It is thought that the name is a borrowing from *warrada*, in the Dharug language west of Sydney. George Raper, watercolour, Sydney region, New South Wales, about 1789. The Ducie Collection of First Fleet Art, nla.pic-vn3579250, National Library of Australia, Canberra.

Often in small and ill-equipped parties, the plant hunters traversed lands still occupied and controlled by indigenous landowners. Although local people often posed a physical threat, the plant hunters were naturally attracted to the indigenous land users for only they possessed detailed knowledge of the properties of plants that grew in their country. Plant hunters garnered much information on the potential uses of plants from non-European peoples, many of them hunter-gatherers. The results of their investigations

Heath-leaved banksia. The *Banksia* genus was named in honour of its discoverer, Joseph Banks, during the voyage round the world with Captain Cook in 1770. Aboriginal foragers sucked nectar from the flower spikes and used the smouldering dried seed cones to carry fire (Stewart & Percival, 1997, p. 13). S. Edwards, watercolour, early nineteenth century. *Curtis's Botanical Magazine*, London. Vol. XIX, 1804.

contributed to the base data for the floras eventually to be published by botanists.[30]

Indigenous peoples were on occasion highly suspicious of the Europeans, particularly when the strangers made inquiries about the plant resources within their territory.[31] For instance, Swedish naturalist Anders Sparrman lived in the Cape Colony between 1772 and 1776, where he recorded the traditional uses of the *goree-bosch* (aloes) – the source of gum aloe applied to wounds.[32] He alleged that it had taken a long time to find this out because the indigenous Africans had contrived to keep the uses of the plant secret from their colonial masters. Possible motivations for keeping plant uses hidden would have included discouraging further colonisation, reducing competition for limited resources, and even the fear that some plants might be used against them through sorcery.

The working life of plant hunter Richard Spruce, who collected plants for Hooker and George Bentham at Kew, provides us with a good example of the perils of plant collecting. He arrived at the mouth of the Amazon in 1849, and while travelling up the river suffered considerable hardship, which included almost starving to death, nearly drowning, and having to overcome a murder plot by disgruntled indigenous porters.[33] Spruce is credited with collecting the seeds of the cinchona tree, the bark of which became the source of quinine to combat malaria. For the material he sent to Bentham, Spruce was paid at the princely rate of £2 per 100 specimens.[34] At the same time he supplied Hooker with a large ethnographic collection of indigenous South American artefacts. After years of living in South America, Spruce returned to England in 1864 as an invalid.

Given the length of the voyage from Australia to England, potted plants suffered considerable stress from being moved through different climate zones, and from the deleterious effects of salt spray. Shipment of actively growing specimens was therefore a risky business, with many failed ventures. Banks had special plant cabins built on the quarterdeck of several ships to help keep alive potted plants en route to Britain.[35] He was experienced at plant transportation, in 1787 having overseen substantial modifications to the deck of the *Bounty* to transport living breadfruit plants from the Pacific region.[36] At their planned destination in the West Indies they were to be grown as a cheap crop to feed plantation slaves. The first attempt failed due to the mutiny of Captain Bligh's crew, who rebelled against the water rationing imposed as a result of having to constantly water the breadfruit plants. The horticultural knowledge of how to keep plants alive on ships improved dramatically during the early nineteenth century, and from 1836 small transportable glasshouses, known as 'Wardian cases', revolutionised the survival rate for commercially important plants on long voyages.[37] The Wardian case remained in use until the twentieth century, replaced by modern packaging, particularly plastics.

George Caley: plant hunter and explorer

Early plant hunters who arrived at the infant Colony of New South Wales were faced with organising arduous collecting expeditions with limited resources. They were entering regions that in many cases had not even been mapped, let alone settled. In spite of the difficulties, however, these men had excellent opportunities to observe indigenous uses for plants. Leading the way at the beginning of the nineteenth century was George Caley.

Caley was born in 1770 at Craven in Yorkshire, the son of a horse dealer.[38] The young Caley gained employment and training first in the gardens of his home town, Manchester, and then in the Chelsea Physic Garden, before moving to Kew. King George III owned the Royal Gardens at that stage, and Banks was the unofficial director. Banks eventually sent Caley to Australia, where he arrived in April 1800 to begin his career as a plant hunter.[39] Here, Banks hoped that Caley would find plants and animals 'of national importance, & lay the foundation of a trade beneficial to the mother country with that hitherto unproductive colony'.[40]

The colonial authorities provided Caley with a house at Parramatta, along with rations. This location would have suited the plant hunter, as it was nearer than Sydney to the inland forests of the mountains, where he would be focusing his collecting. The plant specimens he sent back to Banks were eventually placed in charge of botanist Robert Brown.[41] The identity of Caley as the collector was retained through the distinctiveness of his small handwriting on minute labels.[42] As paper was scarce in the early years of the colony, it was imperative he conserve his supplies. Brown was also in Australia, some of the time collecting with Caley, from 1800 to 1804.[43]

Although Banks has shown through his correspondence that he admired Caley's work ethic and skill in plant collecting, he found his employee to be impetuous and quick-tempered.[44] Lyte summed up Caley's temperament: 'That he was hardworking and enthusiastic was not in doubt, but he had an evil temper and constantly believed that he was being belittled and done down'.[45] Lyte also suggested that Banks maintained his patience with Caley only because of the steady supply of seed packets. It was probably fortunate for Caley that he was not from the upper classes, as Banks remarked in a letter written to placate Governor King that 'had he [Caley] been born a Gentleman he would have been shot long ago in a Duel'.[46]

Gymea lily or giant lily. One of the plant species that George Caley collected in the Sydney Basin of New South Wales, it was part of the 'green treasure' sent to gardens overseas. Aboriginal people roasted the roots and young flower spikes for food (Stewart & Percival, 1997, p. 20). Philip A. Clarke, Adelaide Botanic Gardens, 2006.

Burrawang. British colonists also referred to this cycad species as the 'wild pineapple'. The highly toxic seeds were treated by Aboriginal foragers in a process involving mashing, washing and baking to render them edible. Philip A. Clarke, Kurrajong Heights, Blue Mountains, New South Wales, 1988.

Prudently, Caley established the practice of plant hunters employing Aboriginal guides during their expeditions, stating that 'with giving … bread to the different natives I find I can gain their affections, and get information from them.'[47] In August 1801, Caley wrote to Banks in England, saying, 'I mean to keep a bush native constant soon, as they can trace anything so well in the woods, and can climb trees with such ease, whereby they will be very useful to me.'[48] Caley was specifically looking for an inland Aboriginal person to assist him, for the coastal clans based around Sydney referred to groups living in the mountains as 'climbers of trees.'[49] The ability to climb was essential when obtaining botanical specimens from tall eucalypts, because relying on ground falls of leaves, flowers and seed capsules in a forest of mixed tree species led to inaccuracies. Other plant collectors, caught on fieldtrips without the assistance of a climber, had to cut down the trees to obtain the necessary range of plant material.[50]

Method of Climbing Trees. Aboriginal bands from inland districts around Port Jackson were known as 'waddy men', due to their reliance on this tool in hunting forest animals for food and clothing. European plant hunters employed Aboriginal climbers to obtain specimens of fruits and flowers from the canopy. Thomas Watling, watercolour, Sydney region, New South Wales, about 1792–94. V12075/R, British Museum of Natural History, London.

Caley appreciated the diversity of Aboriginal Australia, recognising that there were significant cultural differences among the clans living in the region surrounding Sydney. In 1804, during an excursion southwest of Sydney in the Stonequarry Creek district, Caley:

> fell in with some natives who had never seen white men before … They seemed to be quite of a different race to those that I am acquainted with, not only by their features but in size … The other natives [from the coast] told me that they eat human flesh, but whether they are cannibals or not I shall not take upon me to say …[51]

On this occasion, the plant hunter met Cannabayagal, reputedly the 'mountain chief' in charge of the Aboriginal band. By Caley's account, the leader must have possessed considerable authority in the region around Sydney, as it was said by coastal Aboriginal people that he was 'invincible and more than mortal.' Each member of the Stonequarry Creek group had 'a frightful countenance', but with Cannabayagal Caley 'found something pleasant in his while I was conversing with him.'

The records of David Collins, made earlier than Caley's, had also asserted that there were major cultural and social differences between the coastal and forest Aboriginal people around Sydney. Collins remarked:

> The natives of the coast, whenever speaking of those of the interior, constantly expressed themselves with contempt and marks of disapprobation. Their language was unknown to each other, and there was not any doubt of their living in a state of mutual distrust and enmity. Those natives, indeed, who frequented the town of Sydney, spoke to and of those who were not so fortunate, in a very superior tone, valuing themselves upon their friendship with the white people, and erecting in themselves an exclusive right to the enjoyment of all the benefits which were to result from that friendship.[52]

Conflict between Aboriginal groups coming into contact in towns such as Sydney would have often have been based upon such major distinctions within their broader community.

Collins shared Caley's awe of the ability of inland Aboriginal people in scaling trees, especially during the hunt:

> The very great labour necessary for taking these [woodland] animals, and the scantiness of the supply, keep the wood natives in as poor a condition as their brethren on the coast. It has been remarked, that the natives who have been met with in the woods had longer arms and legs than those who lived about us. This might proceed from their being compelled to climb trees after honey and the small animals which resort to them, such as the flying squirrel [sugar glider] and opossum, which they effect by cutting with their stone hatchets notches in the bark of the tree of a sufficient depth and size to receive the ball of the great toe. The first notch being cut, the toe is placed in it; and while the left arm embraces the tree, a second is cut at a convenient distance to receive the other foot. By this method they ascend very quick, always cutting with the right hand and clinging with the left, resting the whole weight of the body on the ball of either foot.[53]

Furthermore, Collins claimed that he had seen a 'white gum' about 40 metres tall, which had at least 24 metres of notched trunk before reaching the lowest branch.

In formulating plans to gain an indigenous assistant, Caley took into account the cultural differences he had observed. In a letter to Banks, he described the virtues of the New South Wales Aboriginal people, in particular the inland groups who occupied the forests. Here, Caley said:

> It is the inland or Bush natives that I have been representing, for the water natives have different customs, and are more confined to one place of abode. They know nothing of climbing trees, no farther than in getting up for to strip off bark, for to make canoes, which the inland natives do dexterously. Bennylongs [Bennelong], who was in England, was a water native.[54]

This information on Bennelong was consistent with an account from Collins, who recorded that the Aboriginal man had claimed 'that the island Me-mel (called by us Goat Island) close by Sydney Cove was his own property'.[55] Caley understood that the hunting and gathering skills of Aboriginal foragers were attuned to the types of land they occupied, as was their knowledge of plants. Indigenous people also differentiated between people according to the type of country they occupied, across Aboriginal Australia a major cultural distinction being made between 'saltwater' and 'freshwater' peoples.[56]

When Caley had arrived in New South Wales, much of the forest in the vicinity of Sydney had already been cleared and the Aboriginal-controlled fire regime had been suppressed for at least several years.[57] Caley had the opportunity to go with Flinders around Australia on the *Investigator*, but declined, possibly because his first priority was collecting the flora inland from Sydney.[58] For a decade he roamed the surrounding country and penetrated far into the Blue Mountains in order to maintain a steady supply of different plants to send back to England.[59] In these excursions, it was his early practice to travel with a single packhorse and a convict servant to help carry his gear.[60] As the son of a horse dealer, Caley would naturally have favoured using horses over other pack animals, as he would have possessed the knowledge to look after them. Among the plant specimens he sent back to Banks were packets of waratah seed pods, which were intended for the nursery trade.[61]

Apart from the Blue Mountains surrounding Sydney, Caley's excursions also took him to Western Port (Victoria), Jervis Bay, the Hunter River district, Norfolk Island and Van Diemen's Land.[62] Under instruction from Banks, his collecting was not confined to plants but extended also to zoological specimens. Caley was particularly interested in ornithology, writing to Banks in 1803 that 'I flatter myself that I shall in future gain a better knowledge of this branch of Natural History'.[63] In 1818 the Linnean Society of London acquired 'an extensive and valuable collection of Quadrupeds, Birds and Reptiles, made by Mr. George Caley in New South Wales', for the sum of £200.[64]

In the early nineteenth century, British settlers in Australia possessed mainly obsolescent firearms which gave them only a slight advantage over Aboriginal spears and clubs. In a letter to Banks written in October 1800 Caley requested that he be issued with a gun and a double-barrelled pistol:

> as when I travel into the interior of the country I shall be obliged to carry arms, as well as the person that travels

Waterfall in Australia, by Augustus Earle. The artist and his party visited Wentworth Falls in the Blue Mountains in 1826. Aboriginal people provided settlers with guidance and protection in the Australian bush. Augustus Earle, oil on canvas, Blue Mountains, New South Wales, about 1830. Rex Nan Kivell Collection, nla.pic-an2273848, National Library of Australia, Canberra.

> with me. The arms in this colony are chiefly old muskets, which are very heavy, and swallow too much powder and shot for to load them. A double-barrelled pistol is very useful in the woods when covered with a gun, as many of the natives are well aware that a gun will only go off once; nay, many of them do not mind a single gun now – and I know I shall often be among numbers of the natives both day and night.[65]

Caley was not being overly cautious about the risk of indigenous aggression. In December 1804 Brown wrote to Banks, saying that when he was plant collecting along the Hunter River, 'the unfriendly disposition of the Natives who even attack'd my boat, render'd it unsafe for me to go far from the banks, or to trace any of the branches above where they are navigable'.[66] Better guns, with faster loading mechanisms, became more widely available in Australia later in the nineteenth century.

In what was a deliberate strategy, Caley cultivated relationships with the Aboriginal people he came across by giving them small steel axes and food in the hope of encouraging them to impart some of their environmental knowledge and perhaps to guide him.[67] Historian Rae Else-Mitchell claimed:

> Caley considered it a good policy to keep on friendly terms with the natives, and he soon gained a reputation for fair dealing with them, thus being admitted to many of their secrets and obtaining topographical information which white strangers would not ordinarily be given.[68]

A considerable amount of biological knowledge could be gained from Aboriginal hunters and gatherers, who helped scholars solve a number of zoological puzzles. Caley informed Banks that the 'duck bill animal' (platypus) and 'porcupine ant eater' (echidna) were said by Aboriginal sources to lay eggs.[69] At the time, the existence of egg-laying mammals, eventually classified by scientists as monotremes, was considered so extraordinary that it bordered on fanciful.

Banks, however, recognised the importance of monotremes to the study of zoology. He requested that Caley investigate the matter further, hopefully procuring for him live specimens for sending to Britain.[70] The plant collector eventually bought an echidna from a settler, but it cost him five gallons of rum, a high price back then.[71] At the same time, Banks was corresponding with Brown, who was still in New South Wales after the circumnavigation of Australia with Flinders, and requested that he perform dissections upon the echidna and platypus. Throughout the nineteenth century Australian zoologists continued to study the anatomical mysteries of monotremes, with George Bennett in the 1830s and William Caldwell in the 1880s both employing Aboriginal collectors to gather large numbers of platypus for their studies.[72]

The early collectors of natural history specimens recognised that the involvement of Aboriginal people in the field was essential. After nine years of collecting in New South Wales, Caley recommended in a letter to Banks that all plant hunters recruit Aboriginal guides for their trips:

> There would be no danger of the collectors getting lost or bewildered, which is too oftentimes the case, for the native would trace back their footsteps. He would likewise foresee whether any danger was to be expected from the natives of that part. He would point out the tracks of different animals, which would be unnoticed by a European. Many new birds would be readily obtained, as the natives in general are excellent marksmen and quicker-sighted than our people. The specimens of the different trees might be procured which would otherwise have been missed or left for want of a climber.[73]

Caley was in effect advocating that others form close partnerships with Aboriginal hunter-gatherers, just as he did. Plant collectors arriving after Caley's time also used Aboriginal labour. Prominent land owner Sir John Jamison, who arrived at Sydney in 1814, employed Thomas Jones to collect natural history specimens for him.[74] Jones, like many plant hunters, was in the practice of taking Aboriginal people with him to assist during field trips.

Caley observed the demise of the Aboriginal community in rural areas around Sydney, and his concern for their welfare set him against other colonists. He considered that Europeans were often to blame over conflict with Aboriginal groups. At Parramatta, Caley had several disputes with his neighbour, the powerful Reverend Samuel Marsden, who had gained notoriety as a hard man and was known as the 'flogging parson'.[75] On one occasion, Caley supported his own convict servant in refusing Marsden's request to participate in a raid upon local Aboriginal people.[76] Caley's experiences with indigenous peoples and his acknowledgment of their highly developed bush skills had given him a respect for Aboriginal culture that was missing among other colonists.

Moowattin the tree climber

One of George Caley's main indigenous guides and assistants was Moowattin, whose name was appropriately said to mean 'bush path'.[77] A clue to its proper pronunciation is given by the way it was sometimes written: 'Moowat'tin'. Europeans also called him Mowatty, Daniel or Dan. He was born around 1791 in the area inland from Port Jackson, and as a child was adopted by the family of Richard Partridge, alias 'Rice', who was the colony's hangman. Moowattin was probably about fourteen or fifteen when he took up work with Caley in 1805, and lived in the plant hunter's cottage, next to Government House at Parramatta. Caley appears to have used the name 'Dan' when referring to him.[78]

Caley not only had great faith in the loyalty of his indigenous assistant, but he was also in awe of Moowattin's ability to interpret local Aboriginal dialects. Moowattin was probably a speaker of the Dharug language, which was spoken by people just west of Sydney.[79] In 1808 Caley wrote a letter to Banks in which he outlined the good character of his assistant. Here, prompted by a rumour that Moowattin had been speared and killed, Caley claimed:

> The native I have been speaking of is the most civilized of any that I know who may still be called a savage and the best interpreter of the more inland natives' language of any that I have met with. I can place confidence in him which I cannot in any other – all except him are afraid to go beyond the limits of the space which they inhabit with

> me (or indeed any other) and I know this one would stand by me until I fell, if attacked by strangers.[80]

When Caley was exploring the lands west of Sydney he named one of the rivers he discovered after Moowattin, 'to commemorate the memory of the native to whom I am indebted for the discovery of the Cataract [a large waterfall, Carrung Gurring]'.[81] Regrettably this placename did not become widely recognised and has not survived in common use. Although Matthew Flinders used 'Moowattin River' on his map in the *Atlas of Terra Australis*, it became officially known as Cataract River, and the cataract itself as Appin Falls.[82]

Moowattin accompanied Caley to Norfolk Island in 1805 and to England in 1810, although Banks strongly disapproved of this latter trip on humanitarian grounds.[83] Banks knew well the sad experiences of indigenous people who had been plucked from their homelands and taken to Great Britain. For instance, on Banks' Pacific Ocean voyage Captain Cook had brought the Tahitian named Omai to England. When Omai arrived in 1774 the English treated him as something of a spectacle.[84] The experiences gained in Britain had a deep impact on the Tahitian, to the extent that when he returned home two years later he had become a misfit among his own people. A similar situation had resulted when the Sydney Aboriginal man, Bennelong, was taken to England in 1792 and presented to King George III.[85] He returned to New South Wales three years later and tragically became an alcoholic.

Moowattin spent one year in London, where he was supported by Banks. As with Omai, the Aboriginal man also drew attention, being in the habit of going to the theatre and smoking his pipe at a Chelsea coffee house in the evening, all the time dressed in the 'pink of fashion'.[86] Moowattin may have felt constrained by the urban environment, as he was said to have claimed that London's finest shops and houses were 'not equal to the woods' in his own country. At the age of about twenty, Moowattin became an alcoholic. This must have contributed to the angry exchanges between Moowattin and Caley, who on one occasion broke his own thumb when punching his Aboriginal assistant.

While Caley remained in England, Moowattin was sent back to Australia in 1811 with George Suttor, a former employee of Banks and now a settler in New South Wales.[87] Moowattin worked at Suttor's Baulkham Hills farm for two weeks before suddenly leaving. He sold a 'fowling piece' (shotgun) given to him by Brown, bought some peach cider, and ran off into the bush. Moowattin was said to have 'returned to his countrymen and Native life'.[88] Not much is known about his doings over the next few years, although it was said that he had worked for a while as a farm labourer for James Bellamy at Pennant Hills. Moowattin possibly lived some of the time as a fringe dweller on the outskirts of the rapidly expanding colony and, without Caley's support, may have suffered considerable hardship.

In 1816 Moowattin was arrested for the rape of a convict settler's daughter.[89] During the subsequent trial at the Criminal Court in Sydney, Marsden gave evidence that he had known the defendant for over twenty years and that Moowattin knew right from wrong. The explorer Gregory Blaxland also testified to his Europeanised background. When Moowattin had returned from England, Marsden came across him 'sitting, naked, upon the stump of a tree in the woods ... North of Parramatta'.[90] When Marsden questioned as to why he had thrown away his clothes and gone back to the wilds, he had simply replied 'me like the bush best'. The court found Moowattin guilty, and he was executed on 1 November 1816 at the estimated age of 25, achieving the unfortunate distinction of being the first Aboriginal person legally hanged in Australia.

Recording indigenous plant names

As a plant hunter, Caley had the advantage of access to Moowattin's knowledge of local dialects, and recognised the value of noting indigenous names for the plants he collected. He also recorded the variations when inland and coastal peoples had different words for the same species. Importantly, he often recorded these indigenous words on the labels accompanying herbarium specimens. In 1900 Australian-based botanist Joseph H. Maiden was touring Europe and came across one of Caley's plant collections at the Vienna Herbarium in Austria[91], finding that the preserved specimens were documented with many indigenous plant names from languages in eastern New South Wales. Exactly how the collection came to Vienna remains a mystery although, since Banks maintained connections with a network of botanists across Europe, it can be presumed he sent it there in the early nineteenth century as a duplicate set of specimens. Collections were often freely traded between herbaria, sometimes without the collector's names being retained.

Of the collection in Vienna, Maiden observed: 'Caley gives the aboriginal names of the trees in most cases; we now possess a number of records of the language of extinct tribes which may be of philological as well as botanical interest.'[92] Although a few of Caley's Aboriginal plant names were said to be similar to those listed by colonist William Macarthur in the published catalogues of the New South Wales displays at the Great International Exhibition of Paris in 1855 and that of London in 1862, they nevertheless formed a much more complete record.[93] A comparison between the two sets of word spellings gives a clue to the original pronunciation.[94] For example, for the inland grey box Caley wrote 'barilgora', while Macarthur recorded it as 'barroul gourrah'. Similarly, for the grey iron gum Caley had written the indigenous name as 'mundowey', while Macarthur had it as 'maandowie'.

When it came to the differences between types of plants, Caley possessed keen powers of observation which served him well. Maiden acknowledged Caley as the discoverer of hybridisation amongst eucalypts. One of Caley's specimens of ironbark trees he found in the herbarium at Vienna was labelled 'Burryagro. A hybrid between Barilgora and Derrobarry.'[95] Maiden noted: 'The blacks had but one name [Burryagro] for this, the Ironbark Box [inland grey box], and the Ironbark [grey ironbark] ..., but Caley saw that they were different. Caley's surmise as hybridisation in this case is marvellously shrewd.'[96] The plant hunter often included Aboriginal observations in his records, as in this example: 'Yar'ro and Cum'bora (or Cam'bora), A bastard blue gum ... The natives distinguish them by cutting into the bark, the one being thicker and the colour in one is red and the other white or nearly so. Cum'bora is the small fruited form.'[97] It is likely that many of the words and remarks Caley recorded on herbarium labels came from Moowattin. For example, he recorded 'Mogar'gro. *E. beyeri.* Iron bark [narrow-leaved grey ironbark]. White one? Got by Dan.'[98]

Through following the working life of George Caley we gain a greater understanding of European interests in the Australian flora during the early nineteenth century. In spite of his humble origins, Caley had a successful career as a plant hunter and later as a botanic gardens administrator. Lyte described him as a 'Yorkshire stable-lad turned botanist.'[99] Following his return from Australia in 1810, Caley lived in England until 1816, presumably on the small pension provided by Banks.[100] He would have been engaged mainly in ordering his extensive natural history collections, housed at Kew. In 1816 Caley was appointed superintendent of the botanic gardens at St Vincent in the West Indies. He returned to England in 1823, suffering from a tropical health complaint, and lived in Bayswater, where he died in 1829. Caley's wife predeceased him and he left no children, which helps explain why he was said to have bequeathed an annuity to a pet Australian cockatoo.[101]

Although not an academically trained botanist, it was through the scholarly discipline of botany that Caley became famous. Allan Cunningham, a plant hunter from a later generation, recognised his importance by naming a mountain after him on a journey from Sydney to Bathurst. On 2 June 1817 Cunningham made a note in his journal: 'Mt. Caley. This mount has been named in honour of Mr. George Caley, a most accurate, intelligent and diligent botanist who laboured on the eastern coast of this continent with considerable success a number of years, and who well merits such a mark of distinction.'[102] In the late nineteenth century Maiden honoured Caley's early contribution to Australian botany by naming the New England ironbark *Eucalyptus caleyi.*[103] His name is also celebrated in the duck orchid genus *Caleana.*[104]

5 Allan Cunningham and the Mapping of Australia

Allan Cunningham (1791–1839), another plant hunter sent by Banks to collect in Australia, was born at Wimbledon in Surrey, England, the son of a Scottish gardener also named Allan Cunningham.[1] Although his father wanted him to study law, the younger Allan preferred horticulture and like Caley trained as an apprentice gardener at Kew. In 1814 Banks, the President of the Royal Society and still unofficially running the Royal Gardens, used his influence to send Cunningham to Brazil, with co-worker James Bowie, as plant hunters.[2] Cunningham and Bowie collected plants to be dried as herbarium specimens, and gathered living plants and seeds that were sent to botanical gardens across Europe. Their collections were particularly rich in bromeliads, orchids and bulbs from San Paulo and the Organ Mountains, north of Rio de Janeiro. Afterwards, Bowie went to the Cape of Good Hope to collect, and Cunningham to New South Wales, where he arrived in December 1816.[3] This was over six years after Caley's return to England following his stint in Australia.

Allan Cunningham, King's Botanist for the Colony of New South Wales, accompanied Lieutenant Phillip Parker King on several expeditions to map the Australian coastline (1818–22). Cunningham utilised Aboriginal labour and expertise in collecting plant specimens. Unknown artist, oil on wood, about 1835. nla.pic-an2287723, National Library of Australia, Canberra.

West of the Blue Mountains

By the time Cunningham reached Australia, the British colonists had a firm foothold on the east coast and around Sydney and were no longer hemmed in by the formidable barrier of the Blue Mountains. Explorers Gregory Blaxland, William C. Wentworth and William Lawson had succeeded in crossing the ranges in 1813.[4] Plant hunter Cunningham was given the title of 'King's Botanist', and in this capacity travelled widely throughout the Australian mainland and Tasmania. Away from the towns and settled areas, Cunningham usually travelled with an Aboriginal guide, not always the same person. Cunningham was part of Surveyor-General John Oxley's expedition of 1817 into the interior of New South Wales west from Bathurst.[5] From the outset, the prospect of discovering new plants was considered important enough for this expedition to include another plant collector, the former Scots soldier Charles Fraser, then referred to as 'Colonial Botanist'.[6] Cunningham, with his strong links to Banks, appears to have enjoyed higher status than Fraser but from available accounts of the expedition there does not seem to have been any conflict between the two men. In his published journal, Oxley claimed: 'The botanical productions of the country have however in a great measure been ascertained by Mr. Allan Cunningham, the King's botanist, who accompanied the expedition'.[7]

Local Aboriginal groups mainly kept out of the way of the explorers. An exception occurred on 25 April 1817, when the party reached the Lachlan River and came across a group of thirteen people. Cunningham commented that they were much different in appearance from the coastal groups around Sydney: 'By way of ornament they wear Kangaroo Teeth in their ears, and cockatoo feathers in their hair. Those of them, who were young men, had their beards divided in three divisions, and formed into plaited tails. Their language was different from our Eastern Coast natives'.[8] Four days later, still travelling along the Lachlan, Cunningham collected a new species of *Acacia*. He noted, 'From the circumstance of this tree being the wood of which the natives in the Western Country [west of Blue Mountains] make their spears ... I have called it *A. doratoxylon*'.[9] The name was well chosen, in classical Greek *doryatos* referring to 'spear' and *xylon* to wood.[10] Today, this plant is commonly called spearwood, lancewood or currawang, the last name thought to be derived from an Aboriginal language.[11]

Early explorers penetrating into the remote regions of Australia did not often make face-to-face contact with the indigenous inhabitants, although they frequently found evidence of their presence in the form of footprints, fires, diggings, marks on trees and recently abandoned camps. Quite often the campsites had been deserted just before the explorers' arrival, with food and artefacts left behind and fires still burning. The explorers generally considered signs of Aboriginal foragers being around a good omen. On 5 August 1817, once again in the Lachlan River area, Cunningham wrote: 'The recent marks of natives digging for grubs, and remains of fires, [had] led us to conclude that water could not be far distant'.[12] A landscape full of Aboriginal activity suggested to Europeans that the country had potential for at least some agricultural use.

Cunningham's records provide evidence to suggest that Aboriginal people were responsible for the creation of particular forms of vegetation, through what archaeologist Rhys Jones more recently described as 'fire-stick farming'.[13] During this expedition, on 5 June 1817 Cunningham noted at Peel Range that 'the country at the verge of the horizon southerly is in flames, being fired by the natives'.[14] On 4 August 1817, once more back in the Lachlan River area, he noted:

> Some patches of land that had been formerly fired by the natives producing some good tufts of grass induced us to turn out of our course in the scrub and halt upon it. This [surrounding] scrub continues for some miles with all the sterility imaginable, hence we are extremely fortunate in having an opportunity of turning out of it to a spot where our horses would find good grass, and where we found some water in two native wells ...[15]

Although local Aboriginal groups were often invisible to the European explorers and the first settlers, the various signs of their hunting and gathering activities amounted to indirect evidence of seasonal use of certain areas. Firsthand observations of Aboriginal burning frequencies confirm a pattern of high firing rates through most seasons, to the extent that bushfires could often be used to monitor Aboriginal movements in a district.

Governor Lachlan Macquarie had instructed Oxley to gain as much information as possible about the indigenous inhabitants of inland New South Wales during his expedition. On 29–30 July 1817, Cunningham was involved with the exhumation of the grave of an Aboriginal 'king' along the Lachlan River:

> Near our encampment a native grave of modern construction, from the regular manner and systematical mode in which everything connected with it is disposed, led us to conclude that this mausoleum contained the remains of some person of eminence, either a chief or one who had acquired from his skill in hunting, the respect and awe of his countrymen. It is a mound of earth about 3 feet [92 cm] above the level of the ground ... About

> 6 feet [183 cm] to the west of this mausoleum stood a cypress [common cypress pine] on which was cut out with very considerable labour remarkable characters, the stem having been previously barked and about 30 feet [9.2 m] north west was another having some singular figures deeply cut on its stem – perhaps a description of the man, his age, and cause of death.[16]

While Cunningham was interested in the nearby trees, Oxley's attention was focused on the 'three rows of seats', which probably functioned as decorated earthworks for the benefit of ceremonial dancers.[17]

Oxley was not entirely comfortable about the desecration of the gravesite, but believed that it was required by his official brief to investigate local Aboriginal customs, noting: 'I hope I shall not be considered as either wantonly disturbing the remains of the dead, or needlessly violating the religious rites of an harmless people, in having caused the tomb to be opened, that we might examine its interior construction.'[18] Oxley made a detailed examination of the grave and the body of the large middle-aged man it contained, removing the skull for later examination by a craniologist.[19] This expedition, from April to September 1817, was considered successful from a botanical point of view, with Cunningham having collected specimens representing about 450 plant species, many of them new to botany.

Cunningham was also a member of the second of Oxley's expeditions into the interior of New South Wales in 1818. From Bathurst they traced the Macquarie River northwest before running into a large region of wetlands (the Macquarie Marshes). Changing direction, they discovered more fertile areas to the northeast. As with the previous trip, an Aboriginal presence in the country they passed through was to them a good sign. On 17 September 1818, east of Tamworth, they came across a 'native chief' in a camp of two or three families.[20] The Aboriginal man was injured, probably through a fight or a fall, with broken ribs, a twisted back and apparently no use of his arms. The 'chief's' band had left him behind when they had fled at the sight of Europeans. By Oxley's account, the man was not frightened and 'seemed more astonished than alarmed at the sight of our cavalcade, and expressed his wonder in a singular succession of sounds, resembling snatches of song.'[21] He probably thought the Europeans were spirits of the deceased. In the Hastings River area on 23 September 1818, Oxley declared: 'Numerous smokes arising from natives' fires announced a country well inhabited, and gave the whole picture a cheerful aspect, which reflected itself on our minds; and we returned to the tents with lighter hearts and better prospects.'[22]

As a means of gaining local information, Cunningham paid attention to the activities of Aboriginal foragers he encountered during his expeditions. Because his botanising frequently took him well beyond the settlement frontier, he was able to observe Aboriginal hunting and gathering techniques before they were much altered by the influence of European technologies. On one occasion in 1818, while passing through the Illawarra area, Cunningham's Aboriginal guide was

A Native Chief. In 1818 John Oxley's expedition encountered this man in country east of Tamworth. The Aboriginal man was injured, and was left behind by his family when they fled at the sight of the Europeans. He sang to the expedition members. John W. Lewin, lithograph. Reproduced from John Oxley, *Journals of two expeditions into the interior of New South Wales. Undertaken by order of the British government in the years 1817–18*, 1820.

Grave of Native of Australia. During an expedition in 1817, John Oxley and Allan Cunningham performed an exhumation on an Aboriginal 'king's' grave along the Lachlan River. Cunningham speculated on the meaning of the carvings on the trunks of Australian pine trees nearby. G.H. Evans, lithograph. Reproduced from John Oxley, *Journals of two expeditions into the interior of New South Wales. Undertaken by order of the British government in the years 1817–18*, 1820.

occupied in gathering 'long pieces of the tough stringy bark of the Currajong ... for fishing line'.[23] Another time in 1818, Cunningham met a band of fourteen Aboriginal people from the Shoalhaven area, southwest of Sydney, who were camping near the Merrimarra River Farm, and noticed they had baskets that held water.[24] Upon closer inspection he found that the leaves used to make the containers came from a palm species that in the local language was called *bangla*. With his interest in species that would grow well in Europe, Cunningham was keen to discover as many temperate palms as he could. He used the promise of tobacco, a treasured trade item, to induce one of the Aboriginal people to show him where this plant grew so he could collect specimens. Botanists later gave this plant, commonly called the bangalow palm, the scientific name of *Archontophoenix cunninghamiana* in honour of its botanical discoverer.

The Aboriginal guides who accompanied Cunningham on his land-based trips were generally not named in his journal. From his accounts a number appear to have been involved, as chance permitted. He recorded on 24 October 1818 that he was taking 'an assistant and guide, the nephew of the chief of the Lake Allowree [Illawarra], whose services I purchased for the day, for a small piece of tobacco'.[25] Later during the same trip, on 28 October at Hat Hill to the west of Sydney, he 'induced an intelligent native to accompany me', although this person was reported to have soon feigned sickness in order to return to his wives and children.[26] A day later, Cunningham's journal refers to yet another Aboriginal person being with him.[27] It appears he had most success in obtaining guides from areas where the indigenous people had already some experience with Europeans and could speak at least some English, albeit as pidgin.

Bangalow palms (*above*) and frond bases (*right*). In southeast Queensland, Aboriginal women used the red-brown bases of the frond to make their baskets. Allan Cunningham discovered this plant, which is now widely grown throughout the world as an ornamental, with the help of Aboriginal guides. Philip A. Clarke, Adelaide Plains, South Australia, 2007.

Collecting plants around Australia

Cunningham joined several coastal expeditions led by Lieutenant Phillip Parker King, son of former New South Wales governor Philip Gidley King[28], being referred to as 'Collector to the Royal Gardens at Kew' in King's published accounts of these explorations.[29] Membership of King's exploratory parties had earlier included the Aboriginal guide Bungaree, and later Blundell. Sailing clockwise around the continent in the *Mermaid*, King aimed to map unknown stretches of coastline that had been missed by earlier navigators such as Matthew Flinders and Nicolas Baudin. It was thought that the rivers flowing west from the Blue Mountains might eventually empty into a vast inland sea, which in turn might have a passage somewhere in the northwest of Australia that connected it with the Indian Ocean. The prospect of arable land fed by navigable rivers in the centre of Australia had led to much speculation about the interior of the continent.

On these expeditions, Cunningham was again impressed by the Aboriginal burning practices he observed. On 3 May 1818, King's party was in Van Diemen Gulf in northern Australia (between Melville Island and Arnhem Land) and the plant hunter noted: 'The western horizon was much gloomed by extensive bodies of thick smoke of natives, who appear to be burning off the bush and grass of the country in that direction.'[30] Fires were also seen on 8 May 1818 when the expedition party travelled further east to the Kakadu coast. Here, Cunningham noted, 'The fires of the natives continue to be numerous in various directions; these conflagrations extend over immense tracts of flat country, at intervals bursting into large flames as the wind rises, and continuing until a heavy shower extinguishes them.'[31]

The expeditioners often collected wild plants to supplement their dwindling supplies and to provide fresh food. When they reached Melville Island, north of Darwin, the Tiwi people gave them two baskets, one filled with fresh water and the other 'sago palm'.[32] The 'sago' was probably the pulped nuts of a cycad palm which, in a procedure similar to that used in New South Wales and southwest Western Australia, must be prepared by being roasted, cracked and dehusked, pounded, then soaked in water for several days before cooking.[33] The Tiwi received steel chisels and files in exchange. At Goulburn Island, off the coast of Arnhem Land, Cunningham recorded on 28 March 1818: 'Tracing a beaten path made by the natives, I observed the roots of *Tacca pinnatifida* [Polynesian arrowroot], a plant abundant in low shaded situations had been taken up in quantities, which tempted me to conclude they are eaten by these Australians'.[34]

Cunningham's botanising trips also gave him opportunities to study Aboriginal culture. In January 1819, on an expedition to survey Macquarie Harbour in Van Diemen's Land, he was able to compile a small vocabulary of the local indigenous language.[35] Regrettably, the published version was in the form of a comparative table of indigenous languages, which included lists made by explorers from other parts of Australia, and therefore does not contain plant terms.[36] The word list was 'inserted to shew the great dissimilarity that exists in the languages of the several tribes ...'[37] Cunningham had realised that within Aboriginal Australia there was a large diversity of indigenous language.

Later in 1819, sailing along the coast of western Arnhem Land, King recorded that they brought on board for eating 'a great quantity of the bulbous roots of a *crinum* [possibly the white sand lily] which grows abundantly among the rocks on Sim's Island'.[38] Along the eastern coast of northern Queensland in 1820 they found that cunjevoi tubers and cabbage-palm shoots were also excellent eating.[39] During 1822 in southern Western Australia, Cunningham identified creeping parsley and a species of orach to feed the expedition, the latter being 'used by us every day, boiled with salt provisions, and proved a tolerable substitute for spinach, or greens'.[40]

As he did on his overland trips, on the King expeditions Cunningham acquired the assistance of Aboriginal people when he could. King noted that when the expedition was at King George Sound in the southwest of Western Australia in 1821, a local Aboriginal man whom the expeditioners named Jack 'frequently accompanied Mr. Cunningham in his walks, and not only assisted him in carrying his plants, but occasionally added to the specimens he was collecting'.[41] The plant hunter's botanical findings are presented as an appendix in King's published account.[42] Cunningham's prowess as a collector was enormous – on just one of his voyages with King, he was reported to have collected plant specimens representing 300 different species.[43]

With the King expeditions concluding in 1820, Cunningham continued collecting in eastern New South Wales and went on trips north into a region that was opening up in present-day southeast Queensland. In 1824 he accompanied Oxley to Moreton Bay on a trip that investigated the suitability of the area for the establishment of a convict settlement.[44] It was here that he made the insightful remark that the macadamia-nut tree had value, not just as an ornamental species, but as a food crop for European farmers.[45] On all his trips, Cunningham routinely looked for evidence of how Aboriginal foragers lived off the land. In 1827 at the Severn River in northeast New South Wales, he noted:

> The marks of natives wandering in quest of food were noticed on the timber through which the travellers passed on this day. There were steps on the tree trunks, evidently cut to aid the blacks in climbing, although the bush furnished few opossum and apparently the natives had been seeking larvae or pupae, upon which they must have chiefly lived. These were most often found in the knot at the upper limbs of a straight-growing box.[46]

In the same year, inland from Sydney, Cunningham noted having found trees under which Aboriginal foragers had recently dug to obtain insect larvae from the roots.[47]

Even in the remoter regions, Cunningham encountered indigenous hunter-gatherers who by the 1820s were already using European tools. In 1827 on a trip from Sydney to Moreton Bay (Brisbane), he found chopping marks in the trees that had been made by Aboriginal foragers in possession of steel axes.[48] The tools had probably

An Aboriginal man climbing using a knotted loop of lawyer-cane. To ascend a tree with smooth bark, Aboriginal people chopped notches into the trunk to provide toeholds. Murray Studios, Atherton Tableland, Queensland, about 1890. N.B. Tindale Collection, SAMA 338, South Australian Museum Archives, Adelaide.

been obtained through trading networks, although the blazed trees were well beyond the then frontier of British settlement. Across Australia, indigenous people actively absorbed new technologies from the Europeans, substituting stone with metal and glass, changing from animal skins and bark to cloth, and replacing plant resins with pitch.[49]

Explorers often frightened the groups of indigenous people they met by chance. If they could, the Aboriginal people generally withdrew from contact, which resulted in very little direct communication. Material evidence of their hunting and gathering life nonetheless remained. In 1828 Cunningham was on a trip with Fraser in the mountains near Brisbane in southeast Queensland when they came across an Aboriginal camp at the base of Mount Dunsinane. The occupants had scattered at their arrival, and Cunningham noted:

> At their fire we found the bags and little paraphernalia of the women, showing clearly with what precipitous haste these savages had urged their flight, which had not even afforded them a moment to gather their few articles of economy together. Around were quantities of the large seed of that exceedingly ornamental tree of close woods called chestnut at Moreton Bay ... Upon these nuts the few natives who wander through these lonely regions chiefly subsist. Like the English chestnut they contain some saccharine and much farinaceous matter, and by being well roasted are rendered easy of digestion.[50]

These 'chestnuts' (from the seedpods of the black bean) are toxic without extensive preparation.[51] The nuts also impressed Fraser, who claimed that they tasted like Spanish chestnut and that eating them resulted in 'a slight pain in the bowels, and that only when it was eaten raw'.[52]

In the Bremer River district near Brisbane, Cunningham noted, 'No natives were met with in this stage, although patches of the forest grasses had been very lately fired and the recent traces of these people were noticed on the trunks of the trees, from which they had torn off the outer paper-like bark to roof their huts'.[53] The camps, which were occupied seasonally, were often vacant. One such camp on the banks of the Logan River had a bark hut in it, where it was:

> customary for the tribes, when leaving a district, to deposit in such a situation their kangaroo-nets, dillies [bags], bass mats, chissels [sic], and superfluous implements, until their return. It is considered the greatest breach of faith among these rude nations to touch any of the articles thus placed; a degree of honesty which, it is to be feared, we might look for in vain among their neighbours [convicts at Moreton Bay].[54]

Observations like this provide evidence of how Aboriginal hunter-gatherers then lived.

Although titled 'King's Botanist', Cunningham did not entirely restrict himself to collecting plants. Among the more unusual specimens he sent back to England was the desiccated body of an Aboriginal woman. Fraser, his fellow plant collector, recorded in his journal:

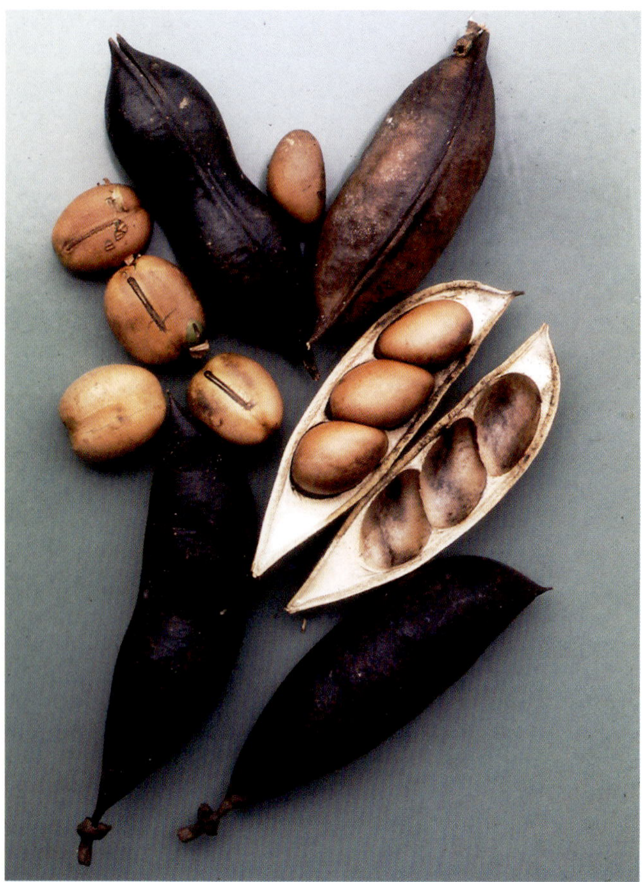

Fruit of the black bean or Australian chestnut, nuts which are poisonous before processing. Plant hunter Allan Cunningham found this species used as an Aboriginal food source at a camp at Moreton Bay. Norman B. Tindale, Tolga, northern Queensland, 1972. N.B. Tindale Collection, AA338/6/34/29, South Australian Museum Archives, Adelaide.

Blue quandong or Brisbane quandong, another of the edible plants that Allan Cunningham encountered in southeast Queensland. Colonists applied the indigenous term 'quandong' to a number of botanically unrelated species, all of which bear edible fruits or nuts. Philip A. Clarke, Mount Coot-tha, southeast Queensland, 2004.

Moreton Bay fig, one of numerous edible plants Allan Cunningham discovered during his expeditions to southeast Queensland in 1827–28. Philip A. Clarke, southeast Queensland, 2004.

> Mr. Cunningham has in his possession the skin of one of the female aborigines, which was procured by Private Platt of the 57th Regiment, from the hut of a native on the banks of the Brisbane River, just above its junction with the Bremer. It consists of only the front of the body, arms, and legs; the fingers and toes have their nails perfect, but the face is wanting, although the ears remain. It had been deposited in the *Dilly* or luggage-bag of a female, and carefully placed within one of their nets.[55]

Across Aboriginal Australia, many communities kept the smoke-dried remains of their deceased kin for a time in huts, awaiting the final ritual of disposal in a cemetery.[56] Cunningham's collection of the woman's remains can only be regarded as theft.

Earlier on this trip to southeast Queensland, Cunningham's party had come across a burial place in a valley northeast of Brisbane. Fraser recorded that they were shown:

> an extraordinary *Cemetery*, if it may be so termed, of the aboriginal natives. It consisted of the hollow trunk of a dead *Eucalyptus*, in which were deposited human bones of all ages, consisting of leg, thigh, and arm bones, vertebrae, and some fragments of crania, all mingled together. I was informed that many of the skulls had been previously carried away by scientific persons.[57]

Throughout the nineteenth century, European collectors displayed a fascination with indigenous skeletons, in particular the skulls.[58] In more recent times, many people have become uncomfortable with the notion of any human remains, whether indigenous or not, being kept for scientific purposes, and since the late 1980s museums around the world that have held such material since colonial times have been involved with handing it back to the appropriate descendants of indigenous communities for reburial.[59]

A dangerous occupation

Field trips in the Australian bush were often dangerous affairs. As expedition leader, King tried to ensure that the interactions between Cunningham and local Aboriginal people were carefully controlled. King was particularly concerned over Cunningham's safety when he was on his own. Although this expedition and others would have been well armed, it was common for plant collectors to wander off alone or stray behind the main party in their absorbing quest to find better specimens or new species. Reporting on his June–July 1819 trip with King to Endeavour River in Queensland, Cunningham said:

> Here was a period of 14 Days, that might have been wholly at my Disposal, had it not been for the annoyances experienced from the 'Prowling' natives, who made a rather determin'd but unsuccessful attack upon the Boat builders and others on shore, while I was at some Distance in pursuit of Flora, which fully occupied my time ... It was a subject of much Regret to me, that in consequence of the Rupture with the Natives, my Walks during the last Week of our stay at Endeavr. River, were either very much circumscribed, or wholly prevented.[60]

There were many other occasions when Aboriginal people threatened expedition members. During King's exploration of coastal Western Australia in 1818, an Aboriginal party in the Dampier area massed to confront them.[61] They acted so menacingly, brandishing spears and throwing stones, that they prevented the explorers landing to obtain drinking water.

In 1819 King's expedition party suffered an Aboriginal attack when visiting Moreton Bay in Queensland.[62] Here, as a result of a dispute over property stolen from the explorers, the Aboriginal people set fire to the surrounding grass, almost burning the party alive. It is likely that they were angered at being denied their presumed right to take items from the Europeans as a form of payment for allowing them to pass through their country.

Cunningham found the Aboriginal practice of burning the bush for hunting and gathering purposes a nuisance when it removed potential plant specimens. At King George Sound in southern Western Australia, he reported in his journal on 24 January 1818 that he was 'avoiding a tract of brushwood on the skirts of the harbour which had lately been fired by the natives and hence could afford me nothing ...'[63]

A number of plant collectors in various parts of the world lost their lives while collecting for their botanist masters. In 1834, for example, North American plant hunter David Douglas, after whom the Douglas fir was named, was found dead in a wild cattle pit in the Hawaiian Islands.[64] Despite the nature of his injuries, it was never conclusively proved that Douglas was gored to death by a bull, with some suggestion that he may have been robbed and his death staged. The Australian frontier also claimed victims. In May 1835, Aboriginal people on the Bogan River in central New South Wales killed government botanist Richard Cunningham, who was exploring with Surveyor-General Major Thomas L. Mitchell.[65] Richard was the only brother of Allan Cunningham. Like his older brother, Richard had been extensively trained at the Royal Gardens at Kew, where he had worked on many of the plant specimens Allan had sent back to England.[66]

Richard Cunningham came to Australia in 1832 to take up the position of Superintendent of the Botanic Gardens in Sydney following the death of Charles Fraser, the previous incumbent. His murder occurred near Dandaloo, west of Dubbo, some time in April 1835 on an expedition to the

Darling River district of inland New South Wales. Cunningham had wandered away from his party one morning to collect plants and become lost. Mitchell returned to camp that evening, but was not at first alarmed, as the plant hunter was in the habit of turning up late.[67] In subsequent days, the party undertook a lengthy search, but found only Cunningham's dead horse, his saddle and bridle, one of his gloves, some straps and a part of a coat. The expedition returned to Sydney with the plant hunter's fate unknown. Months later, mounted policeman Lieutenant Henry Zouch, who was accompanied by 'native black' Sandy, was sent out to investigate Cunningham's disappearance.[68]

In search of information, Zouch captured a 'tribe' of forty 'Myall blacks' (Aboriginal people living beyond the frontier). To gain their release, they offered Zouch the names of four men responsible for Cunningham's murder, three of whom were then taken into custody. Reconstructing the events of the crime in his report, Zouch stated that after Cunningham had become separated from the main party, he approached a group of four Aboriginal men on the Bogan River and made signs to them that he was hungry.[69] At first the Aboriginal men helped him, but while they camped together during the night, Cunningham aroused the suspicions of his companions by constantly getting up. This unusual behaviour led to his being fatally struck on the head from behind with a club the next morning. Zouch did not suggest it, but it is possible that the Aboriginal men believed Cunningham's nocturnal movements and his collecting activities were the ritualised actions of a sorcerer. If that were the case, he would have been seen as posing a threat, and in any case they would have been attracted to his belongings.

When Zouch arrested the alleged murderers, their bags were searched, yielding articles such as a European knife, glove and cigar case. It was assumed that all of these things had belonged to Cunningham. One of the prisoners, Pu-ri-mul (Burreemal), guided Zouch to the place where Cunningham was killed, and here some human bones, a Manilla hat and part of a coat were found. They raised a small mound over the remains and blazed a nearby tree to mark the spot, so that the body could be recovered at a later date. Two of the prisoners escaped, leaving just Pu-ri-mul to be taken to Sydney in the custody of Zouch. The evidence against the remaining Aboriginal man was largely circumstantial.

Awaiting trail before the Criminal Court, Pu-ri-mul was kept in custody at Goat Island, where missionary Lancelot E. Threlkeld visited him at the request of the Attorney-General.[70] Threlkeld was able to communicate with the prisoner through other Aboriginal people acting as translators, and because Pu-ri-mul had learnt the local dialect through conversing with Aboriginal prisoners. It was determined that Cunningham's killers were named Pu-roi-to and Wong-kai-tu-rai, and that they came from a place distant to where the murder had occurred. Pu-ri-mul was reputedly not involved at all with the killing, but he had heard 'that it was about an opossum he [Cunningham] was killed'.[71] Pu-ri-mul knew where Cunningham's grave was because he had been asked by his brother to bury the body, which he did with the help of two companions. He was eventually released, as the Governor of New South Wales considered it would be unlikely that a conviction would be obtained if the matter went to trial.[72] Allan Cunningham was later critical of Mitchell over the handling of his brother's death.[73] As demonstrated by this incident, early plant hunters traversed a landscape that was still dominated by Aboriginal land 'owners', who often had numerical superiority over British exploration parties.

Changing the landscape

Many of the professional plant hunters saw themselves not just as collectors, but also as redistributors of plants around the globe.[74] In the present era of concern over the fragility of Australian ecosystems and the threats posed by invasive organisms, it is hard to comprehend why European explorers wanted to 'improve' the Australian flora through the introduction of exotic plants. The plant hunters actively sought out places that had ecosystems within which they believed that other, potentially more useful, exotic species would thrive. There are plenty of examples concerning the redistribution of the world's flora with the intent of producing favourable outcomes for humanity. In the mid-nineteenth century, with the help of the European botanic gardens, species of cinchona from South America were established in India to help alleviate the increasing costs of importing quinine to combat malaria.[75] Later in that century, various plants from Central and South America were established in plantations in Ceylon (Sri Lanka) to benefit the rubber industry.[76] The deliberate introduction of exotic plants, whether in wild areas or cultivated fields, was at the time recognised as having potential economic benefits for the colonies.

Cunningham was a keen supporter of introducing exotic plants into Australia, once stating of his imported seeds: 'whenever I find a piece of good soil in the wilderness I cause it to be dug up and drop in a few in the hope of providing a meal for some famished European ... or some hungry blackfellow'.[77] On an expedition out from Bathurst in New South Wales, Cunningham buried under a marked tree a bottle containing a parchment that read:

> Due east and west by compass from this tree ... were planted the fresh stones of peaches, brought from the colony in April last [1823], with every good hope that their produce will one day or other afford some refreshment to the weary farmer, whilst on his route beyond the bourne of the desirable country north of Pandora's Pass.[78]

Cunningham appears to have mainly acted out of concern for his fellow travellers, although due to lack of care it is likely that none of his peaches survived. His plant introduction program was extensive and conducted on most of his expeditions, regardless of region. Oxley remarked that Cunningham was sowing acorns, quince seeds, peach and apricot stones on their 1817 expedition across inland New South Wales.[79] During King's 1818 expedition, at King George Sound in southern Western Australia Cunningham had the ground near a waterhole cleared in order to plant European fruits.[80] He provided a list of what he planted later on that trip at Goulburn Island in northern Australia: apricot, lemon, coconut, marrowfat pea, long-podded bean, scarlet runner bean, large-horned carrot, parsley, celery, parsnip, cabbage, lettuce, endives, spinach, broad-leaved Virginian tobacco, sweet and everlasting pea, Spanish broom and milk-vetch.[81]

Cunningham's actions were consistent with those of the first Europeans to reach Australia. Earlier maritime explorers, such as Cook, Bligh, Flinders and Baudin, had introduced into the Australasian and Pacific regions a range of animals, such as fowls, pigs and goats, and were in the habit of leaving behind gardens containing plants such as lemons and leeks.[82] Cunningham was also acting like the famous John Chapman (alias Johnny Appleseed) of early nineteenth century North America, who distributed apple seedlings everywhere he went in order to 'improve' the country.[83] These explorers, plant hunters and botanists all believed that they had a role in helping future stranded visitors, as well as in preparing the land for eventual acquisition by settlers. Early plant hunters found many botanical wonders in the colonies, but considered that these virgin lands needed much improvement through the introduction of more 'useful' plants and animals. In this respect, Europeans acted like other colonising peoples, such as the Austronesians who began their move through Southeast Asia to the southern Pacific Ocean some 5000 years ago, bringing with them to Australia a range of useful organisms like taro and dingoes.[84]

Allan Cunningham had a similar fiery temperament to George Caley, his predecessor. Like Caley, Cunningham also had a falling-out with the governor of New South Wales, in his case Macquarie.[85] His proven redeeming qualities, which he also shared with Caley, were that he was hardworking and fastidious in his recording of specimen data. Not long after Cunningham arrived in New South Wales he gained a reputation as an explorer, in his journals meticulously recording his bearings and drawing detailed maps of the areas beyond the frontier of British settlement. Cunningham's exploration of southeast Queensland in 1828, where he named a number of major topographic features, is particularly significant.[86] His published accounts became useful guides for the squatters and settlers who came later. Although the competitive Mitchell was critical of Cunningham's geographical work, it is likely that these surveys were of immense value to him during his own exploration of inland Australia.[87]

Cunningham went back to England in 1831, but returned to Sydney in February 1837 to take up the position left vacant on the death of his brother of Colonial Botanist and Superintendent of the Botanic Gardens in Sydney.[88] However, finding the job of supervising convict labourers growing fruit and vegetables, in what was generally referred to as the 'Government Kitchen Garden', to be irksome and in conflict with his passion for plant collecting, he resigned after only a few months and made a trip to New Zealand. In October 1838 Cunningham returned in poor health to Sydney, where he died of consumption in June 1839.

Botanists have honoured Allan Cunningham's contribution to Australian botany by the naming of several plant species, such as *Araucaria cunninghamii* (hoop pine) from southeast Queensland and *Nothofagus cunninghamii* (myrtle beech) from southern Victoria and Tasmania.[89] Another species named after him is *Crotalaria cunninghamii* (birdflower-bush), the stem of which, for indigenous toolmakers in the Western Desert, was a major source of the fibre used in the making of sandals and cord.[90] Aboriginal people in the tropics used another plant named after Cunningham, *Bauhinia cunninghamii* (bauhinia-tree), as a source of edible gum and nectar, as well as for making windbreaks.[91] While he was a prodigious hunter of plants of all types, Cunningham's particular interest was orchids.[92]

Birdflower-bush. This species bears the scientific name *Crotalaria cunninghamii*, in honour of plant hunter Allan Cunningham. It was important to Aboriginal toolmakers as a source of stem fibre. Philip A. Clarke, Port Augusta, South Australia, 2007.

Plant hunters following after Cunningham continued his prudent practice of employing Aboriginal guides and assistants. From December 1856 to August 1857, Silesian naturalist William Blandowski made an extensive natural history collection during an expedition to northern Victoria. He reported:

> the specimens were obtained by the assistance of the aborigines, to whom I am indebted for all the information and discoveries I have made, so that I can but claim a small share of the credit of having, with my party, been successfully exploring the desert of Australia for eight months.[93]

Aboriginal guides were silent partners in the records of many early plant collectors. Not only did they help keep the Europeans alive, but their participation greatly enhanced the success of many plant collecting field trips.

6 Resident Plant Collectors and Aboriginal People

The love affair of the botanists of eighteenth century Europe with the unique Australian flora continued through the nineteenth century, with more collecting opportunities coming with the establishment of each new colony. The territory being opened up was far too large for professional plant hunters to cover alone, and although a number of colonists had already been involved in supplying natural history specimens to Europe, there was now a recognised need for more systematic plant collecting to take place. From the 1830s, therefore, European botanists established networks of amateur collectors involving colonists who were well placed on the advancing fringes of settlement to work under their direction via correspondence.

James Drummond and his grandson. Drummond, a prominent plant collector for Kew, was active during the early years of settlement in southwest Western Australia. Ewen Mackintosh, Western Australia, about 1860 (Erickson, 1969).

Plant collecting offered a release from the boredom of frontier life, and was a pursuit that fitted in much better with other activities than zoological collecting. Both involved making precise notes about where, when and how, but the preparation of plant specimens required only pressing and drying, while preparing animal specimens required skinning, drying and the preservation of internal organs in fluid. Being static objects, plants were also much easier to collect than animals, although a working knowledge of botany and an attention to detail when documenting specimens was required. To further guide their activities, botanists often rewarded their colonial collectors for their services with gifts of botanical publications.

James Drummond, Government Naturalist in southwest Western Australia

James Drummond, an early Swan River settler in southwest Western Australia, became a renowned explorer and amateur botanist. He was born in Hawthornden, Scotland.[1] In 1808, at the age of 21, he was employed to look after the Cork Botanical

Gardens in Ireland, an agricultural testing facility run by the government, and in 1810 was elected an associate of the Linnean Society in London. When the botanical gardens facility was scaled down, Drummond decided to try his luck in the Australian colonies. Arriving on the *Parmelia* with Captain James Stirling at the site of the Swan River settlement in 1829, Drummond was appointed as an honorary government naturalist for the fledging colony.[2] Natural history was a family business for the Drummonds, with James' younger brother, Thomas, becoming a collector of botanical and zoological specimens in North America.[3]

For the first nine months of settlement at Swan River, relations between settlers and the indigenous population were peaceful, and Drummond and his family of six children had extensive contact with the local people. Historian Rica Erickson noted: 'Friendly natives showed the Drummonds how to catch the most elusive prey and the boys became intimately acquainted with their language and tribal lore'.[4] Frontier conflicts soon led to Aboriginal raids upon settlers along the Canning and Swan rivers, culminating in May 1832 with the murder of labourer William Gaze.[5] The main offenders were believed to be Yagan and his father, Midgegooroo, who came from the Beeliar district south of Perth, Yagan's resistance against colonisation leading to his being described by a writer of that time as the 'Wallace of the Age', after the late thirteenth century Scottish patriot.[6] Drummond, however, was an apologist for Yagan, claiming that he only killed settlers out of revenge for mistreatment of himself or his companions.[7]

In September 1832 the authorities captured Yagan and sentenced him, along with two other Aboriginal men, to imprisonment on Carnac Island, which is about 13 kilometres west of Fremantle. A few weeks later, Yagan escaped and reorganised his raiding parties, leading to a pattern of conflict with further deaths of both settlers and Aboriginal people. Midgegooroo was eventually captured, tried, and executed by being shot while tied to the door of the Perth Gaol in May 1833.[8] Contact between Yagan and the Drummond family must have occurred soon afterwards when, according to family legend, Yagan brought James Drummond gifts of fish and game, taking the settler to be his own deceased father returned to him.[9] Whether or not Drummond and Yagan ever met, the frontier conflict had an impact upon the lives of everyone living in the region at that time. Now with a price on his head, Yagan was finally killed on 11 July 1833, shot by a young shepherd who was himself killed during the affray. Yagan's head was removed from his body, then smoked, preserved and taken to England as a 'specimen'.[10] It was not returned for reburial by the Nyungar community of southwest Western Australia until 1997. The Western Australian Department of Agriculture recognised his status as an indigenous resistance fighter in 1988 when they named a new early maturing cultivar of barley bred for sandy soils as 'Yagan'.[11]

In 1835 Drummond made contact with the London-based botanist and horticulturalist, Captain James Mangles.[12] For the next few years they were in a commercial relationship, with Drummond sending seeds and specimens to Mangles for distribution across Europe. With collecting materials, especially paper, in short supply in the early years of the colony, Drummond resorted to the practice of drying his fresh botanical specimens between layers of grasstree leaves.[13] In the early years of settlement, most plant collectors would have suffered from a shortage of paper to mount their pressed specimens.[14]

In 1836 Drummond and two other settlers organised an expedition to travel from Guildford near Perth to the east towards York, with an elderly 'Canning River native' called Bobbing to 'see water' for them.[15] The party blazed trees along the way and recorded the Aboriginal placenames, focusing upon waterholes, to help future settlers find their way using Aboriginal guides. Drummond was interested in indigenous cultures and in particular their uses of plants. At a place their guide called 'Duidgee' (Toodyay), Drummond remarked in his journal that there was an abundance of cat's tails (bulrushes) or reed mace, noting that they were important to local Aboriginal bands in 'furnishing a great portion of the food of their women and children for several months of the year'.[16] Finding the area suitable for farming, Drummond relocated his family to Toodyay Valley.[17]

Drummond's close relationships with Aboriginal people continued at Toodyay Valley. On one occasion in 1839, he took the side of an Aboriginal man who was needlessly whipped by a station manager, and reported the incident to the Resident Magistrate.[18] His fourth son, John, associated with Aboriginal people to such a degree that it became a source of tension with his father, particularly when John was accused of taking an Aboriginal man's wife to his bed.[19] In 1840, after Aboriginal people killed settlers near York, John was appointed the first Inspector of Native Police to help keep the peace.[20]

From 1839 settlers in the region had offered to organise an annual feast for local Aboriginal people if they could refrain from burning the fields, as they had formerly done when hunting and gathering.[21] The flour for the occasion was ground at James Drummond's mill, his sympathy undoubtedly generated in part by the deep understanding he had of Aboriginal culture. His letters to Hooker at Kew provide insights into this knowledge, illustrated by detailed descriptions of Aboriginal customs involving sorcery and revenge.[22]

From 1839 Drummond was actively collecting seeds and plants for Hooker, who valued his collector's accounts of the Australian flora and published them in the *Journal of Botany*.[23] Drummond and his youngest son Johnston were also engaged in collecting birds and animals as natural history specimens for the famous author and artist, John Gould.[24] Johnston was particularly well qualified as a zoological collector, being skilled in bushcraft and possessing the ability to converse fluently with inland Aboriginal groups in their own languages. Gould paid them with copies of his books.

In 1839 Drummond went on a botanising trip to Rottnest Island in the Indian Ocean west of Perth, in the company of John Gilbert, who also worked as a zoological collector for Gould.[25] In the early settlement period, conflict between settlers and Aboriginal people meant that mainland collecting was risky. Gilbert said in a letter to Gould:

> I have been in the Interior as far as any Europeans have yet settled, but unfortunately at the time I was there, the Natives committed several frightful murders on the white people, who to punish them killed several of the Blacks in return. This had roused up all their former savage disposition, and it was considered very dangerous to move far away from the settlers houses, at the same time the different Tribes were at war with each other ...[26]

Drummond, his son Johnston and Gilbert often travelled together. In late 1842 Drummond and Gilbert went on an expedition to King George Sound to collect natural history specimens.[27]

During the 1840s and early 1850s, Drummond set off at the beginning of each plant-collecting season from his home at Toodyay Valley with his three ponies and one or two Aboriginal guides.[28] The influences of these guides on his botanical work are apparent, for while he mainly focused on botanical details, Drummond also recorded many interesting accounts of Aboriginal uses and beliefs

Giant kingia (left) and grasstree (right). Plant collector James Drummond described in detail the Aboriginal use of grasstrees in fire-making. He was in the habit of drying his fresh botanical specimens between layers of grasstree leaves, rather than paper, which was scarce. Norman B. Tindale, Crystal Springs, Western Australia, 1972. N.B. Tindale Collection, AA338/6/35/34, South Australian Museum Archives, Adelaide.

of plants. Of the grasstree, he claimed 'The natives are particularly fond of the blackboy [grasstree], and frequently refuse any other nourishment, whilst its sound old flower-stalks furnish them with the means of obtaining a light by friction'.[29] He also gave an account of its use by Aboriginal foragers as a source of edible wood grubs.[30]

Drummond's observations provide deep insights into Aboriginal interactions with the environment. He stated that at King George Sound, in southern Western Australia, Aboriginal foragers subsisted on roots, particular the abundant red root plant, which was available throughout the year.[31] According to his notes, Aboriginal people knew this grass-like plant, with a small onion-sized bulb, as 'mynd'. He described its cultural importance:

> The *mynd*, however, is mostly eaten by the women and children, or very old men, – the young men disdaining it if other food can possibly be procured. Their mode of cooking this bulb is curious, and chiefly performed by the women. It is first well roasted, and then pounded between two stones, together with some *earth* of a reddish colour, nearly free from sand ... This earth is understood to be the production of the white ant [termites], whose hillocks or nests are very common. One measured by Mr Gilbert, the naturalist, was nearly four feet [1.2 m] high, and of considerable girth. The women never travel without a supply of this earth, as in the iron-stone country the *co-kut*, or ants' nests, cease to appear.[32]

The termite mound material neutralised the toxins in the red root plant.[33] Drummond made the further observation that eating this plant food gave the women a purple tongue, which they would poke out to gain sympathy from settlers when asking for food. Anthropologists and archaeologists refer to the practice of eating earth as geophagy; it occurred widely across Aboriginal Australia.[34]

The next most important edible root was said to be the 'tieubuck' (little-kidney), which was found in sheltered places in spring.[35] Drummond said, 'The natives procure it by means of a long pointed stick, which is the only instrument used by the women in obtaining every kind of food from the earth'.[36] His description of the digging-stick as the chief female tool is insightful, given that in the art of Aboriginal Australia this artefact is often used to symbolise women.[37] Although Drummond recorded the Aboriginal use of by-yu nut trees (zamia palms) in southwest Western Australia, the nut did not impress him as a food source. Calling it 'quenine', he claimed the 'nut is poisonous; but the rind, which is of a fine red colour, *after being buried for a month*, forms a chief article of [Aboriginal] food in the autumn. To me it was disgusting, the taste being rancid, and resembling train oil'.[38] In one paper, Drummond listed seventeen food types, mostly roots, and concluded: 'In very dry parts of the country [to the north and east], many other kinds of roots are eaten by the natives; but, as far as can be ascertained, they are otherwise despised, unless under cases of extreme necessity'.[39] Aboriginal foragers generally refer in Aboriginal English to such emergency sources as 'hard time' food.[40]

In Drummond's description of the york gum he commented: 'The Eucalyptus, found on the sandy loam, is called by the settlers York Gum, by the natives Doatta; they use the bark of the root as food in the dry season, chewing it along with the gum of the Manna [manna wattle] ... which produces a large quantity of gum in the dry season'.[41] For wattles in general, he observed that the 'gum of some of these species is used by the natives as food; and the seeds, when ground, give them a tolerable substitute for flour'.[42] He appears to have been impressed by the york gum, stating that Aboriginal foragers would also 'collect a description of manna from the leaves of the York gum-tree, which yield a considerable quantity of saccharine matter'.[43]

Drummond had an interest in collecting fungi. On one occasion, he collected a species that glowed naturally for four or five days before it dried. He had shown the fungus 'to the natives when giving out light ... They called it a *chinga*, their name for spirit, and they were much afraid of it'.[44] In some published notes on his collecting in southwest Western Australia he said: 'The fungi are also palatable to the aborigines: one species belonging to this order, the *Boletus* [genus of mushroom-like fungi], is remarkable for possessing the properties of German tinder when well dried, and for emitting a radiant light in its natural state'.[45]

During his explorations he made detailed observations of Aboriginal hunting and fighting practices.[46] In 1842, during a collecting trip to the present-day Busselton area south of Perth, he came across a concealed set of pit traps surrounding a bare space, designed to catch wallabies and kangaroos.[47] In 1842–43 the *Inquirer* newspaper, based in Perth, published Drummond's expedition reports, as well as a series of his letters on Western Australian botany.[48] These writings contain notes on Aboriginal plant uses. From the indigenous terms Drummond recorded, it appears that his guides were speakers of various Nyungar dialects of southwest Western Australia. When Drummond and Gilbert were part of an expedition travelling north of Perth beyond Moore River to Lake Dalaroo and Wongan Hills in 1842, their Aboriginal guide was Kabinger.[49] On this trip, Gilbert required local Aboriginal assistance in procuring specimens of the mallee-fowl.

In 1844 Drummond was on an expedition to the Victoria Plains region in the company of the 'native' Mangerwart, who was said to have 'owned the land and knows every spring and pool for a hundred miles to the east'.[50] Drummond produced from the trip a detailed account of Aboriginal methods of using game nets to capture small kangaroos and wallabies. In a letter to Hooker dated 30 October 1844, he described how:

> [the] natives break down the shrubs and construct a rough fence, placing their nets in the tracks with the mouths of the nets open towards the thickets. The people then assemble, and disperse themselves in the opposite end of the brushwood from the traps and set up a great noise which induces the terrified creatures, whatever they may be, to rush inwards towards the nets somewhat on the principle of the Highland huntsmen who used to drive the deer into the snare. One or more natives conceal themselves near, to watch and secure such animals as may chance to get entangled. The nets are constructed of the bark of the plant numbered 730 and 731 (*Pimelea argentea* [silvery leaved pimelea]) of my large collection whose valuable properties for that kind of use I pointed out long ago to our Agricultural Society. It gave me no small pleasure to see the poor creatures turning it to account.[51]

At a decisive point during the expedition, Mangerwart cautioned Drummond that if they continued east, for the next four days there would be no water, save that which could be collected from hollow tree trunks.[52]

The presence of the indigenous people was both a blessing and a nuisance for avid plant collectors. While a healthy population of hunter-gatherers suggested fertile country, Aboriginal bands posed a physical threat to small parties of Europeans. In 1844 Drummond wrote to Hooker to complain about the lack of resources to collect plants in remote regions of Australia:

> However well disposed the natives of this country are naturally inclined to be when unmolested, we cannot venture to travel among strange tribes without at least three persons well armed: and the necessary provisions, packhorses etc., which an expedition of the sort requires comes to be very expensive.[53]

He was not overstating the risks. During the winter of 1845 Drummond's son Johnston discovered that their guide Kabinger had been stealing their sheep.[54] Upon being sent away from the station, Kabinger threatened to spear Johnston when he next had the opportunity. On 4 July 1845 Johnston went on a collecting trip to Moore River, taking with him Kabinger's wife as a sleeping partner, and Kabinger's brother Knoberring and his wife. On the night of 12 July, while the party was asleep, Kabinger crept up and ran two glass-tipped spears through Johnston, then retrieved his wife. Although Johnston was able to extract the spears, he died shortly afterwards.

By unpleasant coincidence, Johnston's death occurred two weeks after the fatal spearing of his fellow collector, John Gilbert, in northern Queensland.[55] On hearing of his brother's death, Inspector John Drummond set out with two companions to apprehend the murderer.[56] About a fortnight later they came across Kabinger at his camp in the company of other Aboriginal people. Although he fled, armed with glass-tipped spears, Kabinger was soon spotted hiding in tall scrub nearby, and when he shifted a spear in order to throw it Inspector Drummond killed him with a gunshot to his side. The authorities later considered that Kabinger's death had not occurred lawfully, due to an absence of a warrant, and John Drummond was suspended from duty, eventually to be reappointed at a more junior rank.

James Drummond's grief at the loss of his son caused him to lose interest in collecting for a while, but his passion was reignited in October 1846 when he received an honorarium of £200 from

By-yu nut, a cycad species, here growing amongst bracken. Drummond recorded that in southwest Western Australia by-yu nuts formed a chief article of Aboriginal food in the autumn. Settlers collected nuts and stem starch from cycads, and used Aboriginal techniques to remove the toxins. The settlers also ate bracken rhizomes when other foods were scarce. Philip A. Clarke, near Albany, southern Western Australia, 1987.

the Queen's Bounty for his services to botanical science.[57] This alleviated his poor financial situation and encouraged him to do more botanising trips. During a trip in southwest Western Australia in 1847, Drummond recorded following a 'native path' and on another occasion finding water at a 'native well' in the dry bed of an inland river.[58] Due to the Aboriginal presence, plant collecting remained a dangerous activity in this region. In a letter to Hooker in 1851, Drummond claimed:

> I could have procured many more plants in the north, but for the character of the natives, who were so troublesome that I could only make excursions armed with a double barrelled gun, and in company with mounted police. Both myself and my son [John], who is at the head of the police here, had several narrow escapes with our lives.[59]

In 1855, explorer Augustus C. Gregory offered Drummond the position of botanist on his expedition to northern Australia. Then 86 years old, the plant collector declined due to his advancing age, with Ferdinand von Mueller taking his place.[60]

James Drummond was highly regarded as a botanist in Europe, as evidenced by Charles Darwin's letter to him in 1860 requesting information on the fertilisation of the red leschenaultia.[61] Drummond died at his home in Toodyay Valley in 1863.[62] Von Mueller honoured him by publishing the scientific name of the common nardoo (clover fern) as *Marsilea drummondii*.[63] This was a significant food source for desert Aboriginal foragers in the Cooper Creek to Mulligan River region.[64]

Georgiana Molloy, 'esteemed lady plant collector'

Georgiana Molloy was born into the Kennedy family of Cumberland in England on 23 May 1805.[65] In 1829 she married the much older Captain John Molloy, a seasoned soldier who had fought at Waterloo, and the newly wed couple emigrated to the recently established Swan River Colony (Western Australia). In 1830 they moved south to Augusta, where Captain Molloy was made Resident Magistrate. In 1836, although she had no previous formal training in botany, learning 'on the job', Georgiana Molloy became a collector of seeds and dried plant specimens for Mangles.[66] In 1839 she also sent plants to Dr John Lindley, who was the foundation professor of botany at the newly created University of London.[67] With the benefit of Molloy's herbarium specimens, Lindley published a description of the Swan River flora in 1840.[68]

The frontier region in which Molloy lived experienced much conflict between settlers and local Aboriginal bands. Some of the violence directly involved her husband, but in spite of this Molloy appears to have maintained good relations with local indigenous communities.[69] At Augusta, she claimed, 'The natives are very fond of all the settlers ... and we live on the most peaceful terms. But at the Swan, from the indiscretion of several persons and particularly their servants, they are hostile'.[70] Members of the Aboriginal community at Augusta knew Molloy as 'King-bin', while they referred to her magistrate husband as 'King Kandarung'.[71]

Molloy formed a close social relationship with Gyallipert, an Aboriginal man from King George Sound who had travelled to Swan River in 1832, and then in 1833, to act as a conciliator in the conflict between settlers and Aboriginal people.[72] He apparently spoke 'tolerably intelligible English'. Molloy gave Gyallipert flour for making damper and let him cook game, such as goanna, in her kitchen. In return, he gave her bush food, such as possum and fish, and asked her to look after his spears for him. In correspondence Molloy wrote about the Aboriginal people she encountered in Australia. From Augusta she sent 'some little memento' to a friend in England:

> I send you a bunch of emu feathers given me by Mobin, a native chief. You will perceive that they are covered with a sort of red earth. This they paint themselves with, and, mixed with fat they extract from their food, they besmear the hair, which is turned up 'à la Greque' and confined by many strings of the opossum hair, which the women spin.[73]

The practice of sending indigenous artefacts back home to Europe was common in the early years of settlement, before the formation of colonial museums in the mid-nineteenth century.

Not all interaction between Molloy and the local Aboriginal community was as friendly. On one occasion, an Aboriginal band arrived to dig up potatoes at the Molloy farm while the magistrate was away; shaking their spears, they threatened Molloy and her child when she objected.[74] No one was killed, but the group stole three glass saltcellars, which they probably intended to use to make spearheads. It is likely that the indigenous people of the region considered that they had a traditional right to any plant and animal foods in their hunting and gathering territory, which would explain why their hunters were spearing the settlers' pigs and sheep.[75] The original inhabitants would certainly have felt the pressure of local game becoming scarce due to European hunting, while their access to favoured areas for gathering plant foods was being progressively cut off by the establishment of farms.

Molloy frequently travelled the country with her indigenous companions to check on the flowering of particular plants she needed to collect, such as the Western Australian Christmas tree and giant kingia.[76] The Christmas tree was a special plant for the Nyungar people, who used the roots as a water source, collected its edible gum and believed the souls of the dead rested in the tree before departing to the 'land of the dead'.[77] Molloy recognised the skills of the indigenous plant collectors, remarking:

> The Natives are much greater auxiliaries than white people in Flower and seed Hunting ... They ask no impertinent questions, do not give a sneer at what they do

Western Australian Christmas tree. Plant collector Georgiana Molloy employed indigenous guides to check on the flowering of this parasitic species. It was a special plant for the Nyungar people, who used its roots as a water source, collected its edible gum and believed the souls of the dead rested in the tree before departing to the 'land of the dead'. Norman B. Tindale, Regans Ford area, southern Western Australia, 1968. N.B. Tindale Collection, SAMA 338/6/35/33, South Australian Museum Archives, Adelaide.

Georgiana Molloy lived in southwest Western Australia in the 1830s and early 1840s. She employed Aboriginal people to help her obtain natural history specimens to send to collectors in England. Unknown artist, portrait on ceramic, southwest Western Australia, 1830s. Private collection.

> not comprehend, and above all, are implicitly obedient, and from their erratic habits penetrating every recess, can obtain more novelties. The grand desideratum.[78]

She used food to pay a Wardandi man named Calgood to accompany her on her botanising trips and to collect seeds for her.[79] Molloy also had plans for an Aboriginal man named Banny to shoot bird specimens for sending back to zoological collectors in England.[80]

Molloy found she had to overcome some cultural prohibitions before she could convince Aboriginal people to act as plant collectors. It was Nyungar custom to place wreaths made from plants on graves, thereby leading them to associate plant collecting with death.[81] In my own field work, I have noted resistance amongst Aboriginal people to decorating their houses with flowers, for the same reason. Eventually she was successful in getting 'the native Herdsmen ... also employed bringing in some desired Plant or Fruit' and was able to place on the head of the Aboriginal man Battap 'a large piece of crimson Anterrhinum [*Antirrhinum majus*, snapdragon] in his Wilgied [ochre-rubbed] Locks'.[82] Even before her association with Mangles, Molloy would write down the Wardandi and Bibbulmun names of the plants from which she collected seed.[83] She also documented the Aboriginal uses of the plants, in particular as medicines.[84]

While Drummond had financial motives in collecting plants, Molloy sought no financial reward.[85] This was probably because she was chiefly motivated by her developing interest in botany, and by finding in it a diversion from a harsh and lonely frontier life. She did occasionally receive parcels of books from England for her efforts.[86] Walking around her estate searching for specimens with the aid of Aboriginal people was an activity which she could undertake with the involvement of her young children. Her husband appears to have been generally disinterested in her plant collecting activities. In spite of this, and the lack of resources on the settlement frontier, Molloy achieved much in a few years. She died in 8 April 1843, soon after giving birth to her seventh child.[87] Drummond attempted to have named in her honour a new plant species, the black kangaroo paw, recently discovered by his son Johnston, but it was officially accorded the name *Macropidia fuliginosa*.[88]

Ronald C. Gunn, Van Diemen's Land

Although Ronald C. Gunn had Scottish forebears, he was born at Cape Town in South Africa on 4 April 1808.[89] He spent several years at Bourbon (Réunion) Island before being sent to Aberdeen in Scotland for his education. After a period spent in the West Indies, he emigrated to Van Diemen's Land in 1830. Here he took up various public service jobs, such as Assistant Superintendent of Convicts and Police Magistrate in Launceston, and became the most prolific of the nineteenth century plant collectors working on the island. He died at Newstead, near Launceston in Tasmania, on 13 March 1881.

Gunn's interest in botany appears to have come about as the result of a chance meeting with settler Robert W. Lawrence in late 1830.[90] This was during

Ronald Campbell Gunn, a prominent nineteenth century plant collector in Tasmania, had a particular interest in indigenous foods. Thomas Bock, crayon drawing, 1848. Dixon Galleries, DG471, State Library of New South Wales, Sydney.

a manoeuvre in mountains north of Hobart as part of Colonel George Arthur's 'Black Line' to round up the last of the Aboriginal people living in Tasmania. Since a large number of Europeans were required for there to be any chance of success, both settlers and armed forces were involved.[91] Lawrence's journal entry for 11 November 1830 gives an insight into the punitive exercise. During the morning his party was somewhere in the Nugent district, when:

> we saw some trees barked, and a roving party led by Mr. Massey – Mr M. was engaged in examining some Natives' Huts, they were five in number, and appeared to be those which had been seen several times already, by other parties. Mr Massey had with him a Black who was taken a few days ago, as a guide, to lead him to the haunts of his tribe ... – Found a spear which had been lately made.[92]

Much later on, in correspondence to Hooker, Gunn gave his own account of proceedings:

> The superior knowledge of the Aborigines as far as regarded the bush, and their way of travelling, us being encumbered with heavy knapsacks laden with provisions, gave them an infinite superiority. My four parties, with other similar gangs, had but little success, though we saved many lives and much property constantly moving about.[93]

In praising the efforts of Aboriginal Protector George Augustus Robinson in getting the Tasmanian Aboriginal people together, Gunn noted: 'all tribes being now collected and sent to Iron bound island [Flinders Island] in Bass's straits, where they enjoy their usual occupations, hunting in the bush and so on ...'[94] At the time of the 'Black Line', Lawrence was already collecting for Hooker, and in 1832 he was joined by Gunn.[95]

From the 1830s to 1860s, Gunn sent an enormous volume of material to Britain as scientific specimens, using the Wardian case to send live plants to Kew.[96] In 1850 *The Times* newspaper in London reported that he had sent a living Tasmanian tiger (thylacine) to the British Museum.[97] He collected other animals for the Museum, including the Tasmanian emu, which became extinct sometime between 1870 and 1880.[98] In 1847 he published a paper that argued for the existence of the mythical bunyip, on the basis of an unusual animal skull found along the Murrumbidgee River in New South Wales.[99] (The skull was later identified as having come from a hydrocephalic foal or calf.)

Although Gunn began his collecting work in the same period as Drummond and Molloy, the early settlement of Van Diemen's Land meant that here the frontier was a distant memory. The destruction of Aboriginal life in Tasmania following colonisation had largely removed the opportunity to work directly with active hunter-gatherers.[100] In company of Lieutenant-Governor Sir John Franklin, Gunn visited Flinders Island in 1838, where Robinson ran the Aboriginal mission, Wybalenna.[101] In a letter to Hooker, Gunn reported:

> All the Aborigines of V.D.L. [Van Diemen's Land – Tasmania] having been removed by the Govt to that spot where they are clothed and fed and receive religious instructions. – The change of Life, and perhaps one or two other causes – of which bad water is the most serious has reduced the number from about 400 to 98 ... the race in another season or two will become extinct.[102]

In the same correspondence, Gunn said that the indigenous Tasmanians were as intelligent as any other people, but he laid blame on the benign climate of Tasmania for not having pushed them to developing their culture to a higher level.[103] Although he collected plants on Flinders Island, Gunn does not appear to have taken advantage of assistance from any of the mission folk. He later reported finding the edible climbing apple-berry to be abundant on the island.[104]

In 1842 Gunn published a paper on the edible wild plants of Tasmania.[105] He had to rely on the sparse existing literature, settlers' oral histories and his own field observations of what plants were eaten by settler children, and by wild animals such as cockatoos, bandicoots and kangaroo rats. His work was to some extent an expansion of that published by James Backhouse in 1841.[106] Like most European scholars during the nineteenth century, Backhouse and Gunn considered that Australian food plants had poor potential commercial value. Gunn remarked that Backhouse's publication:

> met with deserved attention in Europe, from the remarkable circumstance that so few of the indigenous plants of these Colonies [in Australia] yield any fruit suitable for human subsistence. In this respect, as has long been noticed by botanists and others, Australia stands singularly apart from every other portion of the known world.[107]

Gunn quoted extensively from Backhouse, in particular the sections involving Aboriginal use of

Greenhood orchids, which plant collector Ronald Gunn listed as a food source. Found in dense clusters, the small edible bulbs were relatively easily gathered by Tasmanian and Victorian Aboriginal foragers. Philip A. Clarke, Bridgewater, Mount Lofty Ranges, South Australia, 1985.

roots and tubers. He listed eight genera of orchids, including those containing greenhoods and potato orchids, which had 'small bulbous roots, which were eaten by the Aborigines'.[108]

Gunn's account of Aboriginal plant uses was drawn entirely from the reminiscences of colonists, such as Backhouse. Of the 'native carrot', or small-leaved geranium, he said, 'I include this plant in my list, as I have been informed that the Aborigines were in the habit of digging up its roots, which are large and fleshy, and roasting them for food.'[109] The 'large white fungus, called in the Colony punk, which grows from the stringy-bark, is said to have been eaten, when fresh, by the Aborigines'.[110] Gunn would not have had access to the journals of Jorgen Jorgenson and George Augustus Robinson, or to other archival sources, all of which were available to more recent researchers compiling accounts of Tasmanian Aboriginal plant use.[111]

In summing up the potentially useful vegetable products from the island, Gunn was sceptical that there was much to offer the wider world, claiming he had listed all species that 'can in any way be rendered available for the sustenance of man. Not one of them is of sufficient value to be worthy of the attention of the agriculturalist or horticulturalist'.[112] Backhouse also played down the value of Australian food plants, in 1843 commenting that it was a 'remarkable circumstance that so few of the indigenous plants of these Colonies [in Australia] yield any fruit suitable for human subsistence. In this respect, the Australian regions stand singularly apart, from every other portion of the known world'.[113] He regarded the wild plants of Tasmania as totally unsuited to agriculture or horticulture.[114]

During 1840–41, Gunn travelled with Joseph Dalton Hooker, the son of Sir William Hooker, both of them collecting plants from across Tasmania.[115] The younger Hooker had arrived on the *Erebus*, as part of the expedition to Antarctica commanded by Captain James Clark Ross. Later in life, Joseph Hooker took over the running of Kew Gardens and became an important figure in the history of science as a champion for Darwin's theories of evolution and natural selection.[116]

Frederick M. Bailey and his South Australian reminiscences

Frederick Manson Bailey was born on 8 March 1827 at Hackney in London.[117] His father, John Bailey, was an experienced horticulturalist with the firm Conrad Loddiges & Sons, nurserymen who specialised in the introduction of new plant species from South Africa and Australia. The Bailey family emigrated to South Australia in 1839, bringing with them a large collection of vines and fruit trees. Although John Bailey had come out to be a farmer, Governor Gawler appointed him Government Botanist and asked him to form a botanic garden instead. As an adult, Frederick Bailey went into partnership with his father and brother to establish the Hackney Nursery, from which a present-day Adelaide suburb derives its name. During the 1850s, he briefly tried mining in the Goldfields district, and in 1858 tried farming in New Zealand. The disturbances of the Maori Wars saw him move to Brisbane in 1861, where he opened a seed store and started collecting plant specimens for sale to overseas herbaria and economic botany museums. In 1875 the Queensland Government appointed Bailey as a botanist to investigate plants suspected of poisoning livestock. He was acting curator of the Queensland Museum in 1880–82 and in 1881 was appointed Colonial Botanist, a position he held until his death in Brisbane on 25 June 1915.

In his official capacity as botanist, Bailey travelled to many parts of Queensland, including the Georgina

River area of western Queensland in 1895 and Cape York Peninsula and the Torres Straits in 1897; he went to New Guinea in 1898.[118] He was the botanist on the 1889 Bellenden-Ker Range expedition, discovering many previously undescribed species.[119] Bailey wrote several botanical monographs and published many papers on topics such as economic plants and timbers, Aboriginal medicinal plants and horticulture.[120] His name was given to more than fifty species of plants.

Although Bailey became prominent as a botanist after he moved to Brisbane in Queensland, it is his experiences of the 1840s that are relevant to this chapter.[121] In later life, Bailey published his reminiscences of the Aboriginal inhabitants of the Adelaide Plains.[122] When he was a child, his family lived so close to Aboriginal camps that he recalled their having to give up experimenting with grinding barley in a coffee grinder 'because the noise startled the natives in their wurlies [bough shelters]'.[123] Bailey's use of the Aboriginal term for shelters reveals the importation during the early nineteenth century of the local Kaurna language term, *wadli* or *wali*, into the South Australian variety of English.[124] With the horticultural interests of his family, it would have been natural for the young Bailey to pay attention to the Aboriginal uses of plants.

His reminiscences provide a good overview of the main indigenous plant foods in the temperate region of southeastern Australia. Bailey noted that Aboriginal women collected vegetables while the menfolk specialised in hunting[125], a social division of labour present throughout Aboriginal Australia.[126] As confirmed by later researchers, Bailey described the importance of roots and tubers to the diets of southern Aboriginal people.[127] He described how Aboriginal cooks gathered the thick and fleshy roots of the flood mallow and baked them in underground ovens. They were described as having the consistency of parsnips.[128]

The tuberous roots of the yam daisy 'were about the size and form of a man's thumbs. They were white, and tasted somewhat like a fresh cocoanut, being milky, and were eaten without any kind of cooking by both whites and blacks'. A note by Bailey on a yam daisy specimen from South Australia in the Queensland Herbarium states that the colonists of South Australia used to eat the roots of this plant, following the practice of the Aborigines who relied on it as food.[129] For Aboriginal people in southeastern Australia, this plant food was perhaps the most important of the roots they gathered for eating.[130]

Bailey provided insights into the preparation of wattle gum as a food additive, recording that:

> larger lumps of gum, formed on the stem of the golden wattle ... were used for food, like we use bread with meat. Especially when they cooked fish, they would give the lumps of gum a little roasting in the embers. This roasting rendered it soft, and prevented it sticking. Particularly I noticed them doing this on the Onkaparinga River, in about 1844.[131]

He commented that the fruit of the quandong (native peach) and ming (bitter quandong) were eaten in the ripe state, without cooking.[132] In terms of artefacts, Bailey remembered:

> meeting a man carrying one of their long shields, full of

Golden wattle gum. Botanist Frederick M. Bailey observed this gum's use by Aboriginal people in the 1840s as a food additive. European bushmen used the wattle bark and the gum, both of which are rich in tannins, as a cure for diarrhoea. Philip A. Clarke, Adelaide Hills, South Australia, 1986.

> what looked like rice, but which, on a closer inspection, proved to be white ants. The natives used the rootstocks of the [scented] sundew Drosera Whittakeri plants for staining. The plant was very abundant on the treeless portions of the parklands.[133]

This record made sense of an earlier reminiscence from Edward Stephens, another colonist, who was not a botanist, that Aboriginal people on the Adelaide Plains gave their bark shields 'a coating of pipeclay or lime, and then ... ornamented with red bands made from the juice of a small tuber which grew in abundance on the virgin soil'.[134]

Bailey listed a number of fruits that were eaten in the 1840s on the Adelaide Plains by settlers' children and Aboriginal people alike, such as the wiry ground berry and native cherry (ballart).[135] Although brief, his accounts from the first decade of settlement in South Australia are important as a record of a flora that was soon to vanish due to vegetation clearance by settlers. The Aboriginal people of the Adelaide Plains, today known as the Kaurna, also suffered a decline due to the arrival of Europeans.[136]

Plant collectors living on the frontier of British colonisation played a vital role in the botanical discovery of Australia. In the cases of Drummond and Molloy, they had considerable interaction with indigenous people and utilised Aboriginal labour in their collecting activities, which makes their records useful sources of Aboriginal plant use data for scholars who are studying Aboriginal relationships to the environment. As a collector in Tasmania from the 1830s, Gunn was not able to employ indigenous guides and did not have Aboriginal informants available to consult on the potential uses of plants, and so the data on indigenous plant uses in Gunn's work is largely anecdotal. Although the working life of Bailey as a botanist was much later than that of the other collectors mentioned here, he was able to draw upon his early life experiences living on the Adelaide Plains, where he had observed a landscape still in active use by Aboriginal hunter-gatherers.

7 Leichhardt and the Riddle of Inland Australia

During the first forty years of British colonisation the maritime surveys undertaken by explorers such as Flinders (1801–02), Baudin (1801–02) and King (1817–22) had resulted in much of the coastline of Australia being mapped. Settlers had spread along the southern coast and penetrated deep into temperate regions along the major river systems. Now, with much of the southern part of the continent known, explorers set themselves goals of finding overland connections between the major cities of the mainland colonies, and began to cross the continent from east to west and from south to north.[1]

The establishing of links between the European outposts on the mainland was important for both strategic and commercial reasons. In 1829–30 explorer Charles Sturt and his party followed the course of the Murray River from its headwaters in eastern New South Wales to its mouth in what was to become South Australia.[2] The prospect of a riverine trade route with links to the sea encouraged pastoralists into the temperate inland areas of the continent. In 1840 drover and explorer Edward J. Eyre set out from Adelaide in South Australia to Albany in Western Australia, travelling along the coastline of the Great Australian Bight.[3] Starting with a party of five, he finally reached his destination with just his Aboriginal guide, Wylie, in 1841.

Although the accomplishments of Sturt and Eyre and the earlier explorers were geographically significant, the arid and subtropical interiors remained as large blanks on the map. In the 1840s colonists were still speculating on whether a large landlocked sea might dominate Central Australia, fed by the rivers discovered by Oxley and Cunningham that flowed westwards from the Great Dividing Range. The theories of the father of modern geography, Prussian scholar Baron Alexander von Humboldt, on the interconnectedness of the topographical elements, suggested that there would be a network of major rivers draining the centre of the continent[4], and the explorers had not yet disapproved the existence of an 'Australian Mississippi'. If such a waterway existed, its potential economic benefits to the mainland colonies that controlled access to it would have been enormous.

Leichhardt the naturalist

One of the most famous land explorers of Australia was the ill-fated naturalist Friedrich W.L. (Ludwig) Leichhardt. He was born in 1813 at Trebatsch in Prussia, the son of a farmer and a royal inspector of peat.[5] During the 1830s, Leichhardt studied at the universities of Berlin and Göttingen, and became a follower of von Humboldt[6], with deep interests in botany, zoology, geology, meteorology and geography.[7] Evading military service, the young Prussian spent some time in England, where in September 1841 botanist Philip B. Webb introduced him to Sir William J. Hooker in the Royal Gardens at Kew, saying to Hooker that this introduction was made on the basis that Leichhardt 'might be of use to your Herbarium'.[8] The meeting does not seem to have gone particularly well, for Hooker preferred to deal with specialist plant collectors rather than generalist natural historians.[9] Besides this, Hooker already had good contacts in Australia, people like Drummond and Gunn, and thought New Zealand

Naturalist and explorer Friedrich W.L. (Ludwig) Leichhardt paid particular attention to Aboriginal plant foods, which he often used in the field to supplement dwindling supplies. Charles Rodius, lithograph, about 1847. nla.pic-an5600270, National Library of Australia, Canberra.

Bunya pine and its seeds (nuts–*facing page, top*). European settlers followed the indigenous practice of eating the nuts of this tree, which Aboriginal people called *bunya-bunya*. It was so important in southeast Queensland that clans owned individual trees (Clarke, 2007, pp. 78–9, 83, 142). Leichhardt observed the Aboriginal use of the bunya pines in 1843. Philip A. Clarke, Great Dividing Range, southeast Queensland, 2004.

would be a more useful place to collect plants. Letters of introduction were not forthcoming.

Leichhardt emigrated to New South Wales in 1842. He was, in his time, probably one of the most highly trained scientists ever to have come to Australia.[10] At the age of 28 he was described as tall and slightly built, with thin facial features, and the people he met must have found him charming, for he appears to have made friends and gathered supporters easily. Seduced by the geographical mysteries of Australia, Leichhardt managed to convince a number of wealthy east coast landowners to finance several expeditions to discover the grazing potential of the interior. The opening up of inland Australia was to become what he called his 'darling scheme'.[11]

Despite lacking the support of Kew, Leichhardt nonetheless started collecting plants and made a number of botanising expeditions between Sydney and Moreton Bay (Brisbane) from 1842 to 1844.[12] He had a particular interest in the food value of the bunya pine, which grows in the Bunya Mountains northwest of Brisbane.[13] At Moreton Bay, he often stayed with German missionaries involved with the local Aboriginal people[14], and took with him from the mission a number of Aboriginal guides during his trips through southeast Queensland.[15] Two young men, Nicky and Gummerigo, climbed trees to collect leaves and fruit specimens that were out of Leichhardt's reach; they were sometimes accompanied by Burbello, an older man who the Archer family of pastoralists had named Abel.

Leichhardt made a study of the indigenous languages around the Moreton Bay and Wide Bay districts, with Nicky's dialect given as 'Karrwa' and Gummerigo said to have spoken 'Karredo'[16], and from Nicky learnt the names and habits of local animals. On a trip along the coast, Leichhardt had with him Nicky and Charley as indigenous guides.[17]

Cabbage tree palm. European explorers, such as Leichhardt, followed the Aboriginal practice of eating the central growth of the palm as greens. Removing the 'cabbage' killed the tree. Philip A. Clarke, Great Dividing Range, southeast Queensland, 2004.

Historian Colin Roderick claimed that 'Leichhardt trusted [his guides] and thought it important for the acquisition of bush lore always to have a blackfellow with him on his excursions. The natives' sharp eyes enabled him to speed up his collection of timbers'.[18] These early experiences with indigenous people would have shaped the way Leichhardt planned his later, more ambitious expeditions.

Leichhardt had available a wealth of publications produced by earlier generations of botanists which he used in pursuit of his botanical studies. From 1793 to 1795, English botanist James Edward Smith had published (in 4 parts) the first detailed study on the Australian flora, *A Specimen of the Botany of New Holland*.[19] Robert Brown, who accompanied

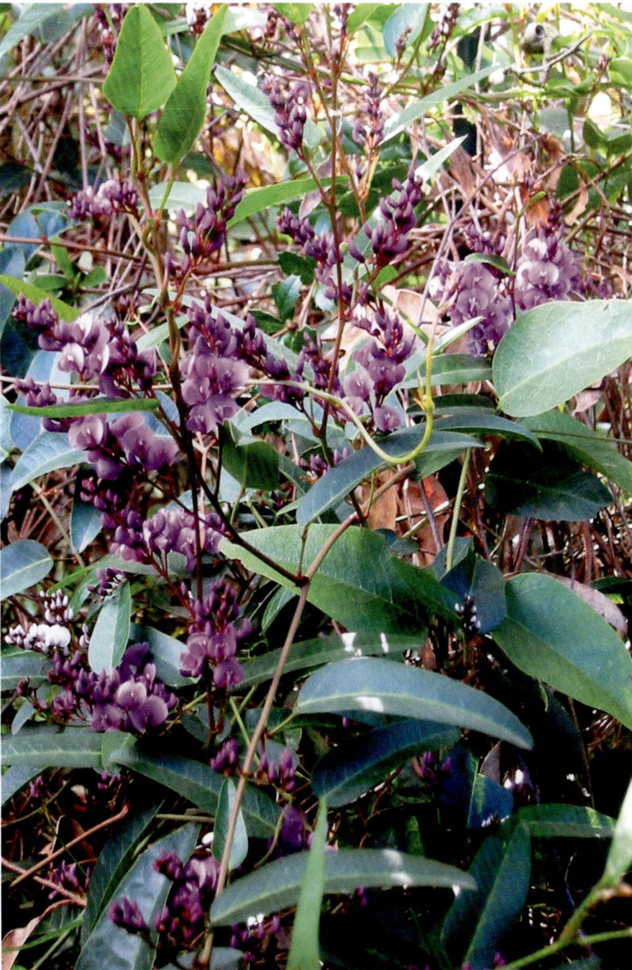

Native lilac or false sarsaparilla. Because of its superficial similarity in appearance to the true sarsaparilla from Europe, the settlers tried making medicinal teas from this creeper. According to Leichhardt, the Aboriginal people around Brisbane in southeast Queensland called it *birri* or *birrwi* and ate the seeds. Philip A. Clarke, Aldgate, Mount Lofty Ranges, South Australia, 2007.

Flinders during the circumnavigation of Australia and was a contemporary of Caley's, had produced *Prodromus* in 1810 and *General Remarks, Geographical and Systematical on the Botany of Terra Australis* in 1814.[20] Hooker had also been active; with the benefit of specimens in the Kew Herbarium supplied by Australian collectors he published the *Journal of Botany* from 1834.[21] These and other published floras provided botanists arriving in Australia with an intellectual framework for the scientific identification of indigenous plants.[22]

With little personal wealth behind him, it was essential for Leichhardt to develop a network of backers to support his geographical and botanical ventures. He was eager to start exploring, drawing on whatever limited resources were available to him. In Sydney he became a friend of the botanist William Phillips, who later helped him prepare his overland field journal for publication[23], and applied unsuccessfully for the job of Superintendent of the Botanic Gardens.[24] Major Mitchell offered him a place on an expedition planned for the north which eventually took in place in 1846, but without Leichhardt, who had plans for an east to west crossing instead.[25]

A key aspect of Leichhardt's exploration plans was his intention to supplement the provisions of his expeditions by drawing upon wild sources, in particular Aboriginal plant foods. His research into Aboriginal plant use was thus not just academic, but of practical use as a bush survival technique. After the success of his Derby to Port Essington trip, he wrote to pastoralist David Archer, with whom he had stayed at Moreton Bay, with a request:

> Do you remember a rich leguminous climber, which the blackfellows called 'Birrwi' and which they eat when ripe – Try to get some seeds of it. If you could employ blackfellows to collect seeds, to which I did not attend so much, when I was in Durrundur [David Archer's property] I should allow you to draw on me for 4–5£ to pay the expense in tobacco, tomahawks, knifes etc. which you might give them. I can send you the catalogue of the native names.[26]

Leichhardt also described the 'Birrwi' (or Breadfruit) as a 'climber, with grapes of violet blossom'.[27] Aboriginal people in the Brisbane area 'call this brush Birri or Birrwi, and eat the fruit on the pod when young'.[28] The plant was the creeper known today as false sarsaparilla or native lilac, and often grown as a garden ornamental.

Across northern Australia

The most successful and famous of Leichhardt's trips commenced in October 1844 at Jimbour Station, near Derby in southeastern Queensland, and finished at Port Essington in the Northern Territory in December 1845.[29] It was an extremely daring and ambitious venture, which took place as Sturt attempted to enter the centre of the continent from Adelaide in the south.[30] Leichhardt was confident that he could establish an overland route from the east coast to the ailing settlement of Port Essington: 'during my recent excursions through the Squatting districts, I had so accustomed myself to a comparatively wild life, and had so closely observed the habits of the aborigines, that I felt assured that the only real difficulties which I could meet with would be of a local character'.[31] While Sturt's expedition failed to reach its goal because of desert harshness and the severity of the seasons, Leichhardt was to have the benefit of travelling through a subtropical landscape that was well watered.

Harry Brown and Charley Fisher

For this expedition, Leichhardt enlisted the services of two Aboriginal guides, Charley Fisher from Bathurst and Harry Brown from Newcastle, who

Charley Fisher and Harry Brown. These Aboriginal men proved themselves essential members of Leichhardt's 1844–45 expedition from southeast Queensland to Port Essington in the Top End of the Northern Territory. Friedrich W.L. (Ludwig) Leichhardt. Reproduced from *Journal of an Overland Expedition in Australia*, 1847.

both possessed broad experience with Europeans. Fisher had previously been employed as a 'native policeman' in Brisbane.[32] Brown's early years were spent on the Aboriginal mission at Lake Macquarie, north of Sydney, where the Reverend Lancelot E. Threlkeld had reported in 1825 seeing him, as a child thought to have been about six or seven years old, at an Aboriginal ceremony, or 'corroboree'.[33] He claimed that he 'spoke a little of broken English'. Settlers knew Brown's father as Moses, although in 1828 his Aboriginal name was written as 'Ngo-ah-ko-ro'.[34] At this time, Brown's own Aboriginal name was not recorded, because to mention it was taboo due to someone sharing his name having recently died.[35] By 1833 Brown had left the mission, and then presumably worked for Europeans.

Threlkeld was full of praise for the usefulness of Aboriginal labour: 'For stockmen the blacks are invaluable, they being exceedingly fond of riding on horseback, and as guides through the bush, none else are so well adapted as they'.[36] He had a particularly high opinion of Brown:

> From childhood, Brown exhibited all the smartness of an intellectual youth, equal to any white lad and when teaching him and other aboriginal youngsters to write the English alphabet, at our establishment on the borders of Lake Macquarie, their ingenuity and quickness of apprehension were exhibited, in chopping with their hatchets on the smooth bark of the standing trees in the wilderness, the Roman characters, thus exhibiting their capacity for learning.[37]

The only sour note was Brown's overindulgence in alcoholic beverages.[38]

The role of the Aboriginal men on the Port Essington expedition was much broader than simply that of guide, for they would have been instrumental in keeping the party supplied with food and water.[39] As had earlier explorers, Leichhardt found that it was easier for one of their guides to make the first approach to the local Aboriginal people they encountered, who often could not bear the 'sight of a white face'.[40] An experienced tracker, Fisher on one occasion was able to find two lost expedition members, the 15-year-old Welsh boy John Murphy, and the 'American negro' Caleb, who went missing in the Condamine River area on 20 October 1844. Leichhardt reported that 'they would certainly have perished, had not Fisher been able to track them'.[41] Leichhardt honoured the Aboriginal men of the expedition party when he named parts of the country he discovered as Charley Creek and Brown Lagoon.[42]

The relationships between Leichhardt and his Aboriginal companions had highs and lows. After the Port Essington expedition, Leichhardt described Brown as 'ill tempered', although he employed him on future expeditions[43], while his relationship with Fisher seems to have been explosive at times. He reported that on one occasion, during an argument, Fisher had threatened 'to stop my jaw'.[44] They came to blows, and Fisher was pushed away by the rest of the party, having punched Leichhardt in the face and loosened two of his lower teeth. The argument came about over Fisher keeping the party waiting while he was out hunting possums and collecting honey. From the journal of zoological collector John Gilbert, who accompanied the expedition for a time, we learn that Gilbert blamed Leichhardt for this particular confrontation, claiming that he would verbally attack the Aboriginal men before giving them a chance to explain any apparent misdeed.[45] Banished for a few days, Fisher was eventually allowed to rejoin the party upon surrendering his axe, a prized possession used for chopping possums and honeycomb out of tree trunks. In spite of this affair, Leichhardt appears to have forgiven Fisher a little after returning from Port Essington, noting that he was 'staying at present at the Clarence [River] collecting birds for Mr Strange and takes to all appearances great care of himself'.[46]

John Gilbert, zoological collector

John Gilbert, who made the unflattering journal entry about Leichhardt's confrontational manner, came to Australia in September 1838 as a collector for naturalist and specialist ornithologist John Gould after meeting him through the Zoological Society of London.[47] Gilbert started collecting in Van Diemen's Land, then went to the Swan River settlement in Western Australia in February 1839, where he worked alongside explorer and plant collector James Drummond during a period of intense frontier conflict.[48] From July 1840 to March 1841, Gilbert was based in Port Essington, north of Darwin, collecting zoological specimens for which he also recorded the Aboriginal names.[49]

The region was then recovering from a cyclone, so settlement conditions were fairly rough. Here, in a letter to Gould, Gilbert wrote:

> I am happy to say the Natives are extremely quiet, and on very friendly terms with the English, so much so that I felt less fear in going about the country already with the imperfect knowledge of them than I did at the Swan [River settlement] after becoming perfectly acquainted with their habits and manners ... They bring me in a good many things, and to-day have brought me three very interesting species of freshwater Fish, which I have preserved.[50]

The Aboriginal people at Port Essington would have had considerable experience with non-Aboriginal visitors already, for the Macassan seafarers who came to northern Australia to collect trepang congregated there in their prahus each April, a practice pre-dating British settlement.[51]

Gilbert returned to England in late 1841, along with his large collection of skins for Gould.[52] He stayed only a few months, then returned to Western Australia for more collecting. In 1844 Gilbert travelled overland from Sydney north to the Darling Downs, where by chance he met up with Leichhardt's overland expedition to Port Essington. Realising an opportunity to obtain more animal specimens, he offered to accompany the party. Before his untimely death in the southern Cape York Peninsula region, he proved himself a good bushman and possibly added to the detailed botanical and anthropological information in Leichhardt's published journal. His own journal was brought back to Sydney by the expeditioners.

Traversing an Aboriginal landscape

Passing through country maintained by Aboriginal hunter-gatherers who frequently burnt the vegetation, Leichhardt noted on 29 November 1844, north of the Boyd River in southeast Queensland, that 'Recent bush fires and still smoking trees betokened the presence of natives; who keep, however, carefully out of sight.'[53] In the Mount Nicholson area on 7 December 1844, he found that 'immense stretches of forest had been lately burned, and no trace of vegetation remained ... The sky was covered by a thin haze, occasioned by extensive bush fires.'[54] The expedition was in the Burdekin River area of northern Queensland on 19 April 1845 when Leichhardt remarked:

> As we were passing over the flats between the creek and the river, we saw a native busily occupied in burning the grass, and eagerly watching its progress: the operation attracted several crows, ready to seize the insects and lizards which might be driven from their hiding places by the fire.[55]

Leichhardt's party also took on the practice of burning grass to rid their camping areas of nuisance 'hornets', or wasps, hiving in surrounding trees.[56]

Leichhardt is generally regarded in the official texts as a poor bushman, with historian Alec H. Chisholm, for example, describing him as a 'blunderer'.[57] Poor bushman or not, it appears from Leichhardt's writings that he was a keen observer who acknowledged local Aboriginal foragers for their detailed knowledge of the environments he was passing through.[58] On many occasions, his expeditions used Aboriginal pathways.[59] Along the Condamine River in Queensland, he observed: 'The well-known tracks [traces] of Blackfellows are everywhere visible; such as trees recently stripped of their bark, the swellings of the apple-tree cut off to make vessels for carrying water, honey cut out, and fresh steps cut in the trees to climb for opossums.'[60] He noted seeing 'heaps of broken muscle-shells [mussels]' left on the banks of lagoons[61], and found holes dug by Aboriginal foragers looking for shells, roots and water.[62] Leichhardt claimed that 'Natives, crows, and kites were always the indications of a good country.'[63]

Living off the land

Leichhardt's exploration party would not have survived the epic journey without being able to take advantage of wild food sources such as roots, gums, fruits, nuts, seeds and greens.[64] At camps abandoned by Aboriginal hunters and gatherers surprised by the sudden appearance of the explorers, Leichhardt often found foodstuffs they had collected. On 31 December 1844 he came across a quickly vacated camp at Comet Creek in southeast Queensland. In conducting a forensic-style investigation of the scene, he noted that 'In their "dillis" (small baskets) were several roots or tubers of an oblong form, about an inch [2.5 cm] in length, and half an inch [1.3 cm] broad, of a sweet taste, and of an agreeable flavour, even when uncooked ...'[65] These may have been tubers of the long yam, a major indigenous food source along the east coast of Australia and in the northern coastal tropics.[66] On 14 January 1845, two Aboriginal women were encountered at Mackenzie River gathering horse beans, which Leichhardt found useful as a coffee substitute after roasting and pounding.[67] He also used the seeds of the sacred lotus in lieu of coffee.[68] Leichhardt would have tentatively tried most of the fruits and seeds he came across during his expeditions, particularly if there was an evidence of indigenous use.[69]

Banyan figs. Leichhardt and his party ate wild fruits such as these. Australian fig species are found in their greatest variety in the northern rainforests. Aboriginal toolmakers collected string fibre from the bark of many species (Clarke, 2007, p. 120). Philip A. Clarke, Adelaide Botanic Gardens, South Australia, 2005.

Lilly pilly or scrub cherry. Leichhardt would have eaten this fruit, an Aboriginal food, as did European settlers along the east coast of Australia and in the coastal Northern Territory. Today the tree is often grown as a garden ornamental. Philip A. Clarke, Adelaide Botanic Gardens, South Australia, 2006.

At Palm Tree Creek in southeast Queensland, Gilbert recorded that they cut down cabbage-tree palms 'for the purpose of obtaining the edible part for vegetables for our evening meal'.[70] Leichhardt was impressed with the quality of the greens available to them in this region:

> Atriplex [saltbush] forms, when young, as we gratefully experienced, an excellent vegetable, as do also the young shoots of Sonchus [thistle]. The tops of the Corypha palm eat well, either baked in hot ashes or raw, and, although very indigestible, did not prove injurious to health when eaten in small quantities.[71]

When Leichhardt's party reached the Kakadu area of the Top End, they found that:

> The natives were remarkably kind and attentive, and offered us the rind of the rose-coloured Eugenia apple, the cabbage of the Seaforthia palm [solitaire palm], a fruit which I did not know, and the nut-like swelling of the rhizoma of either a grass or a sedge [spike rush]. The last had a sweet taste, was very mealy and nourishing, and the best article of the food of the natives we had yet tasted. They called it 'Allamurr' (the natives of Port Essington, 'Murnatt'), and were extremely fond of it. The plant grew in depressions of the plains, where the boys and young men were occupied the whole day in digging for it ... They went to the digging ground, about half a mile [0.8 kilometres] in the plain, where the boys were collecting Allamurr, and brought us a good supply of it; in return for which various presents were made to them. We became very fond of this little tuber ...[72]

In the Nicholson River area of the southern Gulf of Carpentaria, Leichhardt noted that his expedition had:

> collected a great quantity of Terminalia [wild plum] gum, and prepared it in different ways to render it more palatable. The natives, whose tracks we saw everywhere in the scrub, with frequent marks where they had collected gum – seemed to roast it. It dissolved with difficulty in water: added to gelatine soup, it was a great improvement; a little ginger ... and a little salt, would improve it very much. But it acted as a good lenient purgative on all of us.[73]

They also found water lily tubers and seeds, which they had observed Aboriginal people eating, 'extremely nourishing'.[74] Aboriginal gatherers located the tubers by tracing the stems of the heart-shaped leaves floating on the surface of the lagoon down into the mud.[75] The roots were generally eaten after being roasted. At Lynd River in northern Queensland, Leichhardt observed Aboriginal foragers collecting water lily seeds:

> After having blossomed on the surface of the water, the seed-vessel grows larger and heavier, and sinks slowly to the bottom; where it rots until its seeds become free … The natives had consequently to dive for the ripe seed-vessels; and we observed them constantly disappearing and reappearing on the surface of the water … The best way of cooking them was that adopted by the natives, who roast the whole seed-vessel.[76]

It was also Aboriginal practice to eat water lily and lotus lily seeds raw or as small cakes prepared by pounding the seeds between stones into a rough flour, mixing the flour with water, and baking the dough in the ashes of the campfires.[77]

Leichhardt, like others before him in different parts of Australia, found that some foods required special and extensive processing to render them edible. In September 1845 he came across a deserted Aboriginal camp in the Gulf of Carpentaria region, where he discovered:

> [the] seeds of Cycas [zamia palms] were cut into very thin slices, about the size of a shilling, and these were spread out carefully on the ground to dry, after which, (as I saw in another camp a few days later) it seemed that the dry slices are put for several days in water, and, after a good soaking, are closely tied up in tea-tree bark to undergo a peculiar process of fermentation.[78]

Leichhardt was keen to establish what methods the Aboriginal cooks used for processing poisonous foods. He had difficulty working out how they treated the screw palm (pandanus) fruit which he observed in large heaps in their camps.[79] The first few times he ate the fruit, it gave him sore lips, a blistered tongue and on one occasion diarrhoea.

Water lilies. Leichhardt's party found water lily tubers and seeds, which they had observed Aboriginal people eating, to be 'extremely nourishing'. The tubers are located in the mud by tracing the stems of the heart-shaped leaves floating on the surface to their source. Norman B. Tindale, Four Mile Creek, Normanton, northern Queensland, 1963. N.B. Tindale Collection, AA338/6/35/77, South Australian Museum Archives, Adelaide.

Cones and seeds of the screw palm, or pandanus. Because this food requires processing before it is safe to eat, Leichhardt became sick through consuming it in the raw state during his 1844–45 overland trip. Richard T. Maurice, eastern Kimberley, northern Western Australia, 1890s. A1756, South Australian Museum Ethnobotany Collection.

Relations between explorers and indigenous landowners

Observer and observed often traded places. Leichhardt found that Aboriginal people who had kept out of sight as his party passed by often retraced their path and made detailed examinations of the explorers' previous camp.[80] He claimed that direct contact with local Aboriginal groups on this trip was always 'accidental and never sought by them [Aboriginal people]'.[81] The Aboriginal 'cooee' call sometimes indicated the presence of local foragers, even when they were not sighted.[82] Across inland northern Australia the party of Europeans would have seemed extraordinary to Aboriginal inhabitants at the time, as most of them would not have had prior direct experience of Europeans. One large group which Leichhardt met on the Burdekin River appeared to think the expedition's bullocks were the explorers' wives.[83]

Leichhardt respected the Aboriginal ownership of the objects he found. Whenever expedition members took food or artefacts from a deserted camp, he made sure that payment was left in exchange.[84] On 21 November 1844, in the Robinson Creek area of southeast Queensland, his party:

> passed a native camp, which had only lately been vacated, [where] I found, under a few sheets of bark, four fine kangaroo nets, made of the bark of Sterculia [kurrajong]; also several bundles of sticks, which are used to stretch them. As I was in the greatest want of cordage, I took two of these nets; and left, in return, a fine brass-hilted sword, the hilt of which was well polished, four fishing-hooks, and a silk handkerchief, with which, I felt convinced, they would be as well pleased, as I was with the cordage of their nets.[85]

While leaving the sword might have appeared generous, Chisholm thought otherwise: 'The sword was part of Leichhardt's superfluous equipment and he was well rid of it'.[86] Some of the other items left from time to time may have been more useful, things such as spare pieces of iron in the form of bullock nose rings, and on some occasions objects such as brass buttons, a geologist's pick, tin canisters, leather belts and horseshoe nails bent into fishhooks.[87] Leichhardt's appropriation of the useful materials he came across in deserted camps helped replenish his stocks of food and equipment.

It is likely that Leichhardt wanted to establish good relations between Europeans and local Aboriginal groups for the benefit of the settlers who he knew would eventually follow him. In this the Prussian explorer was following Mitchell's common practice of gaining access to country by paying indigenous 'owners' with iron objects.[88] Such was the Aboriginal demand for this exotic material that explorers often had metal objects stolen from them. The indigenous reactions to Leichhardt's gifts or payments are unknown; the metal was probably well received, as northern coastal Aboriginal groups had already developed an interest in iron for their tools through their exposure to the Asian trepangers.[89]

On 12 July 1845 Leichhardt's party was in the southern Gulf of Carpentaria region. Here an Aboriginal group approached them cautiously while they were drying the meat from a slaughtered bullock. Leichhardt held out a branch as a peace sign. The group came in and then:

> They examined Brown's hat, and expressed a great desire to keep it. In order to make them a present, I went to the tents to fetch some broken pieces of iron; and whilst I was away, Brown, wishing to surprise them, mounted his horse, and commenced trotting, which frightened them so much, that they ran away, and did not come again. One of them had a singular weapon, neatly made, and consisting of a long wooden handle, with a sharp piece of iron fixed in at the end, like a lancet. The iron most probably had been obtained from the Malays who annually visit the gulf for trepang. Some of their spears were barbed.[90]

Even though, from Leichhardt's description of the meeting, this group seems to have been totally unfamiliar with Europeans, they had nevertheless been able to obtain metal that must have originated from non-Aboriginal sources.

The 'Malays' mentioned by Leichhardt were most likely Macassarese, who were sea traders based in southern Sulawesi.[91] It is known that they came to Australia with the monsoon winds each year in large fleets, which then split up into smaller divisions to collect trepang in the shallow northern waters. Five days before the above encounter took place, the exploration party had met an Aboriginal man who called them 'Mareka', a term Leichhardt stated meant 'Malay visitors'.[92] The Gulf of Carpentaria was where Matthew Flinders in 1803 had encountered six prahus under the command of a Macassarese man named Pobasso, part of a fleet said to number sixty vessels.[93] If not obtained from visiting Asian trepang-collectors, it is possible that some metallic items could have reached the north through the extensive Aboriginal trade networks then crossing the continent, which may have linked up with European contacts in the south.[94]

Aboriginal people on occasion approached Leichhardt's party for trading purposes. Aboriginal Australia was crisscrossed with trade routes along which were exchanged such things as ochre, narcotics, raw materials for artefact making, finished artefacts, ornaments, sacred items and ritual objects.[95] Dances were also exchanged or passed on at ceremonial gatherings.[96] This trading network established peaceful relationships between groups spread out over large areas.

In some of the regions he passed through, Leichhardt also participated in the local hunter-gatherer economy. In the South Alligator River area of the Top End, Leichhardt noted:

> A little before sunset of the 21st [November 1845] four natives came to our camp; they made us presents of red ochre, which they seemed to value highly, of a spear and a spear's head made of baked sand-stone (grès lustré). In return I gave them a few nails; and as I was under the necessity of parting with every thing heavy which was not of immediate use for our support, I also gave them my geological hammer.[97]

Getting rid of non-essential heavy items was crucial at this point, for the expedition had but one bullock left to use as a pack animal, the others having been consumed.

Not all encounters with indigenous people were as passive as those just mentioned. Gilbert died on 28 June 1845 from the wound caused by a barbed spear lodged in his chest – the weapon had been thrown as he stooped to exit from his tent.[98] This occurred on the Mitchell River in southern Cape York Peninsula when an Aboriginal group attacked during the night, spearing and clubbing several members of the party. The conventional view, as strongly argued by Chisholm, is that the attack was probably the result of the expedition's two guides attempting to interfere with local women a few days earlier. Another view is that the raid was punishment inflicted for the party's removal of lotus seeds at Mosquito Lagoon the week before, which had been done without consent from the traditional owners.[99] A third possibility is that the explorers were attacked because they had trespassed across Aboriginal ceremonial grounds.[100] It is, of course, unlikely that we will ever know the true reason behind the raid. Expedition members buried Gilbert's remains and made a large fire over his grave 'to prevent the natives from detecting and disinterring the body'.[101]

Two other members of the expedition carried serious injuries resulting from the attack for some months, but in spite of their poor condition they had to keep moving. On 9 July 1845, heading south along the coast of the western side of Cape York

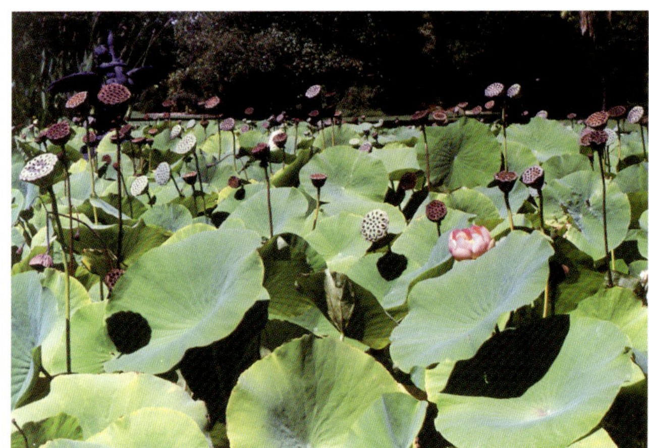

Sacred lotus, seed capsules on stalks. Leichhardt used ground lotus seeds in lieu of coffee. Aboriginal foragers ate the seeds raw, or in small cakes prepared from a flour made by grinding the seeds on a stone. Philip A. Clarke, Adelaide Botanic Gardens, 2006.

Peninsula, Leichhardt found a 'small green looking tree [raspberry jam tree], which smelt like raspberry jam; and, upon burning it, the ashes produced a very strong lye [leached solution], which I used in dressing the wounds of my companions.'[102] Approaching the bottom of the Gulf, on 12 July 1845 he named a river 'the "Gilbert", after my unfortunate companion.'[103]

From east to west

On 17 December 1845 the explorers finally reached Port Essington. They recuperated there for a month before taking ship to Sydney, where they were received as celebrities.[104] Leichhardt presented the Botanic Gardens with 200 different kinds of seeds from his expedition, which were then sown on the grounds.[105] While he had to abandon most of his pressed plant collection during the trip after four packhorses were drowned, he managed to bring back a small number of specimens in boxes covered in greenhide to protect them.[106] His botanical specimens were dispatched to botanists in Germany and France. Encouraged by his initial success, and having received liberal support from the public, the Prussian naturalist explorer began to plan a new expedition.

This next venture was to be the first attempt by a European at making an east–west crossing of the Australian continent, from Sydney to Swan River.[107] It was planned that the party would follow the same parallel of latitude all the way, if possible, to finally settle the question of what lay in the interior of the continent. Leichhardt's party, consisting of seven Europeans and two Aboriginal guides, commenced the journey by travelling overland from the Hunter River to the Darling Downs, bringing with them mules, cattle and goats. On 10 December 1846, they departed from Mr Stephens' station on the Condamine, with their stock built up to forty bullocks, fifteen horses, thirteen mules, two hundred and seventy goats, and one hundred and eighty sheep. These animals, along with their supplies of dried food, were meant to see them out on a trip lasting two years. His confidence

Victoria Square, Port Essington. This was how explorer Leichhardt saw this European outpost at the conclusion of his northern Australian expedition. Friedrich W.L. (Ludwig) Leichhardt. Reproduced from *Journal of an Overland Expedition in Australia*, 1847.

based on the experience of his successful trip to Port Essington, Leichhardt apparently intended to depend on foods found along the way for fresh vegetables and fruit. On this trip, Daniel Bunce was the expedition botanist.

Wommai

Aboriginal guide Harry Brown again accompanied Leichhardt, although not in the best of health, apparently through excessive indulgence in alcohol since his return from Port Essington.[108] A younger man called Wommai, also known as Jemmy, Jimmy or Killali, went with them.[109] Wommai was 'a Stroud Pt Stephens Black' from Newcastle in New South Wales, and was previously in the service of Captain Phillip Parker King.[110] To Leichhardt, he initially seemed 'too slow and uncertain'[111], but he was then only 'a lad and not strong enough in assisting us in loading our mules'.[112] He later said of Wommai:

> Besides being extremely useful he's a good-natured young fellow and by no means as uncivilised as the other one [Brown], who was with me on my first expedition ... The other one is not so useful; but they both have extraordinarily sharp eyesight, which is what makes them exceptionally useful to me.[113]

Wommai appears to have grown into the role of an expedition guide, taking on more responsibility as he gained in experience. It was on this first attempt at an east–west crossing that Brown gave Wommai his chest cuts to make him 'a young man'.[114] This would have given a significant boost to Wommai's standing within his own Worimi community at Port Stephens. Anthropologist A.W. Howitt described the production of prominent raised keloid scars, known as cicatrices, on adult males as being not just for decorative purposes, but also to 'denote the class of the bearer, or his hardihood and prowess'.[115] As significant as it was, the operation would not have required a full-blown ceremony. Wommai's status as a trusted guide amongst the colonists would also have contributed to his status in the broader emerging culture of post-European Aboriginal Australia.

Failure

On this first attempt at an east–west crossing Leichhardt's party was unable to leave the old Port Essington track from the 1844–45 expedition. Heavy rains fell as they struck the brigalow scrubs of the Dawson and Mackenzie rivers. Historian Ernest Favenc, who was scathing in his assessment of Leichhardt as an explorer, described the hardship that followed:

> Fever was the result, and they had no medicines with them – a strange omission. Their only coverings during the wet were two miserable calico tents. Their life, as told by members of the party, consisted of semi-starvation, varied by gorging and feasting on killing days, in which the Doctor apparently set the example; in fact, his character throughout comes out in anything but an amiable light, and one is led to wonder how anyone so destitute of tact and readiness of resource ever achieved the journey to Port Essington, favoured even as he was on that occasion by circumstances and seasons.[116]

Bunce reported that the party ate wild foods, such as boiled 'portulacca' (common pigweed).[117] Having lost first their sheep and goats, then their cattle and most of their horses and mules, the party was forced back east, turning up at the Chauvel station on the Condamine River on 6 July 1847. From start to finish, the whole trip lasted only nine months.

The second attempt

Early in 1848 Leichhardt tried again to cross Australia from east to west, this time leaving from Roma in southeast Queensland.[118] This later expedition, consisting of six Europeans and two Aboriginal guides, was not as well equipped as the previous party had been, taking with them fifty bullocks, thirteen mules, twelve horses, two hundred and seventy goats, and no sheep. The two guides were Wommai and Billy, who had replaced Brown. Billy also came from the Port Stephens area and was noted for having a large bushy beard and whiskers, while Wommai was clean-shaven.[119] Leichhardt was pleased by the expedition's early progress, finding what was then known as Fitzroy Downs, the region between Roma and Mitchell, to be well grassed. At this stage he would have been convinced that his party would find enough food and water to sustain them on their journey. In a letter to a friend in Sydney dated 3 April 1848 and written at Mount Abundance, he mentioned the minor problems of some footsore cattle, the loss of a spade, and 'myriads of flies'.[120] Leichhardt's party left Mount Abundance, the most westerly sheep station on the Darling Downs, the same day, and vanished from European sight.

As the months passed without word, it began to be feared that the exploration party had met with misfortune and perished. In 1855 Threlkeld, lamenting their loss, suggested that if Brown had been present on this last trip, the outcome would have been different:

> I regret that in the last attempt Brown, the black, was not with the heroic explorer. This aborigine might have proved of the greatest use in extricating the party from local difficulties, or in quickly discovering the ambush of hostile blacks. Both are now counted with the dead. Leichhardt in the pursuit of science, Brown in following up his evil propensity for strong drink.[121]

The mysterious fate of Leichhardt's last expedition has entered Australian folklore, with some accounts suggesting that a sole survivor from the missing and 'massacred' party was living within a Central Australian Aboriginal community several years after the party had disappeared.[122] These accounts often name the survivor as August Classen, who was second-in-command of the expedition and the brother of Doris Leichhardt, the wife of Ludwig's brother Hermann.[123] Favenc said:

> For some inexplicable reason, this man, whose name was Classen or Klausen, has always been selected as the hero of the many tales that have been brought in of a solitary survivor of the party living in captivity with the natives; probably, because his was the only name besides Leichhardt's generally known and remembered.[124]

In 1852, in the wake of persistent rumours reaching the authorities about an Aboriginal massacre of Europeans some years earlier, west of the Maranoa River in southern central Queensland, a search party was despatched from Sydney to investigate. Hovenden Hely, who had accompanied Leichhardt on his first east–west attempt, headed the party, which included Brown as guide, a surgeon, a surveyor and five other men.[125] Their findings were inconclusive. Other searches were made over the following decades, but the results of all were equally disappointing.[126]

Around 1900 a prospector was working near Sturt Creek in northern Western Australia when his Aboriginal assistant Jacky found a partly burnt and rusted firearm inside a bottle tree marked on the trunk with an 'L'. Attached to the firearm was a brass nameplate bearing the words 'Ludwig Leichhardt. 1848', which is strong evidence that Leichhardt's party, or at least a remnant of it, had passed through the area on the way south towards the temperate zone.[127] As a follower of Humboldt's theories, the Prussian explorer would have been seeking a river connection between the centre of the continent and Swan River in the south. If Leichhardt really had been in the Sturt Creek area, the last members of his second east–west expedition would probably have perished somewhere in the Great Sandy Desert.

During the decades following his disappearance, botanists recognised Leichhardt's scientific and geographical achievements by naming after him a large number of Australian plants. Many of these names have now been superseded but some plants still bear his name. One such species is the yellow basswood (*Duboisia leichhardtii*), which is today grown commercially as a source of alkaloids for the pharmaceutical industry.[128] A common name that celebrates the explorer is the Leichhardt tree, a northern Australian species which Aboriginal people used as a source of medicine and fish poison.[129] Another is the Leichhardt wattle from southeast Queensland, which also has a botanical name that honours him, *Acacia leichhardtii*.

Down the years in Australian history, Leichhardt has generally been portrayed poorly in terms of his leadership and bushcraft, although in more recent times some scholars have suggested that these alleged faults were magnified by anti-Prussian sentiment and by the jealousies of explorer Major Mitchell and his circle of supporters.[130] Leaving aside the question of his prowess as a leader of men, a reading of Leichhardt's journals and surviving correspondence provides ample proof that he was a skilled botanist who possessed a deep interest in how Aboriginal people interacted with the environment. Such qualities in a researcher today would lead to that person being described as an ethnobotanist.

8 Von Mueller and Australian Botany

The prominent nineteenth century botanist and explorer Baron Ferdinand Jakob Heinrich von Mueller, a prodigious plant collector, became a driving force in Australian botany after the domination of Banks ended.[1] From his Melbourne base as Government Botanist, von Mueller travelled widely across Australia, sometimes relying on indigenous food sources to sustain him during his expeditions. Unlike most botanists based in the Northern Hemisphere, who generally described new Australian species from dried plants and drawings sent to them by collectors, in communicating the results of his plant discoveries to European scholars von Mueller had the advantage of being able to observe many of his plants as living field specimens.

Sir Ferdinand Jakob Heinrich von Mueller (Baron von Mueller) headed a network of amateur botanists across Australia. Von Mueller advocated the employment of Aboriginal people as plant collectors, and possessed an intense interest in European uses of Australian plant species. John Botterill, photograph, about 1867. H4146, State Library of Victoria, Melbourne.

Von Mueller was born at Rostock, Mecklenburg-Schwerin, Germany, in 1825.[2] Following a four-year apprenticeship as an apothecary in Husum, he studied botany at Kiel University, receiving his doctorate at the age of 21 for a thesis on the common shepherd's purse. He then undertook a medical degree but, after losing most of his family to tuberculosis and in the interests of his own poor health, decided not to finish and emigrated to South Australia. He arrived in Adelaide with his surviving sisters in December 1847, where he was initially employed as a pharmacist. The eminent Prussian scholar-explorer Baron Alexander von Humboldt, who had a strong interest in botany, was as much of an inspiration to the young von Mueller[3], another Prussian, as he was to Ludwig Leichhardt, and von Mueller began to explore South Australia, making expeditions to the Flinders Ranges, Lake Torrens and the southeast of the colony. On a trip to the Murray Mallee he nearly drowned when crossing the Murray River. In 1849 von Mueller became a British citizen and anglicised the spelling of his name (formerly Müller). He also briefly tried farming in the Bugle Ranges, east of Adelaide.

His interest in botany becoming ever greater, von Mueller made contact with the Royal Gardens at Kew and, with the help of a recommendation from Hooker, in 1853 secured the appointment as the first Government Botanist in Victoria, a post that he held for forty-three years (until 1896), concurrently with various other positions.[4] He was Director of the Botanic Gardens in Melbourne for many years, from 1857 to 1873. His collecting trips were intended not only to discover and collect new plant species, but also to investigate botanical resources of potential interest to industry. He was also driven by a conviction undoubtedly inspired by his tragic family history that plants could hold cures for diseases such as tuberculosis. In his pursuit of plants worthy of horticultural development, von Mueller was carrying on the plant hunting work initiated by Banks in the previous century, at a time when botanists were striving towards the production of a published flora for the whole Australian continent. Von Mueller's plant collecting trips produced an extraordinary volume of specimens that eventually formed the foundation of the National Herbarium of Victoria.

Recording Aboriginal plant uses

Von Mueller often travelled alone on his field trips, requesting only occasional assistance from local settlers or Aboriginal groups. His interest in Aboriginal people was limited; his concerns were restricted to the plants they used, particularly those which Europeans might develop for horticultural use. Despite his lack of interest in indigenous cultures and their spiritual links with the flora, Aboriginal people nevertheless formed a backdrop to his plant collecting activities, for they were still actively hunting and gathering in the remote parts of Australia that von Mueller visited. It was not until later in his life that a more intense European interest in the former diversity of Aboriginal culture across the continent developed among scholars. The editors of von Mueller's recently published correspondence comment that the botanist showed little sensitivity to the Aboriginal community, considering them a 'barbaric people in need of the civilizing influence of Europeans'.[5] He did, however, consider that Australian hunter-gatherers were physiologically worthy of scientific study, for he sent several fresh Aboriginal skeletons to scholars in Germany via Hermann Beckler, the medical officer on the ill-fated Burke and Wills expedition of 1860–61.[6]

Von Mueller often utilised Aboriginal people's knowledge of plants. In March 1855 he was on a field trip to the Lake Wellington area of Gippsland in Victoria when he wrote a letter to Hooker at Kew:

> Here on the coast and in other parts of Gipps land I observed a Solanum, called by the aborigines Gunyang [kangaroo apple], which promises to become an additional fruit shrub of our gardens. I have not yet obtained the perfect ripe fruit, which is said to be of excellent taste and of which the natives are passionately fond.[7]

Von Mueller was so impressed with the fruit's potential that he hurriedly published an account of the gunyang in Hooker's *Journal of Botany*, based in London, in which he claimed:

> The number of fruits indigenous to this Colony [Victoria] is so limited, that any addition to them cannot fail to attract a far more general attention than even the most important discoveries in the medicinal properties of our plants, or in their geographical distribution or affinity likely would secure. With this view I selected from a series of new plants, which were obtained during my last journey through the eastern parts of this Colony, the "Gunyang," for an early publication. That the natives apply a special name to this production of our Flora warrants its usefulness in their nomadic life; and as, in fact, the Gipps' Land tribes collect this fruit eagerly, and as probably cultivation will improve it so much as to render the plant acceptable for our gardens.[8]

The plant proved to have a fairly wide distribution, and the name *gunyang* from the Ganay language of Gippsland was widely adopted in other parts of Australia.[9] Von Mueller overestimated the willingness of European colonists to adopt Australian species of food plants, however, and his vision of the gunyang becoming a common garden fruit has not yet eventuated.

Von Mueller's most significant account of Australian Aboriginal plant use appeared in Robert Brough Smyth's two-volume work, *The Aborigines of Victoria*, published by the Victorian Government in 1878. It is essentially a list of vegetable food resources utilised by Victorian Aboriginal hunters and gatherers[10], organised into the categories of tubers, roots, young shoots, fruits, seeds, leaves, gums, honey-like secretions, top shoots and truffles. Von Mueller's account bears similarities to those given by Dawson for western Victoria, and by Backhouse and Gunn for Tasmania, although since no references are given it is possible that it was largely based upon his own experiences.[11] More recent research on food plants in southeastern Australia has confirmed that von Mueller's list contains the main species that Aboriginal foragers utilised in the region.[12]

Collecting for the Baron

Von Mueller established a network of plant collectors across Australia whose efforts he rewarded by sending them packets of garden seeds.[13] He honoured the most important of his collectors by using their names when describing new species. For example, *Scaevola brookeana* (heart-leaved fanflower) was so named after Sarah Brooks, a collector in southwest Western Australia.[14] He was always looking for geographically well-placed people to gather plant specimens, whether they were public officials or land owners. Sarah Brooks' brother, John, also one of von Mueller's collectors, employed Aboriginal people on his botanical expeditions. In 1875 John Brooks was exploring inland from Israelite Bay towards Lake Roe in central Western Australia, with an Aboriginal guide known as Black Ben.[15] As was the case for most indigenous assistants, little appears to have been recorded about Black Ben's life. Few, if any, Aboriginal collectors anywhere in Australia have ever been acknowledged in the naming of new plant genera or species. This is in spite of the fact that many collectors would have greatly benefited from their access to indigenous knowledge of the environment.

In his role of coordinating the flow of plant specimens from the field to the herbarium, von Mueller nonetheless recognised the value of his collectors gaining the assistance of local Aboriginal people. In 1883 he appealed to a Western Australian newspaper to 'urge inland and northern and far eastern settlers to induce the natives to bring, in baskets, specimens of all sorts of plants, to be dried at the stations and forwarded to me by post.'[16] Many of his collectors lived far from the capital cities, in areas where they had considerable interaction with local indigenous people in a variety of ways. Sadly, von Mueller's personal drive and enthusiasm was so critical to his collectors' activities that after his death in 1896 most of the network he had established fell apart.[17]

Paul Foelsche

One of von Mueller's most prolific collectors and correspondents was Paul Heinrich Matthias Foelsche, a mounted policeman working in the Northern Territory from 1870 to 1904.[18] He was an

Paul Heinrich Matthias Foelsche. Working as a mounted policeman in the Northern Territory from 1870 to 1904, Foelsche was one of von Mueller's most prolific plant collectors and correspondents. He had a scholarly interest in Aboriginal culture. Paul Foelsche, Palmerston, Northern Territory, late nineteenth century. P. Foelsche Collection, AA96, South Australian Museum Archives, Adelaide.

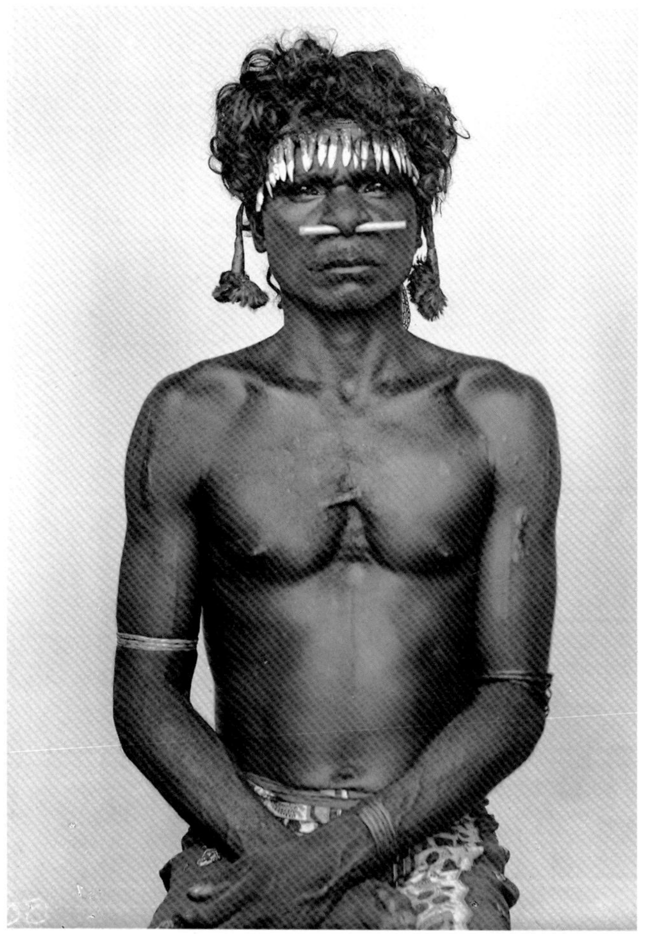

Biliamuk Gapal, pictured at the age of thirty-four. Inspector Foelsche employed this Larrakia man as a tracker. Biliamuk saved the life of plant collector Friedrich Schultz by standing in front of him as other Aboriginal men threatened to spear him. Paul Foelsche, Palmerston, Northern Territory, 1890. P. Foelsche Collection, AA96/200, South Australian Museum Archives, Adelaide.

Aboriginal artefacts collected and photographed by Foelsche. Many of these items were sent to the South Australian Museum and placed on public display. Paul Foelsche, Palmerston, Northern Territory, late nineteenth century. P. Foelsche Collection, AA96/93, South Australian Museum Archives, Adelaide.

educated man living on the frontier and probably became a plant collector as a hobby. Being based in Darwin, Foelsche was able to establish close relations with Aboriginal people of the Top End. One of his close associates was the tracker Biliamuk Gapal, a Larrakia man known by Europeans as 'Billy Muck'.[19] Biliamuk was well known, having in 1869 saved the life of the plant collector Friedrich Schultz by standing in front of him when he was about to be speared.

With such a large population of Aboriginal hunter-gatherers living in Foelsche's policing district, botanist Joseph H. Maiden claimed that the mounted policeman:

> thus possessed unique opportunities for acquiring a knowledge of the aborigines, and being an excellent photographer, he acquired a remarkable collection of negatives of them ... For some photographs of the Northern Territory aborigines he received a 'magnificent gold hunting watch and signed enlarged photograph of the Kaiser'.[20]

The South Australian Museum has acquired large collections of Foelsche's photographs and artefacts, which together form an important record of Aboriginal life in the Top End of the Northern Territory in the late nineteenth century.[21]

Apart from his prowess as a photographer, Foelsche also wrote on the ethnography of the tropics around Darwin. In 1881 he published a major paper, titled 'Notes on the Aborigines of

Aboriginal camp at Palmerston in the Northern Territory. Following the loss of land due to European settlement, indigenous peoples were often forced to live in camps such as this on the fringes of towns. Paul Foelsche, Palmerston, Northern Territory, late nineteenth century. P. Foelsche Collection, AA96/92, South Australian Museum Archives, Adelaide.

North Australia', in the *Transactions of the Royal Society of South Australia*, in which he documented several indigenous plant uses, such as for the caustic-vine (milk bush):

> The remedy the natives apply to cure smallpox is a thick milky-looking juice obtained from a leafless vine found along the shores of mangrove flats. It twines in among other bushes, and is called by Port Darwin native 'Gaoloowurrah.' This juice is put on the sores, and left till it forms a scab, which is washed off so soon as it gets loose, when the sore is found to be healed, the skin is white, and takes about a year to attain its natural colour. This remedy is said to be a sure cure, although some who used it lost their eyesight; but strange to say some patients object to having it applied, but why they cannot explain.[22]

Editor Ralph Tate remarked in a footnote to Foelsche's paper: 'On my recent visit to the Northern Territory, the plant, which was shown to me by a native in company with the author, proved to be *Sarcostemma australe*'. This plant species was widely used in this manner by Aboriginal healers in other parts of Australia.[23]

In another paper, published in 1895 in the form of answers to a survey by the *Journal of the Royal Anthropological Institute of Great Britain and Ireland*, Foelsche outlined the medical practices among Aboriginal people around Darwin. For common ailments, plants featured heavily in his account of their remedies:

> Boils are treated with poultices made with hot water and leaves of certain trees ... Gatherings in the ears are treated with the juice of the fruit of the red Eugenia [eugenia apple?] by squeezing it into the ear after the fruit is roasted. Neuralgia is treated by applying poultices of the same fruit roasted.[24]

In a section of this paper headed 'Magic and Divination', the inspector described the role of Aboriginal 'doctors':

> each tribe has one or more professed doctors who profess to cure disease but cannot inflict it; their only mode of operation with which I am acquainted is to suck the affected part of the body, but they apply no internal remedies. No doctor is called in to treat external injuries or complaints such as wounds, broken limbs, boils, &c.; most natives know how to treat such complaints themselves.[25]

Foelsche was considered to be an anthropological expert on the 'tribes' living around Darwin. The Director of the South Australian Museum, Edward C. Stirling, remarked that 'Mr. Foelsche is a most intelligent and *accurate* observer, knows the natives well, and has great influence'.[26] Von Mueller honoured Foelsche in the botanical naming of the broad-leaved bloodwood, from the

Milk-bush or caustic-vine. Aboriginal healers widely used the white sap of this succulent plant in their treatment of cuts and abrasions. Plant collector Paul Foelsche described its use in the Top End of the Northern Territory. Philip A. Clarke, Flinders Ranges, South Australia, 1984.

Ellis Rowan. The painter employed Aboriginal people to help her collect fresh plant specimens. From her watercolour paintings, botanist von Mueller would later identify the species she had portrayed. John Longstaff, oil on canvas, 1926. nla.pic-an2310713, National Library of Australia, Canberra.

Top End of the Northern Territory and northern Western Australia, as *Eucalyptus foelscheana* (now *Corymbia foelscheana*).[27]

Ellis Rowan

Many amongst von Mueller's network of collectors and illustrators were women, an analysis of his surviving correspondence indicating that he had at least two hundred female correspondents.[28] Artists were needed in the field to record the living plants before they became dried and shrivelled specimens attached to herbarium cards, and at the time painting was considered an acceptable pursuit for women while academic botany was largely left for educated men.[29] In New South Wales, Australian-born Caroline Louisa Atkinson, a keen collector of natural history specimens and a renowned botanical illustrator, was one of von Mueller's associates.[30]

Ellis Rowan from Victoria was one of von Mueller's more famous artist contacts.[31] Rowan specialised in watercolours of flowers, but also painted birds and occasionally insects. The plants and flowers were usually portrayed in groups as handpicked bunches set in a natural context, with the background worked in an impressionist style. On the back of many of Rowan's paintings von Mueller penned the scientific names of the plants in his distinctive large and scrawly cursive.

Ellis Rowan was born in 1848 at Kilfera station, one of her father's properties in the Port Phillip District in Victoria, and started life with the name of Marion Ellis Ryan.[32] In 1873 she married Captain Charles Rowan, a former British army officer who had fought in 1866 as part of the New Zealand forces in the Second Taranaki War. The newly married couple went to Taranaki, where Rowan became a sub-inspector in the armed constabulary, but returned to Victoria in 1877. Charles had an interest in botany and encouraged his wife to become a 'wild flower painter'. Von Mueller already knew Ellis through her father, and he took an active interest in her work. Although she had no formal training, Rowan quickly emerged as a competent and prolific artist, winning major art prizes within Australia and internationally. This gained her some notoriety in the purist art world, which regarded her flower paintings as suited to museums, not art galleries, for they crossed the boundaries between art and natural history illustration.

The desire to collect fresh specimens of exotic plants to paint led Rowan to many remote and dangerous parts of Australia. Small and slight, she became renowned for her determination to find new plant species, as evidenced by accounts of her being lowered by rope over cliffs.[33] From 1887 she made several long trips to northern Queensland. In correspondence to her husband, Rowan gave detailed accounts of the people she met and socialised with, both indigenous and European. Her contact with a broad range of people in such dangerous landscapes makes her unique in her time. While Rowan was often amazed at the customs of the Aboriginal people she met, her experiences with them were largely positive, in 1891 on Bloomfield River, north of Cairns, for example, Aboriginal people saving her from a fall into the water.[34]

Some of Rowan's observations are relevant to the modern academic study of indigenous plant use (ethnobotany). At Bloomfield River, Rowan had the opportunity to observe Aboriginal techniques for preparing cycad nuts:

> This evening we watched the 'gins' [Aboriginal women] preparing their food of zamia (cycad) nuts, which they pound into a soft, pulpy-looking stuff. This is put into bits of bark and water run through it to extract some kind of poison. It was like thick pea-soup when strained afterwards through their dilly bags. I ate some of the nasty, unwholesome-looking stuff, and found it utterly tasteless.[35]

Here, Rowan also commented upon Aboriginal snakebite treatments:

> The natives have remedies for the bites of different kinds. They are more afraid of brown snakes [possibly taipans] than any other, and when bitten, pound up the leaves of a plant, a specimen of which I have painted, putting some on the wound and drinking a concoction of them.

The identity of the plant which was the source of the medicine was later revealed in her journal of a trip to the Torres Strait Islands in 1892. She outlined the uses of various species of trees known as eugenia, claiming that the 'leaves of one of the large varieties is an antidote to the bite of the brown snake'.[36]

Rowan employed Aboriginal people to help her collect fresh plants as painting subjects. On one occasion on the Bloomfield River, she tried with 'tempting accents ("mine give budgery plenty tobacco")' to convince an Aboriginal woman to climb a tree to gather red-flowering mistletoe, eventually succeeding in gaining her cooperation by offering to pay her with clothes, hat and boots.[37] When staying at Mabuiag (Jervis Island) in the western Torres Strait, the local chief ordered village children to collect flowers for her.[38]

It was planned that Rowan would illustrate the book von Mueller intended to write on the whole of the Australian flora, a project abandoned after his death in 1896.[39] She did, however, illustrate three floras in North America, written by botanist Alice Lounsberry and published in 1899, 1900 and 1901.[40] During 1916–18 Rowan made two trips to Papua New Guinea, where she concentrated on illustrating wild flowers and birds of paradise. She died at Macedon, Victoria, in 1922.[41]

Hermann Kempe

From 1877 until 1893 Lutheran missionary Pastor Hermann Kempe was based at Hermannsburg Mission settlement (now known by its Arrernte name Ntaria) in the Macdonnell Ranges of Central Australia. Von Mueller developed a close intellectual relationship with his former countryman, who sent him about 600 plant specimens from the Centre.[42] It is likely that Aboriginal collectors were responsible for many of them. One of the specimens Kempe sent to von Mueller was identified only as 'Eucalyptus no. 8' and with the Aboriginal name 'ilumba [*ilwempe*, ghost gum]'.[43] It was described as a tree growing to 50 metres tall in favourable positions. In return, von Mueller assisted the Mission by sending seeds and cuttings of useful food plants such as figs, arrowroot and Cape gooseberry suited to cultivation in the extreme desert environment.

Under von Mueller's auspices, Kempe published

Flaky-barked Satin Ash. Ellis Rowan painted this species, also known as 'white apple', and which bears edible fruits, during a trip to northern Queensland. Ellis Rowan, watercolour, 1891. nla.pic-an6764596, National Library of Australia, Canberra.

Ghost gum. In 1881 Pastor Hermann Kempe sent von Mueller botanical specimens of this tree species, identified by its Arrernte name of *ilumba*. Philip A. Clarke, Simpson Gap, Macdonnell Ranges, Central Australia, 1997.

several lists of plants collected around the Mission. There is some Aboriginal plant use data, such as for wild currant, about which Kempe remarked, 'Its sweet berries are the favourite eating of the natives here.'[44] Similarly, for the desert pear he wrote, 'the fruits of which are eaten by the aborigines', and for pituri, 'the leaves of this shrub are used by the natives to poison emus'.[45] In 1891, Kempe published a paper on the language spoken around the Mission. The plant use data it contains is slight, possibly because his informants had not lived as hunter-gatherers. He claimed that 'It is only with the help of the boys grown up on the station, and who have become less nomadic than their elders, that the knowledge now gained has been established'.[46]

The botanist-explorer

Von Mueller used his government positions and standing with peak scientific societies to assist in the maintenance of his extensive network. Among his collectors were many prominent explorers, including Charles Sturt, Ernest Giles, John Forrest and William Tietkens.[47] In the case of Giles, von Mueller helped plan, publicise and obtain funding for his expeditions across Central Australia, for which the pay-off was significant.[48] During two expeditions (1872, 1873–74) Giles was able to make a plant collection representing over 250 species, at least forty of which von Mueller eventually scientifically named and described.[49] Desert plants had a severe impact on the explorer's forays into the deserts of Western Australia, with the spines of the hard spinifex grass (porcupine grass) puncturing the legs of his horses, and the camel poison (sandhill corkbark) eaten by his camels rendering them immobile.[50]

Always fascinated by exploring, von Mueller was deeply interested in the fate of Ludwig Leichhardt and his party. The two men had never met, for von Mueller arrived in South Australia less than four months before Leichhardt disappeared into the vastness of the Queensland outback. If von Mueller felt some affinity with Leichhardt, it was probably because both were Prussian scientists living in British colonies and therefore marginalised in some circles, and both professed admirers of their former countryman, Baron von Humboldt. Von Mueller also had a minor role in the Exploration Committee of the Royal Society of Victoria, during the period in which Burke and Wills crossed the continent from south to north (1860–61).[51]

Von Mueller was not an 'armchair botanist', one who let others do all the field work. He was a prodigious plant collector in his own right and as a scholar he was a prolific writer. In 1856 he accompanied explorer Augustus C. Gregory on the North Australian Expedition, which commenced its journey at Point Pearce, north of Brisbane, and proceeded overland to the Victoria River in the Northern Territory.[52] As had happened to Allan Cunningham, the dangers that went with exploration interfered with von Mueller's botanical work. On this northern trip, Aboriginal people

Hard spinifex or porcupine grass (foreground). This spiky plant was the scourge of mounted desert explorers, such as Ernest Giles, because it severely irritated the horses' legs. For Aboriginal people, spinifex was a major source of gum for making artefact cement (Clarke, 2007, p. 121). Philip A. Clarke, Great Victoria Desert, Western Australia, 2006.

Camel poison tree. This species in 1875 poisoned the camels of explorer Ernest Giles when he was travelling northeast of Lake Gairdner in northern South Australia. Aboriginal hunters used its foliage to poison waterholes to catch large game (Latz, 1995, p. 205). Philip A. Clarke, Ooldea sand dunes, western South Australia, 2006.

attacked their party after passing the Leichhardt River, resulting in the 'leader' of the raid being killed. Von Mueller complained: 'Impossible as it was to remain so far behind of the party on account of the hostilities of the natives, who attacked us twice, I have not secured as many kinds of seeds as I might have wished ...'[53]

The North Australian Expedition managed to locate Leichhardt's 1845 camp at the Elsey River near Katherine.[54] In supplementing their provisions, von Mueller recorded in his journal that they ate portulaca in its fresh state, and papery gooseneck once it had been boiled.[55] He commented:

> In this regard, we had almost daily occasion to praise the value of the purslane (*Portulaca oleracea*), which not only occurred in every part of the country explored, but also – principally in the neighbourhood of rivers – often in the greatest abundance. We found it, in sandy and grassy localities, so agreeably acidulous, as to use it for food without any preparation; and I have reason to attribute the continuance of our health, partially to the constant use of this valuable plant. The absence of other anti-scorbutic herbs in the north, and the facility with which it may be gathered, entitle it to particular notice.[56]

Having a botanist as part of the team allowed the expedition to identify fresh edible plants to supplement their supplies of dried food and to forestall the onset of scurvy.

The exploration party consumed the cabbages (leafy growth centres) from at least two species of cabbage palm. They similarly used the screw palm (pandanus), although they found that even after boiling, the cabbage retained some acidity.[57] Amongst the many other Aboriginal foods they consumed in the better-watered northern regions were the roots of water lilies and bulrushes, and fruits from the nonda plum, Leichhardt tree and cluster fig.[58] In the case of water lilies, von Mueller noted that the 'seed-vessels and the roots of these water-lilies form a large proportion of the vegetable food of the northern natives, and the former [giant water lily] particularly will always be regarded as a providential gift in cases of need, by explorers of the North Australian wilderness'.[59] He would have been well aware that water lily tubers were edible, as his preparation for the trip had included reading Leichhardt's published account of the 1844–45 expedition to Port Essington.[60]

Von Mueller's extensive knowledge of wild foods often helped keep him alive on botanising trips. Charles Daley, his biographer, claimed:

> It is known that on some of these excursions Dr. Mueller suffered with uncomplaining fortitude, great privations, and frequently had to supplement scanty or exhausted food supplies by eating edible parts of native plants, or titbits from the natives' larder which hunger and necessity alone could render palatable ... after returning from an arduous outing [to the Victorian Highlands], an examination of the doctor's bag showed, in addition to a few customary handfuls of meal, a number of the large 'Bogong' moths, highly esteemed by the blacks as a delicacy.[61]

For many years von Mueller roamed widely across Australia. From 1853 he made many trips to the Victorian Highlands.[62] In 1867 he was collecting around Albany and the Stirling Ranges in southwest Western Australia.[63] In 1877, then in his early fifties,

Boab or bottle tree. This tree was marked by Augustus C. Gregory's North Australian Exploration Party in 1856. Baron von Mueller was the botanist on the expedition. Paul Foelsche, Victoria River, Northern Territory, 1891. P. Foelsche Collection, AA96, South Australian Museum Archives, Adelaide.

he conducted a plant survey along the Gascoyne and Murchison rivers, inland from Shark Bay in Western Australia, accompanied by an Aboriginal guide and a police trooper.[64]

Such was von Mueller's fame that he was considered to be 'the last explorer in Victoria'.[65] He had been involved with organising many expeditions of exploration, ensuring a steady flow of plant specimens collected from remote parts of the continent. He knew well the dangers that plant collecting presented, particularly in the far north, once warning a colleague contemplating field work in remote regions of the tropics, 'Beware of the saurians [i.e. crocodiles] and aborigines'.[66]

Spreading 'useful' plants

Von Mueller had a keen interest in the introduction of various exotic organisms into Australia, particularly plants and animals that were considered of potential benefit for agriculture.[67] He apparently took blackberry seed with him for dispersal whenever he travelled through country Victoria, claiming it 'deserves to be naturalised on the rivulets of any ranges'.[68] Von Mueller would have known that cultivars like boysenberry and loganberry had come about through the crossing of a group of closely related Northern Hemisphere plants of the *Rubus* genus such as blackberry, raspberry and dewberry.[69] Australia has plant species that are biologically related to these berries, generally known as native raspberry or native bramble.[70]

In his account of Victorian Aboriginal plant foods, von Mueller listed 'two kinds of raspberry (*Rubus parvifolius* [native raspberry] and the rarer *R. rosifolius* [rose-leaf bramble])'.[71] Aboriginal gatherers ate the native raspberry fruits raw.[72] In

New South Wales, Aboriginal healers used the young leaves of the native raspberry to make a tea to be drunk as a treatment for digestive problems referred to as 'bad belly'.[73] Von Mueller outlined the good attributes of the rose-leaf bramble, stating that the 'shrub bears in woody regions an abundance of fruits of large size, and these early and long in the season'.[74] He may well have considered hybridisation between Australian and exotic *Rubus* species a possibility, leading to new forms of fruit suited to the Australian climate.

Indigenous species of fruit in temperate Australia were not highly regarded by most botanists, which was motivation enough for them to experiment with wild hybridisation. Maiden claimed that the mountain raspberry 'yields the best native fruit in Tasmania, ... though perhaps that is not saying much'.[75] He went on to say that he was not overly impressed by the rose-leaf bramble (native raspberry), remarking that it was typical of Australian species of *Rubus*, which 'are for the most part insipid, with a mawkish, granular taste, and with a trace of astringency. They are encouraging to look at, but extremely disappointing to taste'.

Acclimatisation societies, which worked to introduce exotic species of plants and animals to the colonies, both as reliable food supplies and as nostalgic reminders of 'Home', were also heavily involved in the establishment of Australian zoos.[76] Von Mueller was a keen member of the Victorian Acclimatisation Society and his plant introduction activities were not restricted to the settled areas. For instance, he supplied the Trans-Australian Exploration Expedition (1860–61), led by Burke and Wills, with packets of seeds for planting. They were to be used in a fashion similar to the pebbles and breadcrumbs left as a trail by Hansel and Gretel in the Brothers Grimm fairy tale. The expedition's organising committee instructed Burke to 'mark your routes as permanently as possible, by leaving records, sowing seeds, building cairns, and marking trees at as many points as possible consistent with your various duties'[77], clearly wishing to avoid a repeat of Leichhardt's disappearance without trace. Because of problems with establishing the Cooper Creek Depot, the seeds did not reach Central Australia, but were left at Menindie on the Darling River.[78] Ernest Giles was another explorer to whom von Mueller supplied exotic plant seeds to sow during his trips to Central Australia (1872–76).[79]

Fortunately, von Mueller, also a curator of Melbourne's Zoological Gardens, demonstrated more restraint in relation to the introduction of exotic animals, once shooting escaped English magpies to prevent them from decimating the native bird population.[80]

Von Mueller also tried to establish useful Australian species overseas, sending wattle seed to South Africa and eucalypt seed to Europe, for example.[81] One of the Australian wattle species whose virtues he praised was the mulga from arid Australia, which is particularly important as a fodder tree for pasture animals. He described the wood as 'excessively hard, dark-brown, used preferentially by the natives for boomerangs, sticks to lift edible roots, end-shafts of Phragmites [reed]-spears, woomerangs [sic: woomera?], nulla-nullas [clubs] and jagged spear-ends'.[82] Von Mueller's introductions overseas did some good, as he has been credited with the removal of water-borne diseases from parts of the world through the growing of Australian eucalypts to assist drainage.[83]

One Australian plant species that von Mueller did consider worthy of being adapted for cultivation for its food value was the warran yam:

> It is evidently one of the hardiest of the yams, and on that account deserves particularly to be drawn into culture. The tubers are largely consumed by the local aborigines for food. The only plant, on which they bestow any kind of cultivation, crude as it is. Fit for arid situations, but fond of lime.[84]

To give support to his own field observations, von Mueller would have drawn upon the early nineteenth century records of James Backhouse and George Grey from coastal central Western Australia. Backhouse observed Aboriginal use of the warran yam tubers in the Swan River district of southwest Western Australia, where he had:

> examined some holes, where the Natives had been digging for roots of a *Dioscorea*, or Yam, for food. This plant climbs among bushes, in a strongish soil, and the Natives have a tradition, respecting its root having been conferred upon them, in which there are traces of the deluge.[85]

These tubers penetrate about half a metre down into the soil before enlarging into a thick cylindrical structure. The warran yam grounds attracted many Aboriginal bands to them for annual feasts.[86] More recently, archaeologist Sylvia Hallam remarked that for those Aboriginal groups who relied on the warran yam grounds, it was 'not a matter of digging out a root here and there, but of returning regularly to extensively used tracts'.[87] Grey described the warran as tasting 'like a [cultivated] sweet potato'.[88]

Some of the date palms now naturalised around waterholes in Central Australia may well have originated from seeds von Mueller sent to Kempe at the Hermannsburg Mission.[89] The botanist argued that date palms 'should be raised in the oases of the Australian desert million-fold ... Though sugar or palm-wine can be obtained from the sap and hats, mats and similar articles can be manufactured from the leaves, we would utilise this palm beyond scenic garden-ornamentation only for its fruit'.[90] He boasted: 'into Central Australia the date-palm was first introduced by the writer of this work [himself]'.[91] Today, land management authorities regard the spread of the date palm as a serious threat to the indigenous flora of Central Australia.[92]

In spite of von Mueller's interest in altering the flora through deliberate introduction of exotic organisms, he was a prominent opponent to Charles Darwin's theories of evolution and natural selection.[93] Darwin, in his argument about global climate changes during glacial periods, used von Mueller's discovery of 'European species' (such as *Rubus* species) growing naturally in the Australian highlands.[94] He even sent von Mueller a copy of *On the Origin of Species*, but the botanist was not swayed, remaining a firm believer in the immutability of species.[95]

In spite of the early history of poor health that had led to von Mueller's emigration from Germany, he appears to have thrived in Australia, becoming a robust and determined explorer and scientist. A perusal of the index of any Australian flora will demonstrate that his fellow botanists have honoured him with the naming of a great many plant species.[96] One example is the billy goat plum (or Kakadu plum), which was named *Terminalia ferdinandiana*, based on von Mueller's Christian name. This plant has special significance to nutritionists because its pale green fruit, which are the size of large grapes, contain the highest known quantity of vitamin C of any fruit in the world, just one holding the equivalent of up to eight oranges.[97] Its scientific name is well chosen, for during his long career von Mueller had searched for the existence of a panacea among Australian plants. Aboriginal people usually collect the fruit once they have felled the trees, which are tall and slender. In the western Kimberley they make a drink from the billy goat plum, by pounding the fruit and soaking them in water.[98] Another plant in the same genus was also named after the botanist, this time his surname, with Mueller's damson called *Terminalia muelleri*.

During von Mueller's lifetime, Australian botany was still under the control of British botanists, due largely to the importance of collections held at Kew Gardens and in the British Museum. Von Mueller's aim of writing a flora for the whole of Australia was never achieved. Instead, he assisted George Bentham at Kew with the writing of the *Flora Australiensis*.[99] Von Mueller was overlooked as the primary author in spite of his extensive field experience in Australia and the fact that he had provided many specimens to herbaria in Europe. In 1866 he was described by botanist colleague Anthelme Thozet as a 'learned and indefatigable friend ... to whom the scientific and commercial world owe so many valuable discoveries'.[100] In 1871 the King of Württemberg in southwest Germany appointed von Mueller a hereditary baron; he had already been granted the title 'von' in 1867. He died on 10 October 1896 in South Yarra, Melbourne. In recent times, botanist Francis A. Sharr claimed that von Mueller was 'probably the greatest Australian botanist and, in his time, the most famous scientist in the southern hemisphere'.[101]

9 Inland Explorers and Aboriginal Knowledge

By the 1860s naval exploration around the world was approaching a plateau; the globe had been circumnavigated and the oceans and coastlines largely mapped. Land-based explorers still had work to do, such as filling in the empty spaces on several of the continental maps by tracing the courses of major river systems back to their headwaters. After several attempts between 1857 and 1863, John H. Speke and James A. Grant finally succeeded in locating the source of the Nile at Lake Victoria in east Africa, for instance.[1] In the early 1860s another major gap in geographical knowledge lay in the centre of Australia, with Charles Sturt having failed to cross it in 1845, and Ludwig Leichhardt simply vanishing into its vastness in 1848. In the words of historian Alan Moorehead, the interior of Australia in the mid nineteenth century was a 'ghastly blank'.[2]

While explorers such as Eyre, Sturt and Leichhardt paid attention to indigenous cultures when traversing Australia, many others, to their detriment, did not take full advantage of the Aboriginal presence. This flaw, often fatal, was acknowledged in 1866 by Anthelme Thozet, a gardener in the Botanic Gardens in Sydney in 1856–58 before he became von Mueller's plant collector at Rockhampton in Queensland:

> Our pioneer explorers and travellers, in passing through trackless paths previously untrodden by the foot of the white man, in their praiseworthy efforts in the cause of civilisation, often die of hunger although surrounded by abundance of natural vegetable food in the very spot where the aborigines easily find all the luxuries of their primitive method of life, and not a few unacquainted with the preparation which several of the deleterious plants require, lose their lives in venturing to use them.[3]

Though he did not name them, Thozet would have been referring to the tragedy of Burke and Wills, who perished in 1861 when stranded at their Cooper Creek Depot in Central Australia.[4] Their deaths through starvation occurred despite ready access to a wild food source and a permanent water supply. There was also an Aboriginal population living in the area.

Burke and Wills

The race to make the first south to north crossing of mainland Australia was a matter of intercolonial rivalry.[5] Until separation from its parent colony in 1851, Victoria had been known as the Port Phillip District of New South Wales. As a result of the gold rush, however, by the 1850s Victoria had become the richest colony in Australia, with the largest population. Victorians were looking for worthy projects and goals to help support and grow their colonial pride. In the case of the south to north crossing, there was also an element of one-upmanship – a race to beat John McDouall Stuart, who set out from Port Augusta in South Australia with the identical aim. Exploration was not cheap, and promoters of the Victorian expedition had to sell the potential benefits to the public. The gains included the discovery of new species of plants and animals, opening up country for crops and livestock, establishing a route for a telegraph connection between Melbourne and Europe, and the possibility of finding out what had happened to Leichhardt.[6] The changing of the colonial cultural environment was also reflected in increased government funding for the sciences and humanities, as demonstrated by the establishment of universities and museums.[7]

The Exploration Committee of the Royal Society of Victoria appointed Robert O'Hara Burke to lead the Trans-Australian Exploration Expedition (1860–61), with George Landells as his second in command.[8] Several routes were proposed, the Committee finally settling for a crossing of the continent from Melbourne in the south to the Gulf of Carpentaria in the north. The expedition set out from Royal Park in Melbourne on 20 August 1860, where a crowd of about 15 000 had gathered to see them off.[9] They proceeded through northern Victoria, reaching Swan Hill at the Murray River on 6 September and Menindee at the Darling River on 12 October. Burke's initial plan was to establish his main depot at Cooper Creek in Central Australia, and, with summer approaching, wait out the hot weather there before heading north in the autumn of 1861.

From Menindee to Cooper Creek

The expedition was plagued from the outset by poor planning, rash decisions and considerable misfortune. At Menindee, Burke decided to break up the expedition party and leave much of the equipment behind. Some members were already in poor condition, with artist Ludwig Becker suffering from an injured foot – he had been stepped on by a horse.[10] On the trip north to Menindee, Landells had clashed with Burke and resigned, thus placing William John Wills second in command. On 19 October, Burke took an advance party and made a dash for Cooper Creek.[11] Aided by favourable weather and recent rains, they arrived at Cooper Creek little more than three weeks later, on 11 November. William Wright, who had been recruited into the expedition at Menindee, was to follow later with more supplies and camels.[12] Notably, the camel sepoys and Aboriginal guides were not with the advance party. By now Dr Hermann Beckler, the appointed botanist and medical officer on the expedition, wanted to leave the expedition, but in view of the break-up of the party into two groups he agreed to stay on at Menindee, at least until the Exploration Committee could organise his replacement.

On 10 November 1860, trooper Myles Lyons and saddler Alexander McPherson set out from Menindee, accompanied by the Aboriginal tracker Dick, to convey urgent dispatches from the Committee in Melbourne to Burke at Cooper Creek. Included in the dispatches was the most recent intelligence of Stuart's progress on his south–north crossing from Adelaide.[13] When the messengers became lost and desperately short of provisions and water, Dick conveyed them to the care of local Aboriginal people in the Torowoto district.[14] He then returned to Menindee, which he reached on 19 December, having had to walk for eight days as his horse was run into the ground.[15] Dick took Beckler back to rescue the stranded men, who had lived for weeks on about a half a litre of nardoo (*Marsilea*) flour per day. Beckler recorded in his journal on 27 December that they had found:

> Macpherson [McPherson] at a short distance from us, apparently searching for something [nardoo] on the ground … Lyons was at the camp engaged in baking cakes when we came up to him. The seeds of which they prepared a warn [sic: warm?] meal, and out of that either cakes or porridge, is not properly a seed, but the sporangium and the spores of a small plant, the leaves of which are very like clover. It is, I believe, a Marsileana [*Marsilea*], and everywhere to be met with where water stagnates for a time …[16]

These Europeans owed their lives to nardoo, which is a low-growing water fern that produces edible sporecases.

Back at Menindee, expedition artist Ludwig Becker commemorated the heroism displayed by the Aboriginal guide with a portrait of 'Dick, the brave and gallant native guide'.[17] Little has been recorded about this man or of any the other indigenous guides employed during the unfolding of the Burke and Wills saga. It is likely that Dick had earlier become familiar with Europeans at pastoral stations along the Darling River, which was then at the frontier of European expansion towards the north. Dick's cultural affiliations may have been with the Barkindji people who lived along the banks of the Darling River and ranging south. These people had good relations with those who lived in the 'back country', north and west of the river.[18] European settlers regarded the Darling River people highly in terms of their tracking abilities. In

Portrait of Dick. The brave and gallant native guide, Darling Depôt Decb. 21. 60. Dick saved the lives of Myles Lyons and Alexander McPherson during the unfolding of the Burke and Wills drama. Had he accompanied Burke and Wills on the leg from Cooper Creek to the Gulf of Carpentaria and back, the tragedy of their deaths may have been averted. Ludwig Becker, watercolour, 1860. H16486, State Library of Victoria, Melbourne.

reminiscences published in 1889 colonist Simpson Newland claimed:

> Their power of tracking is simply marvellous; they will tell you the track of each horse on the station. They can follow a snake or a rat. It has even been said the most skilled can track a mosquito. I have often known them to follow a small mob of lost sheep through the tracks of others, when to my eyes one as closely resembled the other as grains of wheat.[19]

To the Gulf and back

On 16 December 1860 the two leaders, Burke and Wills, accompanied by John King and Charlie Gray and taking with them six camels and a horse, set off from Cooper Creek to make the crossing to the Gulf of Carpentaria, while William Brahé and three other men were ordered to stay at the Depot for at least three months.[20] Burke fully expected that soon after he had left, Wright would appear at Cooper Creek with more men and supplies from Menindee. Amongst the instructions Burke left Brahé was one authorising him to shoot Aboriginal people if they caused annoyance.[21] After departing Cooper Creek, the Trans-Australian Exploration Expedition was effectively no longer a scientific venture, but a competitor in a desperate race against a rival explorer, Stuart. Considering that summer had just begun, a crossing at this time of the year was a major gamble.

Burke's party of four headed from Cooper Creek directly to the Gulf of Carpentaria. They were fortunate in the route they chose, as well as with the weather initially. On about 9 February 1861, the party reached the Flinders River. Here, the water was salty and showed a strong tidal rise and fall that told the explorers they were near the Gulf. Desperately low on food, however, they started back to Cooper Creek without having seen the open sea. Reaching boggy country, they were saved on one occasion when they came across an

Return of Burke and Wills to Cooper Creek, showing Trans-Australian explorers Robert O'Hara Burke, William John Wills and John King. They returned south from the Gulf of Carpentaria through the Central Australian deserts during the height of summer. Charlie Gray had died somewhere north of the Cooper Creek Depot. Nicholas Chevalier, oil on canvas, 1868. nla.pic-an2265463, National Library of Australia, Canberra.

Aboriginal pathway which took them to a 'nice watercourse', past some 'little pebbly rises where the blacks had been camping', and on to a place where Aboriginal foragers had left behind lots of yams, 'so numerous that they could afford to leave lots of them about, probably having only selected the very best. We were not so particular, but ate many of those that they had rejected, and found them very good'.[22] Although not named, the roots they found may have been pencil yams (small yams).[23]

Farther south, the desert was much harder on the men during the return trip, for it was now late summer. Wills wrote in his journal: 'I am inclined to think that but for the abundance of portulac that we obtained on the journey, we should scarcely have returned to Cooper Creek'.[24] They were not the first explorers to use 'portulac' (portulaca) as greens, James Cook having made use of its antiscorbutic properties when he was travelling along the east coast of Australia in 1770, and Baron von Mueller during Gregory's North Australian Expedition of 1855–57.[25] Desert Aboriginal people relied heavily on portulaca, for seed to make cakes, as well as for thirst-quenching fresh greens and for its foliage cooked in earth ovens.[26] (In 1878 historian Robert Brough Smyth noted: 'The Government Botanist [von Mueller] is to be commended for drawing attention to the properties of this plant. Every explorer and every bushman should make himself acquainted with it'.[27])

The last stint through the southern part of the desert must have been particularly bad, with little

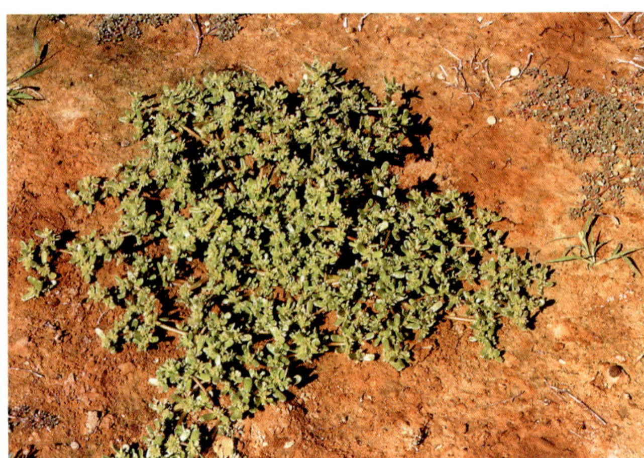

Portulaca, also known as common pigweed, purslane or munyeroo. This succulent creeper bears edible seed which is as fine and black as gunpowder. The leaves and stems were used by desert dwellers as thirst quenchers. Explorers ate this plant to battle scurvy. Philip A. Clarke, Clayton River, Central Australia, 1987.

available food. Gray, who had been ailing for some time, died on 17 April 1861, not far from Cooper Creek.[28] Burke, Wills and King reached the Cooper Creek Depot on the evening of 21 April 1861, to their horror finding the main base deserted. In one of the several near-misses that characterised the Burke and Wills drama, Brahé and his men, who had stayed for as long as supplies allowed, had left Cooper Creek for Menindee earlier that same day. Brahé later defended his actions, claiming that his men had started showing signs of scurvy – sore gums, swollen ankles and feeling weak.[29]

Later explorers of inland Australia referred to the debilitating and life-threatening condition of 'land scurvy' as 'Barcoo rot'.[30] The whole expedition appears to have been very poorly provided for in terms of food. According to Thomas McDonough, one of those who remained at Cooper Creek with Brahé, the daily rations were:

> A pint of rice, raw rice, boiled for breakfast with some sugar, sugar was not scanty, we had plenty of that we brought up more than would do for six months; then we had damper for dinner, and a little salt pork or beef and we usually ate about two biscuits and a pint of tea at night.[31]

On the trip between Cooper Creek and the Gulf, the rations for Burke, Wills and King were restricted to dried meat, pork, rice and flour for damper[32], a diet which would horrify present-day dieticians.

The rations left buried at the Cooper Creek Depot by Brahé were no better nutritionally. The three survivors recovered flour, oatmeal, dried meat, tea and sugar – no dried fruit or vegetables.[33] They ate some of the food that had been left for them, then decided not to attempt to catch up with Brahé. Instead, they would make their way downstream along Cooper Creek before striking overland through the Strzelecki Desert, via Mount Hopeless where there was a police station, and finally towards the settled areas of South Australia.[34] They reasoned that these places were considerably closer than the Menindee Depot. On 23 April, the three men headed downstream along the southern bank. On one occasion they obtained fish to eat from Aboriginal people, paying them with such things as leather straps and matches.[35]

Their attempt to reach Mount Hopeless failing due to lack of water, the explorers were forced to backtrack to Cooper Creek. In their absence they had missed a brief visit by Brahé and Wright, who returned to the Depot on 8 May 1861. The stranded explorers were not completely alone, however, as the local Yantruwanta people showed their generosity by bringing them food, giving them shelter and offering them 'bedgery' (pituri), the highly valued narcotic chewed by Aboriginal men.[36] Incredibly, given their dire circumstances, Burke was for a long while reluctant to accept help from Aboriginal people and on one occasion even angrily refused a gift of fish, ordering King to fire his revolver.[37]

The explorers were given prepared nardoo by the Yantruwanta, but they had no knowledge of its early preparation stages and for a while did not know the identity of the plant species that was its source.[38] Left to fend for themselves, the explorers searched the trees and bushes. In his journal, Wills wrote on 10 May 1861 of his own fruitless search, 'In this I was unsuccessful, not being able to find a single tree of it [nardoo] in the neighbourhood of the camp', although he did find a 'large kind of bean which the blacks call padlu; they boil easily, and when shelled are very sweet, much resembling in taste the French chestnut. They are to be found in large quantities nearly everywhere'.[39] The identity of this plant is a mystery, although it may be the pop saltbush, a forage plant of the region.[40] The

Pop saltbush (*Atriplex holocarpa*). This plant might be the identity of the 'large kind of bean' known as 'padlu', which was one of the wild foods Burke, Wills and King ate on their return to Cooper Creek in 1861. Philip A. Clarke, Kenmore Park, northwest South Australia, 2007.

name Wills recorded might have been an attempt to write *paldru*, which missionary Johann Reuther listed in the early twentieth century as a Diyari word for 'shrub, pods burst open, pop-saltbush'.[41]

After a week of searching, on 17 May the explorers at last located nardoo fern growing on the mudflats.[42] They must have thought their luck had changed for the better, and set about gathering it. From 27 May to Wills' last journal entry on 28 June 1861, the explorers gathered and ate nardoo[43], but inexplicably to them they did not regain their strength. They used a 'pounding stone' taken from a deserted Aboriginal shelter to process what they gathered.[44] Finding such a stone would not have been out of the ordinary; because they were heavy and awkward to carry, Aboriginal people across Australia generally left grinding stones behind at each of their main seasonal camps, which were near water sources.[45]

The health of the three explorers, already poor following the exertion involved in getting to the Gulf and back, continued to deteriorate at Cooper Creek. By late June, Wills was no longer able to move, and Burke was not much better off. On 28 June 1861, Wills noted in his journal that 'starvation on nardoo is by no means very unpleasant, but for the weakness one feels, for as far as appetite is concerned, it gives me the greatest satisfaction'.[46] He recorded that Burke and King were about to go along the creek to obtain help from the Yantruwanta. They had not got far when Burke died.[47] King buried his body and continued his search for help. He was lucky enough to find a store of nardoo in a deserted 'gunyah' (hut), which was enough to feed him for a fortnight.[48] In arid regions, Aboriginal people often kept a surplus of dried food, such as nardoo and seed, stored in wooden containers or in bags of skin or woven string for future use.[49] These were cached

Original gravesite of Robert O'Hara Burke. The explorer lost his life in 1861 at Cooper Creek in spite of a plentiful supply of water. It is likely that his death and that of his companion William John Wills were mainly due to their improper use of nardoo. Philip A. Clarke, Cooper Creek, near Innamincka, South Australia, 1986.

in brush shelters, dry caves or buried in the sand. By the time King could return to Wills at Cooper Creek, his companion had also passed away.

It is clear from Wills' journal, found with his body, that for the most part the Trans-Australian Exploration Expedition had only limited interaction with Aboriginal people.[50] Now the sole survivor, King made the wise decision to join a band of Yantruwanta people.[51] Each day women gave him nardoo, presumably already prepared for eating, while the men sometimes provided him with fish. In return, King would shoot birds for them. In line with Aboriginal custom, as a single man King shared the bough shelter of the unmarried men each night. On one occasion, he noticed that a woman who had just given him a ball of prepared nardoo had a sore arm, preventing her from grinding any more.[52] He treated her by cleaning the wound with a sponge soaked in water he had boiled, then applying silver nitrate. The effect must have been rapid and positive, as from then on this woman and her husband maintained a close relationship with the explorer, which included helping him make camp whenever the group moved.

The Relief Expedition

After weeks of travelling with his adopted band, King saw one day that some Aboriginal people, not of the local group, had come down the creek. They had news of white men on horses, which turned out to be the Relief Expedition led by Alfred W. Howitt which had been despatched from Melbourne in search of the explorers.[53] (It was but one of the five search parties sent from different parts of the country.) The rescue party found King on 15 September 1861 in such a state that he was not at first recognised as being European. The second in command, the surveyor Edwin J. Welch, vividly described the circumstances. Riding at the rear of the main party along the bank of a creek, Welch had noticed a group of Aboriginal people on the opposite bank shouting loudly and waving at him, then pointing further down the creek. In his words:

> The blacks drew hurriedly back to the top of the opposite bank, shouting and gesticulating violently, and leaving one solitary figure, apparently covered with some scarecrow rags, and part of a hat, prominently alone in the sand. Before I could pull up, I had passed it, and as I passed it tottered, threw up its hands in the attitude of prayer, and fell on the ground. The heavy sand helped me to conquer Piggy [his mount] on the level, and when I turned back, the figure had partially risen. Hastily dismounting, I was soon beside it, excitedly asking, 'Who, in the name of wonder, are you?' He answered, 'I am King, sir.'[54]

King returned to Melbourne with the Relief Expedition, where he developed a permanent condition of peripheral neuropathy caused by a prolonged deficiency of thiamine.[55] He never fully regained his strength, in 1872 dying at the age of only thirty-one.[56]

Although the Exploration Committee had probably erred in appointing Burke to lead the Trans-Australian Exploration Expedition, they made a wise choice in Howitt to lead the Relief Expedition. Howitt was an experienced bushman, an accomplished geologist, and an emerging ethnographer and anthropologist.[57] Thanks to their leader's bush skills and the presence of two indigenous guides, Sandy and Frank, the Relief Expedition's party of twelve men largely avoided the health problems that had plagued Burke's party. Howitt made sure that succulent vegetables such as 'portulac', 'wild spinach' and 'mesembryanthemum' (pigface) were frequently eaten.[58] They also travelled during a mild time of the year. Apparently, none of his party developed the symptoms of scurvy. Howitt brought back with him to Melbourne a collection of herbarium specimens made on the trip.[59]

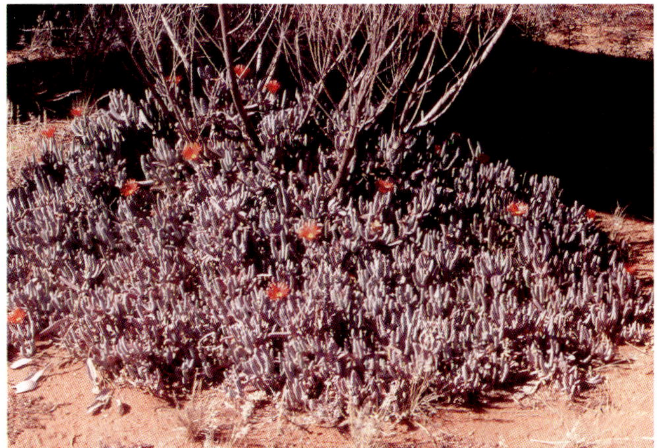

Sarcozona, an inland species of pigface. In 1861 the members of the Burke and Wills Relief Expedition, led by Alfred W. Howitt, ate the leaves of a pigface species to prevent scurvy developing. Philip A. Clarke, Coober Pedy, Central Australia, 2007.

When the Trans-Australian Exploration Expedition was planned, the Exploration Committee was not interested in the study of Aboriginal cultures and Dr Beckler's offer to make a record of the indigenous peoples they encountered had not been taken up.[60] On the Relief Expedition, Howitt corrected this oversight. Like Leichhardt, he made detailed notes from his observations of the Aboriginal people whose country he traversed. At Cooper Creek, Howitt described an Aboriginal practice of storing munyeroo seed as a survival strategy. It was:

> collected in large quantities by the natives after rains. It is even sometimes collected in such quantities as to be preserved for future use. Near Lake Lipson, one of my party found about two bushels contained in a grass case daubed with mud. It looked like a small clay coffin, and was concealed ... The 'Manyoura bowar' [munyeroo] tastes like linseed-meal, and is by no means unpleasant when baked in ashes and eaten hot.[61]

Relations between the Yantruwanta people and the Relief Expedition were good, with presents offered by both sides. A young Aboriginal man offered Howitt a plug of pituri.[62] The Relief Expedition rewarded the band that had sheltered King with gifts such as sugar and old clothes.[63] The Aboriginal people called the flour they were given 'white fellow nardoo'.

Defamation of a food

Media interest in the Burke and Wills saga led to many questions being asked about what had gone wrong, and a royal commission was held in Melbourne. Amongst the commissioners was Charles Sturt's younger brother, Evelyn. The list of witnesses to give evidence included the survivors and certain major players on the Exploration Committee. In its published report, the commission was critical of specific actions by Burke, Brahé and the Committee, although most blame fell upon William Wright, attention being drawn to the impact of his failure to move the base camp from Menindee to Cooper Creek while

Cooper Creek. Like many desert watercourses this creek only flows on the surface during floods; at other times it forms a series of deep lagoons or waterholes. Philip A. Clarke, Cooper Creek, near Innamincka, South Australia, 1986

Burke and Wills were en route to the Gulf.[64] The transcript of evidence contains many questions about the availability of food. There are over fifty references to nardoo in the final published report, making it clear that once the explorers had reached Cooper Creek and eaten the remaining rations it was their main food source. Although the royal commission pointed the finger at those who had made bad decisions leading to the deaths of Burke and Wills, the issue of what blame could be directed at Aboriginal food sources was left for scholars to argue over.

The identity of 'nardoo'

The late nineteenth century saw the beginning of an extended debate about the true identity of the 'nardoo' consumed by Burke's party at Cooper Creek. Botanical writers, including Robert Brough Smyth in 1878, had readily accepted that the nardoo used by the explorers was a species of *Marsilea*, sometimes called clover ferns or water ferns. Botanist Frederick M. Bailey had another view, which was that Burke and Wills had confused the poisonous sporocarps of *Marsilea* for the edible seeds of the *Sesbania*. In 1880 he claimed:

> In North Queensland, according to Mr. T.A. Gulliver, the natives make bread of the seeds [of *Sesbania aculeata*]. I am of the opinion that this is the true Nardoo of the Cooper's Creek natives. The unfortunate explorers, Burke and Wills, might easily have mistaken the sporecases of a *Marsilea* for the shelled-out seeds of *Sesbania*.[65]

Sesbania is a legume plant, with yellow pea-like flowers, originating from the northern tropical areas of Australia and the East Indies, and now widely grown around the world as a food for pasture animals.[66]

Bailey's evidence was circumstantial, largely based upon his correspondence with Thomas A.

Gulliver, who had remarked that 'the natives here (Norman River) do not seem to care about the *Marsilea* seed, and as far as I have seen it does not produce sufficient seed to make it worth collecting; whereas the *Sesbania* is very prolific and can be gathered without any, or rather, with very little trouble.'[67] Bailey's argument was that 'nardoo' should be identified as *Sesbania*, and that Burke and Wills had made a fatal mistake in thinking it was *Marsilea*.

Bailey's theory was attractive, for it explained why Aboriginal people were able to live in the desert region while the European explorers had perished. However, at a talk on indigenous plant foods given at the Royal Society of Queensland in 1884, pharmacologist Joseph Bancroft confirmed that nardoo was *Marsilea*. The published version of his paper noted: 'The native food-plants of the desert are deserving of patient study, and, by a knowledge of them, then lives of explorers may be, at times, preserved.'[68] He believed that research was needed into the correct way of using nardoo. In spite of Bancroft's research, opinion on the identity of nardoo remained divided. Maiden, in his economic botany book published in 1889, had a bet each way when he listed *Sesbania* as 'the "Nardoo" of the aboriginals of the Norman River, Queensland', while he described *Marsilea* as the 'clover-fern' or 'nardoo'.[69]

While Maiden was confused, there were fieldworkers who were certain of nardoo's identity. Scientist Thomas L. Bancroft supported the views of his father Joseph Bancroft. Based on a trip to Lake Kopperamana along Cooper Creek, Bancroft published a paper in 1894 in which he said:

> I learnt that the blacks in that district, and indeed all over the watershed of the Cooper, Diamantina, and Georgina Rivers, still made use of it as in the days of Burke and Wills; and also that the plant is a *Marsilea*, as had been originally stated, but doubted by some [like Bailey], who thought it impossible that sufficient involucres (sporocarps) to serve for food could be obtained from a *Marsilea*, the Nardoo of Burke and Wills being regarded by them as the seed of *Sesbania aculeata*.[70]

According to the younger Bancroft, botanists and explorers had long underestimated the ability of Aboriginal foragers to process such a minute and dispersed food source. In 1898 botanist Max Koch, who had much contact with Aboriginal people while working on pastoral stations in northern South Australia, recorded 'nardoo' as *Marsilea*, noting that the 'natives eat the spore-cases by pounding them up into flour'.[71]

In 1904 Howitt and missionary Otto Siebert of the Bethesda Mission, which overlooked Lake Killalpaninna west of the Birdsville Track, stated that '*Ngardu*, or as it is usually spelled *nardoo*, is the *Marsilia* [sic] *sp.* on which the explorers Burke and Wills endeavoured to support themselves at Cooper's Creek'.[72] In 1910 a commentary in *The Victorian Naturalist* journal quoted Welch as being adamant that nardoo was *Marsilea*.[73] Welch's views and memories should have carried much weight, as he had been a major collector of both Aboriginal and European items from the Cooper Creek region while part of the Relief Expedition.[74] Some doubt expressed by Victorian-based naturalists over the *Marsilea* identity of nardoo was caused by the relatively poor spore-case production in plants of the same or related species growing naturally around Melbourne.

The primary evidence was mounting to support the identification of nardoo as *Marsilea*, but still not all researchers were convinced. In 1915 scholar E.H. Lees claimed that nardoo could not possibly relate to *Sesbania*, as *Sesbania* was not to be found in the area where King had been rescued.[75] But rather than throwing his weight behind *Marsilea* as its identity, he argued that the term 'nardoo' actually related to 'a food obtained from several plants'.[76] For evidence of broad use of the term, Lees gave his field experience in Central Australia, which included one expedition with anthropologist Francis J. Gillen near Charlotte Waters. Here, they had 'partaken of leguminous Nardoo in aboriginal restaurants, where at that time English was unspoken and the white man little known'.[77] But in making the case for 'nardoo' relating to a category of food rather than to a single plant species, Lees overlooked the broad use of 'paua ... the seed of any food plant' recorded in the languages around Cooper Creek.[78]

After considerable field work in Central Australia, as well as the completion of an extensive literature review, biologist and anthropologist Baldwin Spencer was able to confirm in 1918 that 'nardoo' indeed referred to *Marsilea*.[79] He pointed out that in the case of Lees' account of plant use around Charlotte Waters, the local Arrernte people lived well outside the range where nardoo food was used and were therefore unlikely to have recognised the word at all prior to European settlement. In other words, some Aboriginal groups could have adopted the term from Europeans. Spencer had other doubts too, and in relation to Gillen's field experiences remarked:

> Over many a camp-fire between Lake Eyre, in the south, and Borroloola, on the Gulf of Carpentaria, we have discussed most things connected with the natives, but he [Gillen] never suggested and we never found any evidence to show that the word nardoo was used in connection with anything except Marsilea, its fruit, and the cakes made from this.[80]

It was not possible for Spencer to check with Gillen himself, as he had died in 1912. During his investigation Spencer obtained specimens of the nardoo that the Burke and Wills party had actually collected around Cooper Creek, as well as samples obtained later by Howitt.[81] All were botanically identified as *Marsilea*.

Twentieth century field work has confirmed that the name nardoo can be applied to any of a group of small clover-like species of fern that grow on the mudflats or in the water between dunes, and in claypans and around waterholes.[82] From my own field experience in the Diamantina River to Cooper Creek region, nardoo grows in dense olive-green mats on the mudflats after rain, gradually turning red-brown as conditions become drier. After the plants have withered away, the darkened sporocarps can be found embedded in the dry mud, where they must remain viable for many years before taking advantage of the next infrequent heavy rainfalls in the desert. The common Australian English name was reputedly derived from *ngardu*, which is a word for the plant from Aboriginal languages spoken across the inland region intersected by the borders of Queensland, New South Wales and South Australia.[83] The plants are today sold in many native plant nurseries as an attractive water plant with leaves forming a dense floating mat.

The sporocarps of the nardoo water fern are pea-shaped growths that bear the spores. These are usually referred to simply as 'nardoo', and Aboriginal foragers generally gathered them up from the dry mud after the plant had died.[84] Even in this dried form, nardoo is only rendered edible after extensive preparation. Because of its abundance and availability during drought, it was a major 'hard time' food in parts of arid Australia although Aboriginal hunter-gatherers in temperate and tropical Australia appeared to have universally ignored it. For desert foragers, Howitt claimed that nardoo 'may be called their "stand-by" when other food is scarce. In many places, miles of the clay flats are thickly sprinkled with the dry seeds [sic: sporocarps]'.[85] For desert dwellers, the availability of nardoo, and the ability to grind it, enabled them to remain in country that experienced long dry periods.

Nardoo preparation

Nardoo requires a process of pounding, sluicing and baking into cakes to render it edible.[86] Chemical analysis has shown that it contains thiaminase, an enzyme that blocks the absorption of thiamine (vitamin B_1) in human bodies.[87] The highest levels of enzyme activity are found in vigorously growing plants, in other words plants with green material. While probably having some worthwhile nutrients and a high level of fibre, nardoo has not attracted the attention of nutritional analysts. It is thought to be low in protein and starch and is probably best eaten along with other plant and animal foods.[88]

In northeast South Australia, medical doctor George Horne and former outback policeman George (Poddy) Aiston described how the nardoo sporocarps were:

> collected by [Aboriginal women] sweeping them up into a *pirrha* or wooden bowl with a bunch of twigs. They are flattened [in appearance] from side to side and about .5 cm in diameter. If the thick, smooth casing is removed they are seen to be full of a yellowish powder which has a rather bitter taste.[89]

Howitt recorded in the journal he kept while on the Relief Expedition that (at Cooper Creek) nardoo sporocarps are 'gathered by the native women, and, after being cleaned from the sand, are pounded between two stones and baked as cakes'.[90] Mounted constable Samuel T. Gason, stationed at Lake Hope on the western side of the Strzelecki Desert in South Australia in 1865–71 to help pacify relationships between graziers and Aboriginal groups, observed the use of nardoo. He described it as:

> a very hard fruit [sic: sporocarps], a flat oval, of about the size of a split pea; it is crushed or pounded, and the husk winnowed. In bad seasons this is the mainstay of the natives' sustenance; but it is the worst food possible, possessing very little nourishment and being difficult to digest.[91]

Surveyor C. Twisden Bedford, working in the Georgina River district of southwest Queensland in 1886, was amazed at the skill required in winnowing nardoo, observing:

> The nardoo seed is extensively used by the aboriginals as an article of food. They grind it up between two stones, sifting the husks from the grain with great dexterity by a peculiar trembling motion of their coolimans or wooden vessels containing it, which method I have seen several white men try and imitate without success. The flour or meal they thus obtain is then baked into a kind of bread, which is not at all unpalatable.[92]

Thomas Bancroft made more detailed observations of the preparation of nardoo at Annandale in southwestern Queensland:

> The involucres [sporocarps], which are very hard, are pounded between two stones; a handful of them is held in the left hand and fed to a stone on the ground, a few grains being allowed to drop from the hand by separating, abducting the little finger, a smart blow being struck with a stone in the right hand, which effectively pulverises every grain at once; it is surprising with what rapidity they can do this work. The flour is mixed with water, kneaded to a dough, and baked in the ashes.[93]

Although not stressed in any of the above accounts, it is likely that sluicing was an important stage of the process, as the enzyme action of thiaminase is greatly reduced by water.[94] Given the importance of water, it makes sense that grindstones used for pounding are typically found at waterholes located near mudflats where the nardoo grows.[95] The large grindstones that Aboriginal foragers used for processing a variety of seeds and for nardoo are today typically found from western New South Wales, across Central Australia, through the Western Desert to the Pilbara in central Western Australia.[96]

Wangkangurru woman (*above and detail*) grinding nardoo. Long-term survival for Aboriginal desert dwellers relied upon the use of grinding stones to prepare meal from nardoo spore-cases gathered from the dry mud. Norman B. Tindale, Pandi Pandi, northeast of South Australia, 1934. N.B. Tindale Collection, AA338/5/12/12 & 14, South Australian Museum Archives, Adelaide.

Grindstones. For desert dwellers these stone slabs were essential for removing the spore-cases from nardoo and the husks from grass seeds. The resulting meal was cooked into damper. Being heavy, large grindstones were typically left at campsites near waterholes. H. Basedow, Yelta, western New South Wales, early 20th century. A22601, South Australian Museum Archaeology Collection, Adelaide.

The process of producing flour through grinding must have been laborious; Howitt recalled that: 'the "tap-tap" of the process may be heard in the camp far into the night at times.'[97] The sound was also said to have carried for some considerable distance from camp.[98] In the Diyari language of northeast South Australia, *pita-ru* was the lament of hardworking women, meaning 'always-pounding'.[99] It referred to their inevitable reliance upon nardoo sporocarps during drought time, picking them out of the dried mud and pounding the hours away.

Although nardoo is widespread across Australia, only Aboriginal bands living in the most extreme desert environments used it. In the semi-arid Darling River district of western New South Wales, the Barkindji appear to have completely ignored nardoo in favour of grass seed.[100] Although it is present in Victoria, Robert Brough Smyth suggested that Aboriginal people here 'seem to have been unacquainted, generally, with the use, as a food, of the clover-fern, *Nardoo*, though the natives of the north-western parts of Victoria must have had intercourse with the tribes who use it'.[101]

Nardoo: good or bad?

In response to the initial reports from those members of the Trans-Australian Exploration Expedition who had remained at Menindee, von Mueller reported in *The Gardener's Chronicle and Agricultural Gazette* of 29 April 1861:

> MELBOURNE:- New Kind of Food. – Two of the party [Lyons and McPherson] of the Victorian expedition who lost their horses, and were only rescued after a lapse of several weeks, saved their lives by learning from the aboriginals how to pound up the seed vessels (sporangia) of the little Cryptogamic plant called *Marsilia* [sic.] *hirsuta* by R. Brown, and bake them into bread; they also made a porridge of the same. Both kinds of food were considered by Dr. Beckler nutritious by no means unwholesome, and free from any unpleasant taste.[102]

Beckler, as expedition botanist and senior member left in charge of the contingent that stayed at the Menindee base camp on the Darling River[103], was the only member of the party to both collect plants and make botanical notes.[104]

A vastly different opinion of the nutritive value of nardoo emerged in the aftermath of the Burke and Wills expedition. In spite of von Mueller's early enthusiasm for nardoo, he claimed a year later that 'The nutritive properties of the *Marsilea* fruit are evidently scanty'.[105] Joseph Bancroft did not wholly support Mueller's negative conclusions, claiming in 1884 that 'though Nardoo is pronounced by Baron von Mueller to be a "miserable article of food," the great value of it to starving travellers should not be lost sight of'.[106]

Botanist Joseph Maiden favoured von Mueller's opinion of nardoo being unsuitable as human food, stating in 1889 that 'This plant is much relished by stock ... It is, however, better known as yielding an unsatisfactory human food in its spore-cases'.[107] Anthropologist Northcote W. Thomas stated in 1906: 'When other food is scarce *nardoo* is the stand-by of the natives in the centre of Australia, but its nutritive properties are small'.[108] In 1931, naturalist Charles Daley claimed that Burke and Wills had found the nardoo meal 'innutritious and hard to digest'.[109]

Looked at through the eyes of these nineteenth and twentieth century European Australians, and with its notorious connection with the deaths of the inland explorers, the reputation of nardoo as a food source looked as though it would never recover. Nonetheless, it has attracted considerable attention from botanists. In a meeting of the Victorian Field Naturalists Club in 1918 it was reported that the Economic Museum in the Melbourne Botanic Gardens had 'two grinding stones reputed to have been used by Burke and Wills at Cooper's Creek, also portion of a cake made of Nardoo meal'.[110] Exhibited at the same meeting were nardoo sporocarps collected at Cooper Creek in 1861 by Howitt's Relief Expedition. The public's fascination with nardoo has led to the South Australian Museum receiving large quantities as specimens, collected by travellers and missionaries passing through the Cooper Creek region from the late nineteenth century. In 1946 an Australian poet lamenting the loss of Burke and Wills wrote that 'Nardoo is no fit food for white men'.[111] Similarly, in a magazine report in 1951, Douglas Kemsley remarked: 'The nutritive qualities of the nardoo are very low, and the explorers Burke and Wills could not sustain life on this food'.[112]

Nardoo plant. This small arid-zone fern grows in dense mats on the mudflats after rain. Note the dried dark and split pea-shaped nardoo spore cases on the dry mud. Philip A. Clarke, south of Birdsville, Central Australia, 1986.

Nardoo specimens. Because of the connection between this plant food and the explorers Burke and Wills, there are many samples of *ngardu* (nardoo) in Australian museum collections. Ifould, western Queensland, 1902. South Australian Museum Ethnobotany Collection, Adelaide.

The indigenous users of nardoo fared equally poorly at the hands of European Australian writers. In spite of ample evidence that Aboriginal hunter-gatherers possessed the landscape knowledge and technology to gain a living from the challenging environs of Cooper Creek, Alan Moorehead in 1963 maintained that Aboriginal people involved in the Burke and Wills drama were 'the most retarded people on earth'.[113] Such a view of Aboriginal culture is unlikely to lead one to an appreciation of indigenous foods.

Right up to the late twentieth century, many scholars wrongly used the nature of a supposedly impoverished flora to help explain the apparent 'primitive' condition of Australian hunter-gatherers.[114] In 1861, Carl Wilhelmi published an account of the Aboriginal 'manners and customs' in the Port Lincoln district of South Australia. He pointed out that 'although to Europeans the country offers scarcely any kind of eatable fruit, it yields a pretty good variety of such as affords valuable food to the blacks'.[115] In 1867 John Crawfurd read out a paper to the Ethnological Society of London, which was titled 'On the vegetable and animal food of the natives of Australia in reference to social position, with a comparison between Australians and some other races of man'.[116] It was chiefly a commentary on Thozet's 1866 list of the 'esculent native plants of the colony of Queensland'.[117] Crawfurd had asked Bentham from the Linnean Society to comment upon Thozet's plant list, and he responded that:

> There is probably not one of the 'roots, tubers, bulbs, and fruits' [listed by Thozet] that would ever be touched by those who could get a supply of wholesome grain or cultivated food-roots, and perhaps none either that could by any process of cultivation be brought up to the standard of those which have already been modified for human use.[118]

After consulting with John Gould, Crawfurd considered that 'the natives of Australia were far better provided with animal than with vegetable food'.[119] He found that Thozet's account 'a curious record of the vegetable poverty of Australia, in so far as human food is concerned'.[120]

The extent to which nineteenth century scholars believed Aboriginal Australia suffered from a paucity of natural resources was not restricted to the flora. In 1889 naturalist J.J. East stated that on this arid continent 'The absence of animals suitable for beasts of burden is not compensated by the watercourses of the interior, [making] 'the history of any race of inhabitants in such a country ... one of monotonous uniformity'.[121] This view essentially remained in place until the late twentieth century, with most scholars wrongly using the lack of suitable plants and animals for agricultural development to help explain the apparent 'primitive' condition of Australian hunter-gatherers.[122]

An informed reading of nardoo's contribution to the deaths of Burke and Wills suggests that they were trying to live off green nardoo, which was heavily loaded with the poisonous thiaminase. Nor were they preparing it properly – sluicing and cooking were crucial parts of its preparation.[123] Writer Edith Coleman suggested also that the explorers' preparation methods may not have included adequate winnowing, which would have separated the indigestible husks from the more nutritious contents.[124] The resulting contamination would have had serious consequences, as the leaves and fronds of nardoo contain much more thiaminase than the sporocarps.[125] In the case of Wills, his consumption of 'shellfish' at Cooper Creek on at least one occasion may have exacerbated his illness if he had eaten them raw, as mussels also contain high levels of thiaminase.[126]

Although they initially tried a few other wild foods, the explorers desperately waiting for their rescue at Cooper Creek were compelled to largely live on the nardoo that was available close by. Welch confirmed this in an article titled 'The Explorer – Dietary Experiences', published in *The Australasian* on 12 February 1910, in which he commented that fish, pigeons, bush rats, mussels and yabbies were plentiful around Cooper Creek, but the explorers were too weak to collect them and therefore had to be content with just nardoo.[127] He went on to say that 'nardoo alone meant a lingering fight with death'. The explorers were consuming it while suffering from severe malnutrition and battling vitamin B deficiency, a condition generally known as beri-beri. Their physical strength was weakened initially through exhaustion from the trip to Gulf and back, then through their poor preparation of the nardoo. By the time they had returned for the last time to Cooper Creek, they had lost the capacity to range further afield in search of other foods.

The tragedy of the Burke and Wills expedition highlights the importance of the proper transfer of landscape-based knowledge from the indigenous occupants to the newcomers. If the explorers had gained greater intelligence of local plant foods and their preparation, they may well have survived their ordeal. This argument is supported by the example given above involving the Aboriginal guide Dick, who worked with local bands in the Torowoto district to save McPherson and Lyons. In spite of his alleged failings as a leader, Leichhardt demonstrated on his 1844–45 trip that, in comparison to Burke and Wills, he was much more attentive to what Aboriginal foragers collected, ate and how they prepared their food. If the final leg of the Trans-Australian Exploration Expedition had included experienced Aboriginal guides in their party, even if recruited from the south, their chances of survival would have significantly increased. Aboriginal expedition members with bush skills would have been better placed to communicate with local Aboriginal groups, leading to a more effective use of wild food by the explorers.

Forrest and Windiitj

The Trans-Australian Exploration Expedition is an extreme example of poor planning and resource management. Other exploration teams fared much better in equally challenging country. The account of the partnerships between Perth-based explorer John Forrest and his Aboriginal guides provides a good example of how expedition parties can live off the land. Aboriginal expedition members have rarely been credited with having crucial roles in the official records of exploration, but an exception is provided by Forrest's published journal, which contains an extensive record of his involvement with Aboriginal people in the Western Australian deserts.[128]

On his first major expedition in 1869, which was to the desert area west of Leonora in central Western Australia, Forrest took with him Jemmy Mungaro and Tommy Windiitj (Windich), chosen to act in the broad role of guides. Not much is recorded about Mungaro, although it is known that he claimed to have previously seen a spot in the desert where there were remains of white men, possibly Leichhardt's lost party.[129] Windiitj already had extensive experience working with Europeans. Born among the Kokar people about 1840 near Mount Stirling in southwest Western Australia, he spoke the Njaki-Njaki tongue, but would probably have also understood Kalaamaya to the north, and the Nyungar dialects to the south.[130] While still young, Windiitj is thought to have gone to live with Europeans in the Bunbury district, after an epidemic had decimated the local Aboriginal population. In the early 1860s he was working as a 'native assistant' in the police force based at York; as an expert horseman and tracker, he had played a prominent role in several highly publicised arrests. Windiitj was also a seasoned explorer, having in 1865 been part of the third expedition of road surveyor Charles Cooke Hunt, who returned to the arid country east of York to deepen the wells he had previously sunk.[131]

Aboriginal guide Windiitj (kneeling) with explorers Alexander Forrest (left) and John Forrest (right). Windiitj accompanied John Forrest on his trips (1869–74) into the central deserts of Western Australia. He located water and supplied expedition parties with wild plant and animal foods. Photographer unknown, about 1874. 24261p RWAHS. Courtesy of the Battye Library, Perth.

Forrest had no intention of making any of the mistakes suffered by earlier desert expeditions. He was keenly aware of tragedies such as that of Burke and Wills – in fact, part of the reason for his 1869 expedition was to search for another famous missing expedition. In his published journal, Forrest said:

> Early in 1869, Dr. Von Mueller, of the Melbourne Botanic Gardens, a botanist of high attainments, proposed to the Government of Western Australia that an expedition should be undertaken from the colony for the purpose of ascertaining, if possible, the fate of the lost explorer, Leichardt [sic.]. Reports had reached Perth of natives met with in the eastern districts, who had stated that, about twenty years before (a date corresponding with that of the last authentic intelligence received from Leichardt [sic]), a party of white men had been murdered.[132]

Although as a surveyor Forrest was accustomed to navigation by astronomical observation, he relied heavily on his Aboriginal scouts in the daily quest for water and horse feed. His guides provided the exploration party with kangaroo, emu, turkey, possum, ducks and 'sugar bag' (honey) to live on.[133] During this trip a local Aboriginal man called Dunbatch, and later nine others from another group, were attached to the expedition and helped find 'native wells'.[134] The botanical collections made by Forrest's third in command, Malcolm Hamersley, were later sent to von Mueller in Victoria.[135]

During his 1870 journey from Perth to the Great Australian Bight, which traced some of the paths taken by Eyre back in 1840, Forrest again took his reliable companion Windiitj, and another Aboriginal man, Billy Noongale.[136] On his 1874 expedition into Central Australia from the western coast of Western Australia, Windiitj and Tommy Pierre were the Aboriginal party members who provided for him.[137] Such was his respect for Windiitj that on the 1874 trip Forrest named after him a permanent spring north of Wiluna.[138] In his published journal, Forrest recorded on 27 May 1874 that his party:

> Followed up the Kennedy Creek ... passing a number of shallow pools, when we came to some splendid springs,

> which I named the Windich Springs, after my old and well-tried companion Tommy Windich, who has now been on three exploring expeditions with me. They are the best springs I have ever seen ...[139]

In 1876, while working as a guide with the construction party erecting the overland telegraph line from Perth to Adelaide, Windiitj died of pneumonia and was buried at Dempster Head near Esperance.[140] He was remembered in the naming of several streets in various southern Western Australian towns.[141] Over a hundred years after his death, the Aboriginal man was still honoured, with a cultivar of barley named 'Windich' in 1989.[142]

On the 1874 trip Forrest collected botanical specimens and later published a description of the plants he found.[143] Wild plants such as the desert fig were welcome sources of fresh food for the expedition.[144] In spite of Forrest's recorded use of the desert fig, Maiden suggested that the food value of this species was marginal, commenting: 'the appetites of explorers frequently become voracious, and not too discriminating'. For desert Aboriginal people, however, this is a favourite food.[145] Such discrepant views demonstrate that the definition of 'edible' is highly subjective.

Forrest had many encounters with Aboriginal people during his trips, particularly around the precious desert rockholes. Indigenous traditional landowners met by the explorers were occasionally openly hostile, but mostly their initial reactions were fear and submission. Aboriginal inhabitants were generally aware of the explorers' presence, but kept their distance. On a few occasions they were caught by surprise. Forrest recorded the following on his 1874 expedition into Central Australia: 'When near the spring, saw natives' tracks, and shortly afterwards a fire with a whole kangaroo roasting in it. The natives had made off when they saw us, leaving their game cooking'.[146]

In areas where their Aboriginal guides were unable to understand the local languages, such physical evidence was relied upon by explorers to inform them of how people lived on the land.

The use of Aboriginal landscape-based knowledge was fundamental to the survival of many explorers. The ambitions of expedition leaders needed to be balanced by a concern for the health of the men in the party, as people such as Caley, Cunningham and Leichhardt had established long before the Burke and Wills saga. Although many explorers utilised indigenous knowledge of the land, not all Aboriginal assistance was freely given. Some of the desert explorers, such as Peter E. Warburton and Robert Austin, would capture Aboriginal people and make them lead the party to water.[147] Similarly, in 1896 in central Western Australia, David W. Carnegie seized an Aboriginal man for the purpose of locating water.[148] The unfortunate captive was initially reluctant to help, or perhaps because of language difficulties did not understand what he was being asked to do. Carnegie coerced him into cooperating by making him eat salted beef to cause a thirst. In 1897 the Calvert Search Expedition in the northern Western Australian deserts also forced Aboriginal people to act as guides.[149] They chained up their guides in camp each night, the chains looped around their necks and padlocked to a sturdy tree to prevent escape.

Europeans engaged with the Australian landscape first as explorers who mapped the land, then as settlers who occupied it and began to transform it. The degree of understanding that explorers had of hunter-gatherer interaction with the landscape was often a factor in the ultimate success of their expeditions. By following Aboriginal tracks, they could pass more easily through the country and access crucial sites for food and water. The use of wild plants provided explorers with the opportunity, if they knew how to collect and prepare them, to supplement expedition supplies. Eating fresh fruits and vegetables, even if unpleasant tasting, prevented scurvy. The combined experiences of European explorers and plant hunters from the late eighteenth century showed the prudence of keeping indigenous guides with them, best demonstrated by the relationship between John Forrest and Windiitj. In the case of Burke and Wills, their lack of knowledge about the Aboriginal cultural landscape, and their apparent unwillingness to learn, contributed greatly to their deaths.

10 The Study of Aboriginal Plant Use

The British discovered Australia's plants at the same time as they first encountered the continent's indigenous cultures, but the general absence of botanical expertise amongst the settlers impeded the making of detailed records showing how indigenous people interacted with the flora and made documentation of these relationships a slow process. Added to the shortage of intellectual 'tools' for describing plants was an initial lack of scholarly enthusiasm for the study of 'primitive' hunting and gathering cultures.[1] Only when many Aboriginal groups were suffering from the impact of British settlement did either trained or amateur botanists and their collectors begin the immense task of describing the unique Australian flora.[2]

In the latter decades of the nineteenth century, the study of Aboriginal plant use was encouraged by developments in a broad range of scholarly disciplines, in particular pharmacology, linguistics and anthropology. By the beginning of the twentieth century, the academic investigation of the plants indigenous people used had been recognised as a specific area of interest, the new field of ethnobotany.

Pharmacological analysis

In addition to food and manufacturing materials, the world's flora has provided humans with a wide range of complex organic chemical substances as medicines. Interested colonists began to determine the properties of Australian plants through a combination of experimentation with various wild species and their observations of indigenous people gaining a living from the environment, and in doing so found new ways of obtaining commercially useful substances such as medicinal oils.[3] Regardless of the extent to which they drew on Aboriginal medicinal plant knowledge, any newly discovered remedies were generally utilised within the European medical system – a conservatism which meant that throughout the nineteenth century the majority of medicines dispensed by Australian apothecaries (pharmacists) were concocted from recipe books published in Great Britain.[4] Towards the end of the century, European pharmacologists and medical practitioners began to investigate indigenous sources of medicines, narcotics, stimulants and poisons from around the world in attempts to discover chemical substances for a range of new applications.

Joseph Bancroft

Brisbane-based Dr Joseph Bancroft pioneered the pharmacological analysis of the Australian flora.[5] In 1872 he studied the properties of the Aboriginal narcotic pituri, and discovered its nicotine alkaloid contents.[6] Alkaloids are an important class of nitrogen-based organic compounds occurring in plants that includes commonly used medicines such as atropine, morphine and quinine. Amongst the Aboriginal people of eastern Central Australia, pituri was highly prized, with the best leaves and sticks coming in trading pouches from the Mulligan River area of southwest Queensland.[7] This species, which botanists know as *Duboisia hopwoodii*, occurs as a large bush and grows widely across the Australian arid zone.[8]

The physiological action of pituri is complex.

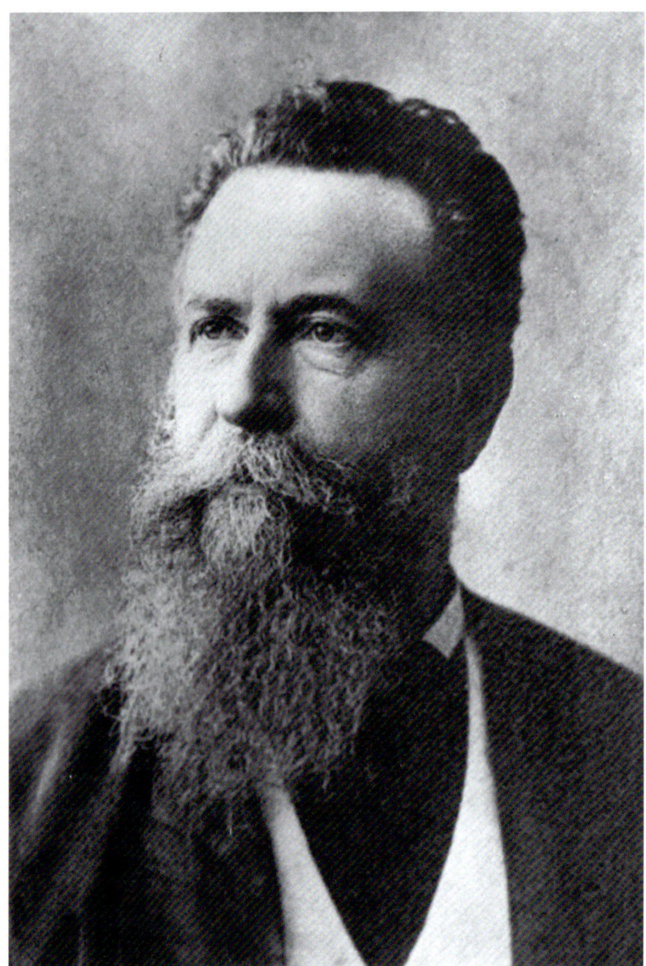

Joseph Bancroft, Australian pharmacologist. He studied the chemistry of the Aboriginal narcotic pituri, which is rich in alkaloids, in the 1870s. He also discovered the medical properties of a botanically related species, the poison corkwood tree. Photographer unknown, late nineteenth century. 57297, John Oxley Library, Brisbane.

Pituri. Desert Aboriginal people used the foliage of this small tree to produce a powerful narcotic, as well as to poison game animals. Several pharmacologists other than Bancroft studied the chemistry of this plant. Philip A. Clarke, Port Augusta, South Australia, 2007.

During Bancroft's experiments, he administered doses of pituri extract to animals such as dogs, cats, rats and frogs.[9] As a stimulant it creates a sensation of wellbeing, while as a narcotic it suppresses hunger and thirst, and was used by Aboriginal people for that purpose on long hunting expeditions. Early European observers noted the effect of pituri upon its users. In 1861, the three beleaguered members of the Burke and Wills expedition at Cooper Creek used pituri supplied to them by local Aboriginal people.[10] King later described its effect on him: 'After chewing it for a few minutes I felt quite happy and perfectly indifferent about my position, in fact much the same effect as might be produced by two pretty stiff nobblers of brandy'.[11]

Bancroft also discovered the medicinal properties of the botanically related poison corkwood tree, *Duboisia myoporoides*.[12] This plant appears to have been used as a narcotic, water reportedly being poured into holes carved into its trunk, left overnight and drunk the next morning.[13] In eastern New South Wales, Aboriginal fishers placed poison corkwood foliage in pools to bring eels to the surface.[14] Some of Bancroft's *Duboisia* research specimens were eventually sent to von Mueller, who remarked:

> Now an interesting field opens to Dr. Bancroft for further research. Let the Doctor try the foliage of Duboisia myoporoides, as he could easily, for a little payment, get a blackfellow to administer small doses of that plant ... There could be no danger in the experiment if the quantity is given cautiously.[15]

By the late 1870s a medicine derived from this plant

Pituri bag. This woven Aboriginal bag contains prepared narcotic for trade. Pituri was once the basis of a major trading economy in eastern Central Australia. Collector unknown, northeast of South Australia, 1890s. A1808, South Australian Museum Aboriginal Ethnography Collection, Adelaide.

Poison corkwood tree. Its medicinal properties have been studied since the 1870s. Von Mueller suggested to Bancroft that during his experiments on human subjects he should employ an Aboriginal person to administer this medicine. Philip A. Clarke, Adelaide Botanic Gardens, 2006.

was being widely used as a mydriatic (to control pupil dilation) in ophthalmic surgery in Australia, America and Europe.[16] Today, a hybrid between the poison corkwood and the yellow basswood (*Duboisia leichhardtii*) is commercially cultivated for alkaloid production.[17]

The publication of Bancroft's initial work motivated other researchers to begin investigation into the chemical properties of Australian plants. Chemist Archibald Liversidge conducted further experiments with the alkaloids contained in pituri, although he had problems in obtaining enough for analysis from his European source, who lived on the Barcoo River in eastern Central Australia:

> Mr Wilson had considerable difficulty in procuring a sufficient supply for my purposes; he states that the blacks prize it very highly; so much do they value it that it can only be obtained from them in very small quantities at a time, hence it involves the expenditure of much time and trouble to collect together a few pounds weight of the substance. The blacks in his district on the Barcoo obtain it from the Diamantina blacks who trade yearly with the Mulligan or Kykockodilla tribe, in whose country the piturie [pituri] grows.[18]

Liversidge paid close attention to how Aboriginal people prepared pituri for use, deducing that their addition of wood ash when chewing the green leaf was for the same reason that Indian people mixed lime with betel nut, which was to help release the alkaloids and thereby increase potency.[19]

Bancroft prepared a pamphlet titled *Contribution to Pharmacy from Queensland* and summarising the properties of a number of Australian plants, for the Colonial and Indian Exhibition held in London in 1886.[20] Bancroft's son, Thomas Lane Bancroft, who in his turn became a renowned pharmacologist[21], outlined the steps of his research methods in a paper titled *Research into the Pharmacology of Some Queensland plants,* published in 1888. Of his preliminary testing he noted:

> The method which I followed in selecting plants for examination was based on the fact that most of the active known drugs are bitter, and consisted in tasting the bark, or occasionally other parts of the plant; when, if these were found bitter, acrid, or pungent, a certain amount of the plant was collected for experiment. If the plant was sweet, sour, astringent or tasteless, no notice was taken of it further ....[22]

Relying on this technique undoubtedly led to his missing a number of plants that were potential sources of drugs.

Many other researchers gained further clues from indigenous plant use, as Joseph Bancroft and Liversidge had done with their analyses of pituri and poison corkwood.[23] Historical uses of Australian plants have also guided pharmacologists. During the 1880s Edward H. Rennie investigated the chemistry of two species that early British colonists used to combat scurvy – the sweet tea and the sour currant-bush.[24] In the 1890s Henry G. Smith commenced detailed chemical studies of eucalypts and Australian conifers.[25] Other researchers interested in the chemistry of Australian plants included Frederick M. Bailey, Joseph H. Maiden and Ferdinand von Mueller.[26]

Surveys of 'primitive' language and culture

By the time that scholars in the late nineteenth century were beginning to appreciate the former high diversity of Australian Aboriginal cultures, and to realise that they possessed a deep understanding of the environment, they found the indigenous population in decline, with many languages and cultures believed to be 'going extinct'. In many regions the generation born prior to colonisation was rapidly passing away. Edward Palmer in northern Queensland urged the recording of plant uses of Aboriginal people before they died out: 'This knowledge is likely to die out with them, unless some means are taken to place on record such information as can be gathered in the present day'.[27] Similarly, of the outback regions of Central Australia anthropologists Howitt and Siebert expressed a fear of knowledge loss: 'the older blacks, who lived in the times before the occupation of their country by the whites, are now rapidly dying out, and with them the old beliefs and customs are being lost'.[28] Many concerned researchers feared that the chance for recording knowledge of Australian hunter-gatherer cultures was fast being lost to them.

In 1883 Palmer read a paper before the Royal Society of New South Wales titled *On Plants Used by the Natives of North Queensland, Flinders and Mitchell Rivers, for Food, Medicine, etc.*[29] In the published version he wrote:

> The aboriginals appear to be possessed of considerable knowledge of indigenous plants and their uses in their several districts, as well as the periods of their flowering and fruiting; they also use many for their supposed medicinal qualities; and, considering that nearly half of their daily food consists of roots and fruits, it is no matter for surprise that they should possess some knowledge of plants. But apart from their interest in them for food purposes, they have names for a great many of plants which they do not use, and are familiar with the habits of nearly all the vegetation of their particular district.[30]

Palmer's knowledge had been gained from his own observation of Aboriginal hunter-gatherers, principally in the Bourke and Cook districts, who he said were 'still in their primitive or original state'.[31] The value of his plant use records was increased through having the species concerned identified by botanists such as Bailey, von Mueller and the Reverend William Woolls.

The increasing academic and antiquarian interest in Aboriginal cultures led to a growth in colonial reminiscences being published and many earlier historical works being reprinted.[32] In the latter decades of the century, ethnologists brought together the findings of early publications, as well as from their own fieldwork, as they delved into specific areas of indigenous practice and knowledge.[33] While ethnology focused upon drawing conclusions from comparisons between many human societies, early anthropologists began the study of specific cultures. In the temperate regions they began recording what would later be termed 'memory culture', based on verbal accounts from elderly indigenous people who could recall their lives as hunters-gatherers before the deleterious impacts of European settlement.[34]

Robert Brough Smyth

The recording of vanishing Aboriginal knowledge of plants became a specific concern in the late nineteenth century. In 1878 the multi-talented mining engineer, historian and meteorologist Robert Brough Smyth published a detailed account of Aboriginal hunting and gathering in *The Aborigines of Victoria*. While primarily focusing on Victoria, he also drew heavily from accounts based in other parts of Australia for comparative purposes.[35] From Smyth's account, it appears that some hunting and gathering activities were still practised in parts of eastern Victoria, but in the western districts had largely ceased. He provided Reverend Bulmer's list of 'vegetables commonly eaten by the natives of Gippsland', while describing Mr Hogan's list from Lake Condah as 'vegetables formerly gathered for food by the Aborigines of the Western district'.[36] It was generally believed that the impact of European settlement had by now decreased the opportunities to conduct actual field work with hunter-gatherers in southern temperate Australia.

Smyth included a separate record of Aboriginal plant foods from government botanist von Mueller in his book[37], and also published the results of his requests to the managers of the Coranderrk, Lake Hindmarsh and Lake Condah missions to send him fresh plant specimens, along with their Aboriginal plant use information, to enable von Mueller to identify the precise species. Although Smyth was not an academically trained linguist, the word lists he garnered form a unique record of the Victorian Aboriginal plant nomenclature.[38] After noticing apparent discrepancies in the collated data, he made the following astute observation concerning the complexity of indigenous taxonomic systems:

> It will be observed that in some cases the natives have given different names to the same plant. Great care was taken by my correspondence, and I cannot believe that any error has crept in. It is not improbable that there are, in the same tribe, more names than one for a plant, and that a plant in one stage of development may have a different name when it is in another stage ... I point out these apparent discrepancies, not to lessen the value of the work, but to increase it. Probably Mr. and Mrs. Green [European sources of the Coranderrk data] had not with them on every occasion the same native; and a native might on one occasion give the name of the foliage, on another that of the fruit, and on another perhaps that of the root.[39]

Edward M. Curr

The task of surveying the records of the languages and customs of Australian Aboriginal groups as they were at first European settlement was immense. Edward M. Curr (1820–89), an enthusiastic amateur linguist, applied himself to the job and in 1886–87 published his results in four large volumes.[40] Curr was a former squatter who had extensive experience with the Aboriginal people living on his sheep runs in the western Goulburn Valley of Victoria in the 1840s.[41] Drawing on his early life as a colonist, he was able to compare firsthand the impact that Aboriginal foragers once had on the landscape through burning the bush, and the consequences of their later being prevented from doing so. Discussing the Australian hunter-gatherer, Curr claimed in his reminiscences that:

> Living principally on wild roots and animals, he tilled his land and cultivated his pastures with fire, and we shall not, perhaps, be far from the truth if we conclude that almost every part of New Holland [Australia] was swept over by a fierce fire, on an average, once in every five years. That such constant and extensive conflagrations could have occurred without something more than temporary consequences seems impossible, and I am disposed to attribute to them many important features of Nature here ... it may perhaps be doubted whether any section of the human race has exercised a greater influence on the physical condition of any large portion of the globe than the wandering savages of Australia.[42]

He naively attributed the continent's generally poor soils and thin covering of vegetation to the frequent Aboriginal use of fire, although he was correct in portraying Aboriginal people as active resource managers.

With his base knowledge of Aboriginal culture, Curr was in a good position to plot the diversity of language and custom across the continent. To augment his historical data, he sent out questionnaires to former settlers, explorers, pastoralists and government officials who had been in contact with local Aboriginal groups and compiled their answers for publication. (Missionary George Taplin had already successfully used this technique when gathering data on Aboriginal culture and tradition in South Australia.[43]) Although Curr's word lists are brief, many contain unique records of Aboriginal plant nomenclature.

John M. Black

The majority of the nineteenth century botanists in Australia were skilled amateurs recruited from a class of well-educated men with careers and expertise in diverse areas. A good example was the South Australian botanist, John M. Black[44], whose additional interest in Aboriginal languages led him to publish small vocabularies for a number of languages that were in decline.[45] He was a skilful language recorder, paying particular attention to the type of sounds used by indigenous speakers. Recording indigenous terms in the southern Flinders Ranges of South Australia on 26 March 1880, he wrote:

> Today, Good Friday, some natives came to look for something to eat. We asked them how they called in their language sandalwood [Australian sandalwood] – 'baru,' with the Somerset 'r'. Peppermint gum [peppermint box] = wita, mallee – wira.'[46]

Other recorders of Aboriginal words were less discriminating between the different sounds.

During his plant collecting trips to remote locations, Black had numerous opportunities to meet Aboriginal people and discuss their relationships with the flora. Words for plants feature prominently in some, but not all, of his word lists. The Lower Murray vocabulary he obtained at Port Elliot in 1892, from Ngarrindjeri man Karammi, had 118 words in total, with only one for a plant species.[47] In 1915 he obtained a Wirangu (Wirrung) vocabulary from a group camped at Murat Bay on the western coast of South Australia. Within the set of 165 Wirangu terms are fifteen relating to particular plant species.[48] During a trip to Yorke Peninsula in 1919 Black recorded 'burkiana' as 'the country around Point Pearce Mission Station, so-called from the number of "oil-bushes"'; 'burko' meant 'oil-bush'.[49]

Black's major botanical writings, such as the published *Flora of South Australia,* unfortunately did not include much Aboriginal plant use information. For example, in the description in his flora of the monterry (native apple), which was a major Aboriginal food source, he simply said, 'The berries, called "muntries" in Victoria and in our South-East, are used for making tarts'.[50] On one occasion he appropriated an indigenous place-name from western South Australia when naming what he considered was a new plant genus. For the creeping carrot, *Uldinia mercurialis* (now called *Trachymene ceratocarpa*), he wrote that the 'name of the new genus is derived from "uldilnga gabi," the native name of "Ooldea Water," more generally known as the Ooldea Soak, and about three miles from the Ooldea Railway Station'.[51] Ooldea is famous as the site of camp of Daisy M. Bates, journalist and an early authority on Aboriginal society, on the edge of the Nullarbor Plain.[52] Black assisted Bates in the publication of her list of the Wirangu vocabulary spoken on the west coast of South Australia.[53]

Johann G. Reuther

Missionaries were involved in some of the earliest attempts at documenting Aboriginal languages and customs across Australia.[54] This is in part explained by their need to understand indigenous cultures as an aid to religious conversion and to enable the translation of the Bible into local languages. Some missionaries developed an intense scholarly interest in indigenous cultures, resulting in the compilation of records that, in terms of depth and scope, go far beyond their practical concerns in managing the affairs of their congregations.

In the 1890s, at the remote site of what became the Bethesda Mission on the banks of Lake Killalpaninna, east of Lake Eyre in Central Australia, German missionary Johann G. Reuther made a detailed study of Diyari language and culture.[55] The data he amassed included Aboriginal site-based mythology and plant uses. Accompanying his thirteen volumes of manuscripts was a large collection of Aboriginal artefacts and the desert plants they used, which is now housed in the South Australian Museum.[56] Each of the botanical specimens, although not originally identified scientifically, has a label with its Aboriginal name and a catalogue number linking it to records in the language volume.

Across Australia, missionaries often made unique accounts of the 'primitive' cultures they were striving to mould into Christian societies. They tended to work with communities that were suffering from the effects of colonisation as the political and social landscapes changed around them, rather than with communities yet untouched by European influences.

Pressed plants. Reverend Johann G. Reuther and his Diyari informants collected and prepared these museum specimens, along with recording their indigenous names. Johann G. Reuther, Lake Killalpaninna, Central Australia, about 1900. J.G. Reuther Collection, AA266, South Australian Museum Archives, Adelaide.

Lake Killalpaninna channel. To the right is the sand dune on which a German Lutheran Mission was established in 1867, where during the late nineteenth and early twentieth centuries the missionaries Otto Siebert and Johann Georg Reuther studied Diyari language and culture. Little remains of the buildings that once stood on the site. Philip A. Clarke, northeast of South Australia, 1986.

Anthropology and the study of plants

In the latter half of the nineteenth century, scientists began the detailed study of the diversity of cultures across the world. The publication of Charles Darwin's *On the Origin of Species* in 1859, and *The Descent of Man* in 1871, stimulated scholarly interest in human evolution.[57] It was thought that the investigation of modern 'stone age' peoples would offer insights into the 'primitive' origins of Western-style European culture.[58] With its societies of Aboriginal hunter-gatherers, Australia was considered one of the last places in the world where theories of human evolution could be tested in a living laboratory, and in 1890 James G. Frazer, a scholar of world religions, wrote in *The Golden Bough:* 'The aborigines of Australia have totemism in the most primitive form known to us.'[59] In this period, comparisons were made between the beliefs and practices of different societies with the goal of reconstructing human history and formulating laws of culture. With an emphasis on how people extracted a living from the environment, scholars perceived significant differences between hunter-gatherers, nomadic pastoralists and horticulturalists. As a legitimate subject for scientific investigation, indigenous hunter-gatherers sat alongside animals, plants and minerals.

In the late nineteenth century scholars commenced the systematic collecting of indigenous artefacts from around the world. As specimens, rather than trophies, they were used in books and in museums to demonstrate human evolution.[60]

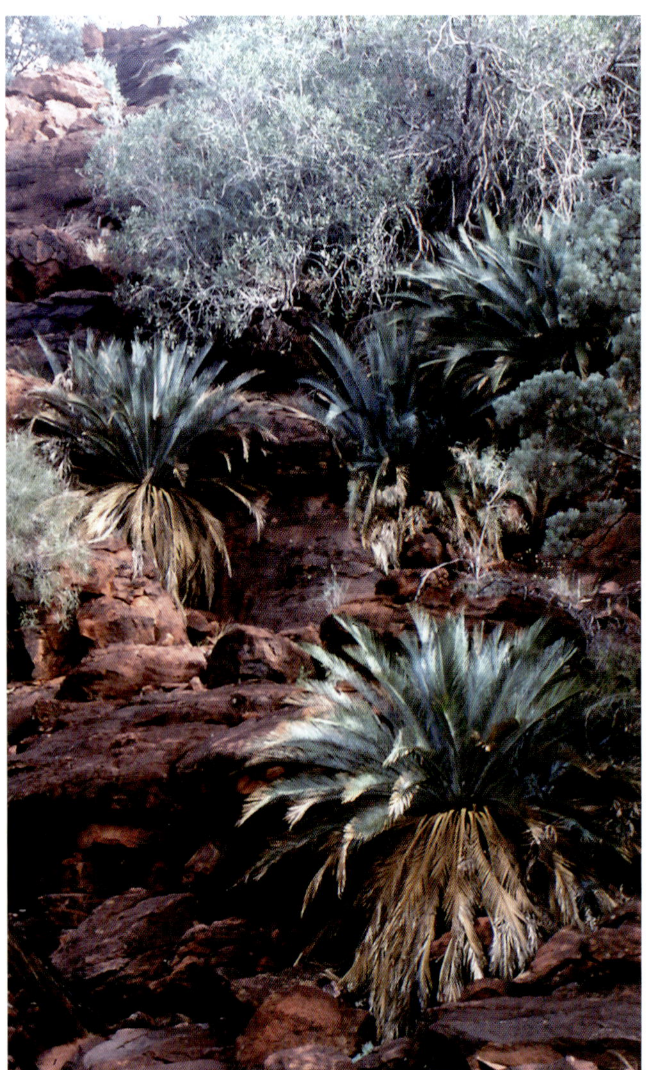

Rock fig trees (top) and Macdonnell Ranges cycads (middle and bottom). The mountain ranges in Central Australia have higher rainfalls than the surrounding desert plains, a fact which has led botanists to look for relic plant species from wetter times, such as these cycads, and palms. Philip A. Clarke, Palm Valley, western Macdonnell Ranges, Central Australia, 1997.

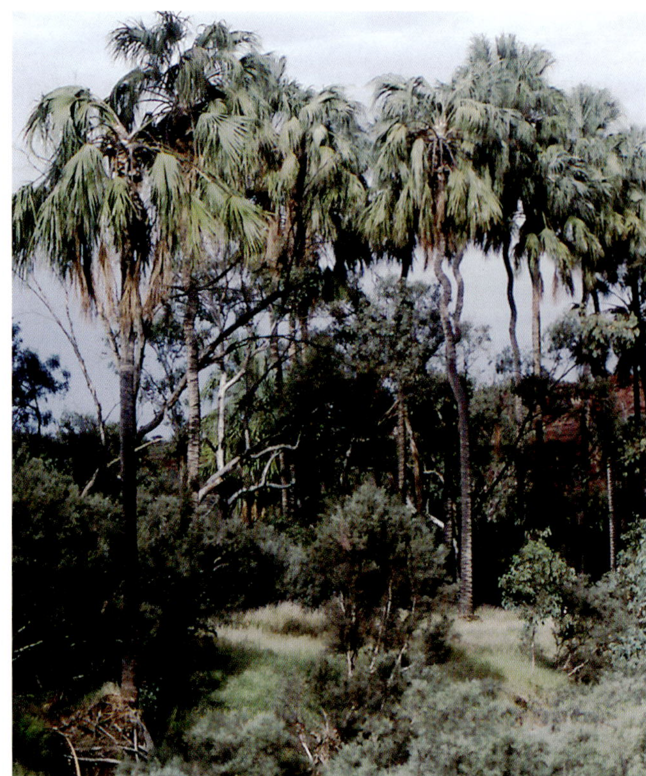

Central Australian cabbage palms. In 1894 the Horn Scientific Exploring Expedition travelled through the mountainous regions of Central Australia to investigate the sheltered biological 'islands' and 'primitive' human cultures. Philip A. Clarke, Palm Valley, western Macdonnell Ranges, Central Australia, 1997.

Professional collectors of natural history specimens catered for scholars interested in indigenous peoples. In Australia, German collector Amalie Dietrich spent several years in Queensland (1863–72), where she actively sought fresh Aboriginal skeletons for her European clients.[61] In spite of her gruesome interests, Dietrich owed her life to the Aboriginal people, for a group saved her from drowning when she was trying to collect specimens of rare water lilies.[62] Norwegian anthropologist and zoologist Carl S. Lumholtz collected natural history specimens in northern Queensland from 1880 to 1884, at one stage attaching himself to an Aboriginal band and travelling with them for fourteen months.[63] With Aboriginal assistance, Lumholtz was able to capture a new species of tree kangaroo which today bears his name. In 1889 he published his Australian experiences in the unfortunately titled *Among Cannibals*.[64]

By the end of the century, the botany of temperate Australia was well known, while the tropical and arid zones still had many species waiting to be collected and described. In this period, scientists had a primary role in expeditions into remote Australia. Frederick Bailey, as the Colonial Botanist of Queensland, was part of Archibald Meston's 1889 Bellenden-Ker Range Expedition, inland from Cairns in northern Queensland. From this trip, Bailey named a species of scleroglossum as *Vittaria wooroonoorani* (now called *Scleroglossum wooroonooran*), the latter part being the Aboriginal name for the Bellenden-Ker Range.[65] Meston, who was the Aboriginal Protector in southern Queensland from 1898 to 1903, made another trip to the Bellenden-Ker Range in 1905, this time without Bailey.[66] On all his expeditions Meston, himself a keen botanist, relied heavily upon Aboriginal people as guides and plant collectors.

The Horn Scientific Exploring Expedition and Baldwin Spencer

Central Australia was recognised as a particularly good place to study Australian hunter-gatherers, for in the late nineteenth century the region was still at the frontier of European expansion. The Elder Scientific Exploration Expedition took place here in 1891, but due to being badly planned and poorly equipped was widely regarded as a failure in terms of its scientific output.[67] In 1894 the Horn Scientific Exploring Expedition, organised by wealthy pastoralist and mining magnate William A. Horn, fared much better. This party, formed chiefly of scientists, travelled through the mountainous regions of the Centre to investigate the sheltered biological 'islands' that form desert refuges for ancient forms of plant and animal life that have vanished from the surrounding plains.[68] Palm Valley, where isolated populations of Central Australian cabbage palms and Macdonnell Ranges cycads grow, was one of their destinations. The Horn Expedition was also interested in the manners and customs of the Aboriginal inhabitants of the Centre. Assembled at Oodnadatta, the expedition moved through the Finke River Basin and across the Macdonnell and James ranges, to the south and west of Alice Springs.

The Aboriginal people employed by the Horn Expedition as collectors of natural history specimens were highly regarded for their powers of observation[69], South Australian Museum Director Edward C. Stirling remarking that the people at Finke River demonstrated an:

> acuteness of observation which at once enabled them to distinguish animals or plants which very closely resembled one another. As Mr Schulze [a missionary at Hermannsburg] informs us they have separate names for twenty-two kinds of snakes, the distinction between some of which is a matter of difficulty even for a trained zoologist.[70]

His experience on this expedition influenced one of its members, Baldwin Spencer, Professor of Biology at the University of Melbourne at the time, to make a career shift to the study of Aboriginal culture.[71] He had already cultivated an interest in anthropology at Oxford University, having been a research

Indigenous plant collectors, with Archibald Meston and his son Jack, bringing plant specimens into camp on the Second Government Scientific Expedition to the Bellenden-Ker Range. Indigenous people possess a wealth of environmental knowledge that many European plant collectors have utilised. Photographer unknown, Bellenden-Ker Range, south of Cairns, northern Queensland, 1905. H.G. Stokes Collection, AA312, South Australian Museum Archives, Adelaide.

Members of the Horn Expedition and their camels taking a break. Aboriginal guide Harry is seated at the water's edge. The expedition employed a number of Aboriginal hunter-gatherers to collect natural history specimens for them. W. Baldwin Spencer, Abminga Waterhole, south of the Finke River, Central Australia, 1894. F.J. Gillen Collection, AA108/Album F/AP5660, South Australian Museum Archives, Adelaide.

assistant to Sir Edward Tylor, a founder of academic anthropology in Great Britain, and combining his interests in biology and anthropology made sense. Unlike many of his predecessors in Australia, Spencer was a keen supporter of Darwin's theories of evolution and natural selection, and aimed to compile as a complete an account as possible of one of the last surviving Stone Age cultures, 'that affords as much insight as we are now ever likely to gain into the manner of life of men and women who have long since disappeared in other parts of the world'.[72]

Spencer edited the four-volume official report of the Horn Expedition, for which Stirling wrote the anthropological section.[73] Maiden inspected the gums and resins collected to provide an appendix.[74] Stirling was not impressed with the eating qualities of many Aboriginal plant foods, finding that even when the fruit of the climbing shrub known as the desert pear (or desert banana) was ripe, it was 'about as palatable and nutritious as sawdust'.[75] Aboriginal people eat the fruit either raw or cooked.[76] Stirling's comments suggest that this species' common names derive from the shape of the fruit rather than its taste or texture.

The Horn Expedition set a high standard for the future conduct of focused investigation in remote parts of Australia. Spencer was inspired to return to Central Australia for further research which led him to establish an important anthropological partnership with the Alice Springs telegraph operator, Francis (Frank) J. Gillen.[77]

Spencer and Gillen Expedition, 1901–02. From left to right: Purunda (Warwick), Francis Gillen, Harry Chance, Baldwin Spencer and Erlikiliaka (Jim Kite). Francis J. Gillen, Alice Springs, Central Australia, 1901. F.J. Gillen Collection, AA108, South Australian Museum Archives, Adelaide.

From Melbourne, Spencer directed Gillen's field research, receiving from him copious amounts of anthropological data and biological specimens. Gillen's main informants were Aboriginal people, but he also utilised a number of European field collectors.[78] In a 1901–02 expedition across the Northern Territory, Spencer and Gillen employed two Aboriginal men, Purunda (Warwick) and Erlikiliaka (Jim Kite), as guides.[79] German missionary Carl Strehlow, based at Hermannsburg Mission in the western Macdonnell Ranges, also studied Central Australian Aboriginal cultures.[80] Although dealing with some of the same cultures as Spencer and Gillen, Strehlow's work was published in German, making the products of his research less accessible to the British anthropological community.[81] From the late nineteenth century, researchers in Australia and elsewhere were aided by advances in the technology of photography, which enabled the accurate visual recording of both natural history subjects and hunter-gatherers in the field.[82]

Gillen (standing second from left) and Spencer (standing second from right) with a group of Arrernte elders. The anthropologists worked at a brush shelter, erected at the Alice Springs Telegraph Station, for the purpose of recording ethnographic data. The information they gathered was used in several major anthropological publications. Francis J. Gillen, Alice Springs, Central Australia, 1896. F.J. Gillen Collection, AA108/Album L/AP5844, South Australian Museum Archives, Adelaide.

Walter E. Roth

Spencer was not the only anthropologist of the time with biological training. Walter E. Roth, a contemporary of his at Oxford, had also studied biology before coming to Australia.[83] While Spencer was based in temperate Victoria, Roth moved to Queensland, where his employment gave him the means to engage in a detailed study of Aboriginal culture in the subtropical and arid regions: 'Since 1894 my tenure of office as Surgeon to the Boulia, Cloncurry, and Normanton Hospitals, respectively, has afforded unrivalled opportunities for making inquiry into the language, customs, and habits of the North-West-Central Queensland aboriginals.'[84] Roth was essentially an ethnographer, recording the cultures he observed rather than theorising about them. His scientific background is apparent in his eye for detail when recording his observations.

Ethnological Studies Among North-West Central Queensland Aborigines, Roth's highly acclaimed monograph, was published in 1897 and is regarded as the first of its type concerning Australia.[85] In it he provided a detailed account of Aboriginal plant use, commenting on vegetables and fruits as food: 'Indeed, it is difficult under this heading to know what is refused.'[86] Roth recorded trading practices involving pituri, describing how in southwest Queensland Aboriginal 'messengers are sent direct to the Ulaolinya tribes at Carlo with spears and boomerangs, "Government" and other blankets, nets, and especially red-coloured cloths, ribbons, and handkerchiefs to exchange and barter for large supplies of the drug.'[87] Roth followed this work with eighteen ethnographic bulletins in the period 1901–10, based on official reports written in his capacity of Aboriginal Protector and covering wide-ranging aspects of indigenous culture.

All of Roth's bulletins include Aboriginal plant uses, although the most detailed account appears in his treatment of *Food: Its Search, Capture, and Preparation*, which has a 'list, with details of preparation where necessary, [and] comprises a fair proportion – some 240 – of the edible plants used by the North Queensland aboriginal.'[88] For most of the listed species he provided preparation information, area of use and Aboriginal terms. Of portulaca, for example:

> In the Boulia district this may be eaten raw in its entirety, or only the seed used. The latter is obtained by taking a goodly-sized bunch and rubbing it between the two hands more or less horizontally, the seeds dropping through the inter-digital spaces into a wooden bowl; it is subsequently washed, ground, and eaten raw.[89]

He gave its name in the Pitta Pitta language of southwest Queensland as *kuni*.

Roth's descriptions of Aboriginal traditions in *Superstition, Magic and Medicine* provided insights into beliefs concerning plants, ancestors, spirit beings and the Creation. For example, the people at Hann River in east Cape York Peninsula believed that the bamboo, which is important for making spears, was first brought to their country by an Ancestor:

> A finch came and stole the rain from the country traversed by the Hann, and took it away north with him to his home in the neighbourhood of Port Stewart. A small red-bird, resident on the Hann, revenged himself by going up to the Port, stealing the roots of as many bamboos as he could, and planting them on his return around his own camp.[90]

Roth used beliefs in plants as a window into aspects of Aboriginal culture. He remarked that for the people in the Bloomfield River area of southeast Cape York Peninsula, 'Double-fruits in bananas, nuts &c., are believed to be made by certain invisible beings for the sake of amusement – "sports" in a double sense.'[91] Roth's publications today provide a valuable record of Aboriginal traditions, customs and languages in northern Queensland. Later in his career, Roth conducted similar anthropological work in South America.[92]

Economic botany

The Royal Gardens at Kew had experienced a decline resulting from poor management after the death of Banks in 1820.[93] To save them, the Government brought the Gardens under public control in 1841, with William J. Hooker becoming the first official Director.[94] Hooker, who had been a protégé of Banks, wanted to create a herbarium 'collection that would render great service, not only to the scientific botanist, but to the merchant, the manufacturer, [and] the physician ...'[95] In 1847 he opened the first public Museum of Economic Botany, which had on display items such as fruits (dried and preserved in spirits), seeds, wood specimens, stems of palms and tree ferns, plant-based drugs, gums, resins, fungi and drawings.[96]

Joseph H. Maiden

In Australia, botanist Joseph H. Maiden became the champion of economic botany, writing a large number of papers in the field.[97] In 1881 he became the first curator of the Technological Museum in Sydney, where he established a herbarium while holding a number of concurrent positions, including Consulting Botanist to the Department of Agriculture from 1890, Superintendent of Technical Education from 1894, and Government Botanist and Director of the Botanic Gardens in Sydney from 1896.[98] Taking advantage of his status, Maiden was instrumental in persuading British institutions to release early herbarium specimens back to Australia.[99]

In 1889 Maiden published a book on economic botany, *The Useful Native Plants of Australia*, which incorporated many Aboriginal plant uses into his discussion of each potentially useful species.[100] The chapter titles are in keeping with the interests of economic botany, starting with 'Human Foods and Food Adjuncts' and ranging through 'Forage Plants' to 'Drugs', 'Gums, Resins, and Kinos', 'Oils', 'Perfumes', 'Dyes', 'Tans', 'Timbers' and 'Fibres'. His plant descriptions outlined their physical properties and potential European uses. In 1900 Maiden wrote a paper on 'native food-plants' in which he argued:

> Knowledge in regard to the indigenous vegetable food resources of these colonies should be considered an absolute necessity by those whose avocations take them out of beaten tracks, especially in the dry country, while the ordinary citizen may find himself occasionally in a position in which an acquaintance with the scanty vegetable food products of the bush would be useful to him … The Australian aborigine has bestowed no cultivation on food-plants (with perhaps a trivial exception), but there is no doubt that many of the fruits, roots, &c., could have been improved by culture and selection extending over lengthened periods.[101]

Maiden's compendium and 'native food-plant' paper brought together much information on Aboriginal plant use, most of it gathered by other botanists and field observers.

Their lack of understanding of Aboriginal culture was an impediment for a number of botanists. An even bigger hurdle was not knowing how to engage Aboriginal informants in useful conversation. Maiden commented:

> The poor aboriginal chiefly takes interest in the vegetation as supplying him with his scanty food, or as affording him fibre useful in securing fish and other animal sustenance. As far as we know, the Materia Medica of the blacks is of a very meagre description … Civilised or semi-civilised blacks frequently know but little about their native Materia Medica, and the difficulty of obtaining reliable information is enhanced (as I have experienced to a slight extent) through the extreme willingness of town blacks to impart information in regard to any plant which may be shown them, which impresses one with the thought that they are too willing to oblige. But perhaps this is mainly owing to asking them leading questions.[102]

Thus, because of difficulties with research in the field, and their primary focus on the flora, economic botanists in the late nineteenth century restricted their concerns to determining the physical attributes of each species, rather than looking at its cultural role and documenting how Aboriginal hunter-gatherers living off the land interacted with the environment.

Max Koch

Amongst the botanists and collectors studying the economic benefits of the flora, it was inevitable that at some point a researcher's primary interest would light on indigenous plant uses. An example of this can be seen in the way Max Koch practised botany. In the 1890s the German-born Koch was one of Maiden's collectors, working chiefly in arid Australia. He came to Australia in 1878, ending up managing the Mount Lyndhurst sheep station in the northern Flinders Ranges of South Australia.[103] In his spare time, Koch collected plant specimens and seeds which he sold to herbaria and botanic gardens across Australia and overseas, making the inclusion of Aboriginal plant use data on herbarium specimens a selling point. In 1899 Koch wrote to the Royal Gardens at Kew, successfully offering for sale various plant specimens and seeds: '[the] botanic names are well authenticated, remarks as to the uses of plants are made and aboriginal names as far as known are recorded on each ticket'.[104]

In the published form of a paper Koch read to

the Royal Society of South Australia in 1898 he described his personal interest in the flora:

> being intimately associated with the rearing of stock, I have made it my business to investigate which plants are most suitable and valuable for pasture, and notes of the economic value of each plant, as far as known to me by personal observation, are herewith given. I have also ascertained a few aboriginal names for various plants, as well as the uses they are put to by the natives, and I trust these additions to my list will be of some interest.[105]

Koch's publications were essentially lists of specimens he collected, arranged by plant family, and annotated with notes about uses by indigenous peoples and pastoralists. This style was later taken up by naturalist John B. Cleland and biologist T. Harvey Johnston in their extensive Aboriginal plant use studies of the early twentieth century.[106]

Amongst Koch's botanical writings are many references to Aboriginal plant uses. For instance, in 1898 he described the methods that desert dwellers used to extract drinking water from the needle-bush or water tree:

> Aboriginal name, *Kooloova* [*kuluwa*]. The aboriginals when hard-pressed for a drink extract water from the running roots, called *Nappa-koparee.* They first burn down the bush, thus driving all the moisture into the roots, which they dig out. One end of the root is exposed to the heat of a fire, and the water trickles out from the other end into a receptacle, often consisting of a wallaby skin turned inside out.[107]

Rather than recording just one Aboriginal name, Koch listed alternatives for each plant he collected. For the onion grass or yalka, which was then not yet properly identified, he stated the 'small bulbs have a nut-like taste, and are eagerly dug after by the natives, who call them *Kudnamurra, Ala, Yower, Tharaka*, and the name *Yower* being mostly used by Mt. Lyndhurst blacks'.[108] Koch realised that Aboriginal people and Europeans had different ways of discriminating between types of plants, which sometimes led to indigenous people classifying different species together. He wrote in 1900 that '*Acacia cibaria* [grey mulga] is the species the seeds of which the blacks gather for food, and it is called by them *Mulka*, or by another tribe *Wodnera*, the same as *Acacia aneura*'.[109]

Ethnobotany

At a time when much of the world was under the control of European colonial superpowers and the description of floras well underway, scholars in the emerging field of anthropology became interested in the study of hunter-gatherers and their culture. The relationships that indigenous peoples had with plants became another determinant of cultural identity, hunter-gatherers being seen as more 'primitive' than pastoralists and horticulturalists. While economic botanists regarded plants that possessed only symbolic value as of minor interest, anthropologists recognised them as culturally significant. A growing number of scholars began to focus upon the ways human culture influenced the use and perception of the flora.

The term 'ethnobotany' was coined in 1895 to better describe the study of 'plants used by primitive and aboriginal people', 'ethno' referring to the study of people.[110] Its author was American botanist John W. Harshberger, who first used it in addressing a university archaeological association. Harshberger's description of ethnobotany had a strong cultural flavour, for he argued that it could serve as a tool for determining the evolutionary position of Native American tribes.[111] He also suggested that ethnobotany would help determine past distributions of people and their trade routes. To the anthropologists working in North America a continent formerly shared by indigenous horticulturalists and hunter-gatherers, this must have had considerable appeal. Around the world ethnobotany became the successor of academic studies previously referred to as 'applied botany', 'aboriginal botany' and 'botanical ethnography', and has attracted researchers from a wide range of disciplines.

A rough distinction between economic botany and ethnobotany is that the former concerns European-style agricultural and industrial uses of plants, while the latter is focused on indigenous (non-European) interactions with the flora. In 1978 ethnobotanist Richard I. Ford stated that economic botany 'emphasizes the uses of plants, their potential for incorporation into another (usually Western) culture, and ... their benefactors have indirect contact with the plants through their by-products', whereas ethnobotany is 'concerned with the totality of the place of plants in a culture and the direct interaction by the people with the plants'.[112] Some more recent definitions of economic botany appear to incorporate the interests of ethnobotany, Gerald E. Wickens, for instance, claiming in 1993 that the economic botany field is concerned with 'those plants utilized whether directly or indirectly for the benefit of Man. Indirect usage includes the needs of Man's livestock and the maintenance of the environment; the benefits may be domestic, commercial, environmental or aesthetic'.[113]

Through the twentieth and early twenty-first centuries, Australian ethnobotany has maintained a common area of interest for a wide range of specialists, from whom it has gained many flavours.[114] Ethnobotanists with scientific, medical or economic interests in the indigenous use of the environment have treated the field as a study of the physical properties of the plants and their potential for use outside the Aboriginal arena. In contrast, anthropologists, linguists and cultural geographers have used ethnobotany as a vehicle to highlight the cultural importance of the flora. Both approaches are valid and, when expertise allows, can often be used in conjunction, particularly in ecological studies. Increasingly, indigenous peoples in Australia are engaging in ethnobotany as part of their social history studies. The definition of ethnobotany that I prefer is that its primary concern is the study of how specific cultures perceive and utilise the flora.

This book has covered the breadth of European–Aboriginal interaction on the frontier of the botanical discovery of Australia. While the first British explorers and settlers believed that Aboriginal people were too 'primitive' to own the land they hunted and gathered on, it was eventually recognised that Australian hunter-gatherers possessed plant use information that was of economic and scientific importance to the Empire. Aboriginal people were not always directly encountered during early expeditions into remote Australia, although their abandoned campsites provided evidence of how they lived off the land. When travelling through unfamiliar landscapes, explorers and professional plant collectors alike acquired the services of indigenous guides. For the European settlers, some of the plants formerly used by Aboriginal hunter-gatherers had major roles as emergency resources on the settlement frontier. In terms of European-style horticulture, however, the Australian flora has had only a minor long-term impact. Today, the main reminders of the close relationships between colonists and indigenous people on the frontiers of settlement are the Aboriginal plant names that have entered Australian English.

The expansion of British settlement during the nineteenth century transformed Aboriginal Australia, resulting in the cessation of hunter-gatherer lifestyles in many regions. With botanists and plant collectors continuing to utilise indigenous knowledge and labour, however, the range of interest in Aboriginal relationships with the flora broadened to include scholars variously described as pharmacologists, nutritionists, historians, ethnologists, linguists and anthropologists. By the beginning of the twentieth century, intellectual interests in Aboriginal plant use were being embraced by the emerging field of ethnobotany.

Since the commencement of European colonisation around the world, scholars concerned with the environment have greatly benefited through their association with indigenous peoples and their cultures. In Australia, further study of Aboriginal plant use will undoubtedly yield much more 'green treasure'.

Endnotes

Introduction

1 Australian Plant Name Index (http://www.anbg.gov.au).

Chapter 1 Early Explorers and Aboriginal Guides

1 Flannery (1994) and White (1994) have provided a palaeontological view of the unique Australian fauna, while Hill (2004) described the origins of the Australian flora.
2 Dampier (1697 [1998, p. 218]).
3 Orchard (1999, p. 12), Morley & Toelken (1983, p. 13) and Nelson (1990a, p. 285). Van Lohuizen (1967) provides an outline of the trip.
4 The species, which do not have widely recognised common names, are *Acacia truncata* and *Synaphea spinulosa*.
5 Rolls (2002, p. 6) and Thieret (1958, p. 23).
6 Zamia palms are cycads. Beaton (1982, pp. 55–6), T.W. Edgeworth David (in Wyndham, 1890, p. 119), Lawrence (1968, pp. 149–50), Thieret (1958, pp. 14–7) and Whiting (1963, Table 2, pp. 276–8) summed up preparation methods for Australian cycad nuts.
7 Dampier (1697–1703), Finney (1984, pp. 9–13), Moore (2001, p. 3) and Nelson (1990a, p. 286). See Bach (1966) for an outline of Dampier's trips.
8 Webb (2003, pp. 3–5).
9 Dampier (1703, p. 122).
10 Symon & Jusaitis (2007, pp. 11, 16–7, 20, 24–6, 28–9). Dampier called the species 'Colutea Novae Hollandiae'.
11 Finney (1984), Moyal (1986) and B.W. Smith (1992).
12 Aitken (2006, pp. 148–52), Barker & Barker (1990, pp. 40–2), Cook (1768–1779), Crosby (2004, pp. 297–8), Field (1993, p. 143), Lyte (1980, chapter 6), McGillivray (1969), Stearn (1969, 1974), Vallance et al (2001), Whitehead (1969) and Wickens (1993, p. 85).
13 Field (1993, p. 141).
14 Gilbert (1966), Moyal (1976, pp. 10–11; 1986, pp. 23–4, 28) and Webb (2003, p. 5). Maiden (1909) wrote a book titled *Sir Joseph Banks: The 'Father of Australia'.*
15 Cook (1893, p. 247). See also Aitken (2006, p. 148) and Whitley (1969, p. 248).
16 Moyal (1976, p. 11).
17 Cook (1768–71 [1968, pp. 393–4]). See also discussion of Cook's fruits by Hunter (1793, p. 478).
18 Banks (cited Maiden, 1900, p. 130).
19 Banks (1768–71 [1962, p. 115]). Probably the tree zamia (*Cycas media*).
20 Finney (1984, pp. 15–6) and Nelson (1990a, p. 286).
21 For a discussion of the problems sailors had with scurvy and how its effects were ameliorated by plant foods see Hughes (1975), Kodicek & Young (1969) and Pearn (1983, pp. 51–2).
22 Field (1993, p. 142), Moyal (1976, p. 10), Stearn (1974, p. 8) and Whitley (1969).
23 Finney (1984, pp. 108–9).
24 Clark (1987, pp. 19–20), Reynolds (1989, pp. 67–8, 94, 98) and Williams (1985).
25 Finney (1984, p. 41).
26 Blainey (1977), Clark (1957, chapters 1–2; 1987, chapters 1–2), Day (2001, chapters 4–5), Denoon & Mein-Smith (2000, pp. 86–94), Macintyre (1999, pp. 17–35), Reynolds (1989, pp. 67–8, 94, 98), Southgate (1967, pp. 103–31) and Williams (1985).
27 James F. Bennett (1843, cited in Rolls, 2002, p. 171).
28 For example, see John Bussell (cited Lines, 1994, pp. 149–51). Donaldson (1985) discussed how early colonists heard indigenous Australian languages.
29 Hiatt (1996, chapter 1)
30 Lines (1994, p. 151).
31 For references on the cultural geography concept of the cultural landscape in Aboriginal Australia, refer to Anderson & Gale (1992), Baker (1999), Clarke (1994), David (1998; 2002), Head (2000a&b) and Young (1992).
32 Berndt & Berndt (1989), Charlesworth (1998), Charlesworth et al (1984), Clarke (2003a, chapter 2) Hiatt (1975), Kolig (1989) and Stanner (1953 [1979]).
33 Day (2001, p. 258). See also Clarke (2003a, chapter 13) and Moorehead (1968, 62–75).
34 Low (1999) and Rolls (1997).
35 Clark (1987, chapters 2–5), Clarke (2003a, chapters 12–14), Flannery (1999, 2002), Griffin & McCaskill (1986, p. 5) and Walker (1889).
36 Baker (1919), Maiden (1889, chapter 9) and Powell (1990, pp. 94–5).
37 Blainey (1977, pp. 99, 106), Cooper (1952, p. 173), Cumpston (1970, pp. 3, 29, 104), Day (2001, p. 51), Moore (1924, pp. 83–86), Nunn (1989, pp. 19, 22) and Peron (1802).
38 Blainey (1977, p. 101), Collins (1798–1802, vol.2, pp. 119–20) and Finney (1984, pp. 58, 101). Starbuck (1878 [1964]) provided tables showing returns of whaling vessels sailing from American ports.
39 Troy (1990, p. 141).
40 Bauer (1969, 1983), Day (2001, chapters 14–15) and Kalma & McAlpine (1983).
41 Berndt & Berndt (1954, pp. 40–122), Clarke (2003a, chapter 11) and Macknight (1976).
42 Clarke (1994, pp. 227–36), Flannery (1997), Williams (1974; 1997) and Woolmington (1972, p. 26).
43 Clarke (2003a, chapters 12–14). Ruhe (1986) described the 'older Australian landscape'.
44 Rowley (1972c, pp. 361, 384, 406, 414, 421–2, 442).
45 Abbie (1976, p. 46), Attenbrow (2003, chapter 3), Butlin (1993), Couper Black (1966, p. 97), Lourandos (1997, pp. 35–8), Mulvaney & Kamminga (1999, p. 68) and Smith (1980, pp. 67–77).
46 Butlin (1993, part 5), Clarke (2003a, chapter 12), Lines (1994, chapter 12 & p. 328), Mulvaney (2002), Rowley (1972a) and Smith (1980). Smyth (1878, vol.1, pp. 253–69) considered the origin of diseases inflicting Aboriginal people.
47 Jones (1990).
48 Clarke (2003a, chapter 14).
49 Rowley (1972c), Smith (1980) and Tindale (1941b).
50 Moyal (1986, p. 65).
51 Finney (1984, p. 1).
52 Clarke (1996; 2003a, pp. 198–200, 204–5, 218) and Reynolds (1990). Flannery (1998) has provided numerous examples of the interaction between explorers and indigenous peoples.
53 Troy (1990, pp. 46–8).
54 Clarke (2003a, pp. 204–5) and Lowe (2002, chapters 10–11).
55 Hunter (1793, chapter 21) and Tench (1788–92 [1996, pp. 185–98]).
56 Hunter (1793, chapter 21) wrote the Aboriginal names for the men as

'Colebe' and 'Ballederry'.
57 Tench (1788–92 [1996, pp. 188–9]). Attenbrow (2003, pp. 23–4) listed the 'Boorooberongal' as the Buruberongal people, and the 'Càdigal' as Gadigal people.
58 Tench (1788–92 [1996, p. 194]). For a description of tooth removal rituals in eastern Australia, refer to Collins (1798–1802, vol.1, pp. 579–80), Howitt (1904, pp. 569, 574–6, 586–92) and S. Leigh (1821, cited Threlkeld, 1824–1859 [1974, pp. 333–4]).
59 For accounts of Aboriginal ritual knowledge, refer to Clarke (2003a, chapters 2 &3) and Keen (1994).
60 Cooper & McLaren (1997). Barrallier (1802) gave an account of the early attempts to cross the Blue Mountains. Flannery (1998, pp. 90–3) provided an account of Barrallier's 1802 expedition.
61 McLaren (1996, pp. 244–7).
62 Clarke (2003a, pp. 199–200), Jones (1974, p. 321), Lines (1994, p. 194) and Tindale (1974, p. 148).
63 Mitchell (1838, vol.1, p. 31).
64 Clarke (2003a, part 3), Davis (1989), Jones et al (1997), Reid (1995a&b) and Thomson (1939).
65 Robinson (1830 [Plomley, 1966, p. 300]).
66 Holden (1966).
67 Bunce (1859a, p. 67). Language data published in Bunce (1859b).
68 McCarthy (1966), K.V. Smith (1992, pp. 30–8) and Troy (1990, pp. 137). Boongaree also spelled as 'Bungaree'. See K.V. Smith (1992, pp. 168–8) for the many variations of his name.
69 Flinders (cited Flannery, 2000, p.xiv).
70 Brown (1801–05 [2001]), Flinders (1814a & b), K.V. Smith (1992, chapter 4) and Vallance et al (2001, pp. 232, 238, 260, 352).
71 R. Brown correspondence to J. Banks (30 May 1802, Banks papers, Mitchell Library, Sydney).
72 Brown (1801–05 [2001, pp. 231, 303, 352, 354, 361]), Flinders (1814a, vol.2, pp. 205–6, 208–9, 238–9) and K.V. Smith (1992, pp. 53–60).
73 Flinders (cited Flannery, 2000, p. 127).
74 Good (1801–03 [1981, pp. 82–3]).
75 Davidson (1934, 1936) provided an overview of Aboriginal weapon styles across Australia.
76 Flannery (2000).
77 Cunningham (1817, cited Lee, 1925, p. 310) and K.V. Smith (1992, chapter 7).
78 King (1827, vol.1, p. xxxix).
79 The name for this explorer has also been written as Captain Faddei Faddeevich (Fabian) Bellingshausen. Day (2001, p. 57), Finney (1984, pp. 159, 164–5), K.V. Smith (1992, chapter 8) and Whitley (1933, p. 310).
80 Bellingshausen (1819–21 [1945, p. 162]).
81 Cleary (1993) and Troy (1993b).
82 Threlkeld (1824–1859 [1974, p. 316]).
83 Bellingshausen (1819–21 [1945, p. 163]).
84 Bellingshausen (1819–21 [1945, p. 163]). See also Iredale (1955, p. 35).
85 Bellingshausen (1819–21 [1945, pp. 348–9]) and Froggatt (1932, p. 115).
86 Bellingshausen (1819–21 [1945, p. 188; see plate 15]). Threlkeld (1824–1859 [1974, p. 316]) gave the name of Boongaree's wife as 'Gooseberry', while Smith (2005a) described her as 'Queen Gooseberry'.
87 Bellingshausen (1819–21 [1945, pp. 188–90, 329 39; see plate 14]). This practice was mentioned earlier by Collins (1798–1802, vol.1, pp. 552–3) and Tench (1788–92 [1996, p. 194).
88 S. Leigh (1821, cited Threlkeld, 1824–1859 [1974, p. 333]).
89 Threlkeld (1824–1859 [1974, p. 32]).
90 Bellingshausen (1819–21 [1945, pp. 337–8]).
91 Bellingshausen (1819–21 [1945, p. 190; see plate 14]).
92 Bellingshausen (1819–21 [1945, p. 184]).
93 Bellingshausen (1819–21 [1945, p. 188]).
94 Bellingshausen (1819–21 [1945, pp. 184–5]).
95 Brown (1801–05 [2001, p. 232]) and Threlkeld (1824–1859 [1974, p. 316]).
96 K.V. Smith (1992, pp. 21–2, 51, 53–5, 65) and Vallance et al (2001, pp. 238, 259, 287). Nanbaree was also recorded as Nan.bar.ree, Nanberry, Nan-bar-ray and Nanbury.
97 Attenbrow (2003, p. 14), Bradley (1786–92, pp. 162–3, 183, 187), Collins (1798–1802, vol.1, pp. 65–6, 86, 597–8), Hunter (1793, pp. 134–5, 166–8, 406, 543), Tench (1788–92 [1996, pp. 104–6, 117, 134, 136, 141, 163, 171, 177, 262]) and Threlkeld (1824–1859 [1974, p. 10]). See also J. Hunter, *Extract from Govr Philips [sic] Dispatches, 1 January 1791* (Series 35.15, Papers of Sir Joseph Banks, State Library of New South Wales, Sydney).
98 Troy (1990, pp. 41–3, 110, 112, 135) provided biographies of Arabanoo, Abooroo (Abaroo) and Nanbarry (Nanbaree).
99 Tench (1788–92 [1996, pp. 102–3]).
100 Troy (1994, p. 62).
101 Troy (1990, pp. 112–13).
102 Threlkeld (1824–1859 [1974, p. 10]).
103 Collins (1798–1802, vol.1, pp. 563–83).
104 Collins (1798–1802, vol.2, p. 67).
105 Collins (1798–1802, vol.2, p. 124).
106 Tench (1788–92 [1996, p. 136]).
107 Collins (1798–1802, vol.1, pp. 133–6) and Tench (1788–92 [1996, pp. 138–9]).
108 Troy (1990, pp. 42–3).
109 Collins (1798–1802, vol.1, p. 594). Collins called her 'Boo-roong', which appears to have been more commonly used for Abaroo in her later life.
110 Collins (1798–1802, vol.1, p. 561).
111 Collins (1798–1802, vol.1, pp. 602–4). Collins referred to Boladeree as 'Ba-loo-der-ry'.
112 Dawes (1790–91) listed 'Nan-bar-re' as one of the 'Names of Native Men'. Troy (1990, p. 43) discussed Dawes' informants.
113 Hunter (1793, p. 134).
114 See chapter three.
115 Threlkeld (1824–1859 [1974, pp. 10, 32]).
116 King (1827, vol.2, p. 5), Lee (1925, p. 459) and K.V. Smith (1992, p. 116).
117 K.V. Smith (1992, p. 116) and Troy (1990, p. 46). Blundell is also written as 'Bundell', 'Bundle' and 'Bondel'.
118 Collins (1798–1802, vol.1, p. 177). See also Troy (1990, pp. 46, 111). Collins referred to Blundell as 'Bondel, a native boy'.
119 Cumpston (1970, p. 42).
120 King (1827, vol.2, p. 17). See also King (1827, vol.1, p. 45; vol.2, pp. 62–3).
121 C. Fraser (cited Hooker, 1830, p. 238).
122 See overview by Clarke (2003a, chapter 12).
123 Lines (1994, pp. 159–60).
124 For example, Good (1801–03 [1981, p. 48]) and King (1827, vol.1, p. 47, 202, 213; vol.2, p. 141).
125 Collins (1798–1802, vol.1, p. 544).
126 larke (2003a, chapter 3), Hiatt (1996, chapter 5) and Malinowski (1913).
127 William Wills, on a Sunday in February 1861 (cited by Burke & Wills Commission, 1862, p. 229).
128 lack (1917, p. 9; 1920, p. 92), Meyer (1846 [1879, p. 198]) and Taplin (1879, pp. 37, 128). *Grinkari* has also been written as 'gringkari', 'kiringkari' and 'krinkeri'.
129 Bates (1912 [1992, p. 16]), Lines (1994, pp. 178, 328), Tilbrook (1983, p. 9) and von Brandenstein (1988, pp. 108, 111–12). *Djanga* is written in many ways, such as 'jengar', 'junga' and 'tjanka'.
130 Boyce (1970, pp. 21–3).
131 Blackburn (1979).
132 Berndt & Berndt (1954), Clarke (2003a, chapter 11), Thomson (1949, pp. 82–94) and Urry & Walsh (1981).
133 Balint (2005, p. 20), Berndt & Berndt (1954, chapters 3–10, 12), Earl (1846, pp. 244–5, 249–50), Favenc (1888, pp. 331–2, 344–7, 382–4), Goodale (1971, pp. 5–7), Macknight (1976), Urry & Walsh (1981, pp. 90–5) and Warner (1958, pp. 456–60).
134 Berndt & Berndt (1954, pp. 15, 75–6), Clarke (2007, pp. 128–30) and Macknight (1976, pp. 43–44, 46–7).
135 Macknight (1976, p. 43) discussed use of cheesefruit-tree bark as a dye. Maiden (1889, p. 298) claimed this bark produces 'tints of yellow'. Worsnop (1897, p. 10) recorded the Asian use of Australian nutmeg.
136 For Torres Strait Islander ethnography and history, refer to Beckett (1987), Moore (1979) and Sharp (1993). For influences of the Papuan and Torres Strait Islander cultures upon Cape York Peninsula Aboriginal people, see Clarke (2003a, pp. 183–6), Haddon (1904–35, p. 300), McCarthy (1938–40) and McConnel (1953).
137 Baker (1999, pp. 65–74), McCarthy (1940) and Turner (1974, p. 197).
138 Clarke (2007, chapters 2 & 8).
139 Wightman, Dixon et al (1992, p. 30).
140 Blake et al (1998, p. 85).
141 Clarke (1998a, p. 15), Flannery (2002, p. 54), Plomley & Henley (1990, p. 65) and Walker (1900, p. 4).
142 McLaren (1996, pp. 34, 36–7, 123–4, 133, 148, 171–2, 210, 247–8, 259).

Chapter 2 Settlers and Australian Plants

1 Collins (1798–1802, vol. 1, introduction), Day (2001, pp. 23–5) and Phillip (1789).
2 Collins (1798–1802, vol. 1, p. 7), Day (2001, p. 35), Low (1988, pp. 19–23) and White (1788 cited Flannery, 1999, pp. 52–3).
3 Collins (1798–1802, vol. 1, p. 544).
4 Delbridge et al (1997, pp. 298–9) and Ramson (1988, p. 117). For a more detailed definition of 'bushfoods' (= 'bush tucker') see the editorial comments in the *Australian Bushfoods* magazine, issue 7, May-June 1978, p. 19.
5 Arthur (1996, p. 16).
6 Foster et al (2003, p. 76).
7 William A. Cawthorne (1843, cited Foster et al, 2003, p. 76).
8 *The Observer*, South Australian newspaper, 27 July 1901.
9 Refer to the title of Bruneteau (1996).
10 Brand Miller & Maggiore (1993) and Maggiore (1993) provided an analysis of the nutritional values of 'bush foods'.
11 Grimshaw et al (1994, p. 23). See also Gandevia (1981, pp. 256–8), Hagger (1979, p. 6), Hunter (1793, pp. 78–80), Hutchinson (1958, pp. 1–2) and Tench (1788–92 [1996, p. 151]).
12 R. Atkins (1792, cited Rolls, 2002, p. 27]).
13 Phillip (cited Campbell, 1932, p. 77).
14 Bradley (1786–92, pp. 92, 95–7, 107), Collins (1798–1802, vol. 1, pp. 23–4, 30–1, 40, 43, 57–8), Hunter (1793, p. 78), Tench (1788–92, pp. 55–6) and White (1790, pp. 214, 216). The spearing of convicts was reported in a letter received by J.B. Banks from Arthur Phillip, 2 July 1788 (Series 37.04, Papers of Sir Joseph Banks, State Library of New South Wales, Sydney).
15 White (1790, p. 166). Gandevia (1981) described White's search for wild foods and medicines. Dan (1983, pp. 4–6) gave an account of White's role on the First Fleet.
16 White (1790, p. 187).
17 White (1790, p. 191). See also Bradley (1786–92, p. 76), for a remark about finding a fig cake.
18 Bradley (1786–92, pp. 135–6). The 'celery' may well have been the sea celery or wild parsley described by Hardwick (2001, p. 43)
19 Bunce (1859a, p. 8). Bunce referred to the plant as *Atriplex halimus* (coast saltbush), which is now called *A. cinerea*. Bunce (1859a, p. 39) also mentioned other greens.
20 See the reference to the economic impact of new mainland colonies in the Tasmanian statistical report for 1839–40, provided by Ewing (1846, p. 143).
21 Thomas (1925, p. 69).
22 *Adelaide Observer*, 4 July 1914.
23 Low (1988, p. 112).
24 Low (1988, p. 140). Clarke (1988, p. 69) gave an account of Aboriginal and colonial use of bracken in southeastern Australia. Plomley & Cameron (1993, pp. 10, 17, 21, 25) provided an overview of Aboriginal use in Tasmania. Bracken was also a major food source for the Maori of New Zealand (Brooker et al, 1987, pp. 74–7).
25 L. de Rougemont (cited Maiden, 1900, p. 125). See also Maiden (1889, p. 60). De Rougemont's real name was Henri Louis Grin (Barrett, 1948, chapter 10).
26 Maiden (1900, p. 124).
27 A. Thozet (cited Maiden, 1900, p. 125).
28 Thieret (1958, p. 17).
29 Smith (1982, 1996).
30 Von Mueller (1883, cited Thieret, 1958, p. 17) claimed that Aboriginal people did not use stem starch from cycads before European settlement.
31 Jones (1993, p. 50) stated that the species used was *Macrozamia communis*, while Thieret (1958, pp. 17–18, 20) claimed it was *M. spiralis*. Aboriginal cooks also processed burrawang nuts for eating (Hardwick, 2001, pp. 29–31; Low, 1989, pp. 83–4; Wreck Bay Community & Renwick, 2000, p. 19).
32 Moore (1884, p. 117).
33 Maiden (1889, p. 20).
34 Cribb & Cribb (1982, p. 148) and Pate & Dixon (1982, pp. 36, 98, 191, 221, 236, 239). Latz (1995, p. 152) provided the account from Central Australia.
35 Taplin (1859–79, pp. 57, 151). Taplin wrote Ngarrindjeri as 'Narrinyeri'. See Clarke (1988, p. 70).
36 Clarke (1988, pp. 68, 70), Gott (1983, p. 8) and Low (1988, p. 135). See F.M. Bailey in the *Adelaide Observer*, 13 June 1914, p. 41.
37 Campbell (1932, p. 83), Powell (1990, p. 93) and White (1790, pp. 153, 158). The wild figs were possibly from the Illawarra fig tree.
38 Backhouse (1843, p. 371). Formerly known as *Myrtus tenuifolia*. See discussion by Low (1989, p. 53) and Cherikoff & Isaacs (1989, p. 52). Barker & Barker (1990, p. 56) provided an overview of Backhouse as a plant collector.
39 Maiden (1900, p. 127). Maiden used an obsolete plant name, *Pleiococca wilcoxiana*.
40 Maiden (1889, p. 8).
41 For the wild quince, see Maiden (1900, p. 130), who used an obsolete plant name, *Nephelium leiocarpum*, for it. In the case of the geebung, refer to Maiden (1889, p. 51).
42 Maiden (1889, p. 29), Powell (1990, p. 93) and Smith & Smith (1999, p. 47). Species of lilly pilly and wild apples are mostly in the *Syzygium* genus, although some were formerly put in *Eugenia*. These plant foods are probably those referred to by Hodgkinson (1845, p. 4) for similar uses.
43 Maiden (1900, p. 122). Maiden referred to this tree as the 'Australian Baobab'.
44 Maiden (1889, p. 49; 1900, p. 128).
45 Hunter (1793, pp. 478–9).
46 Maiden (1900, p. 127). Maiden used an obsolete name, *Atalantia glauca*.
47 Maiden (1889, p. 46). Maiden used an obsolete name, *Muehlenbeckia adpressa* var. *hastifolia*.
48 Maiden (1900, p. 129). Maiden used the now obsolete name, *Diploglottis cunninghamii*.
49 Gunn (1842, p. 39) and Maiden (1889, p. 19). Gunn used the now obsolete name *Coprosma microphylla*, while Maiden used another former name, *C. billardieri*.
50 A.F. Richards (cited Tate, 1882, p. 137). This source is presumed to be the same Mrs Richards at Fowler Bay who sent ethnographic specimens to the South Australian Museum sometime around the 1880s and 1890s.
51 Maiden (1900, p. 130).
52 Bunce (1859a, p. 50) referred to this plant as *Tasmania fragrans*. See also Cribb & Cribb (1982, p. 60).
53 Bradley (1786–92, p. 92). See discussion by Attenbrow (2003, p. 77).
54 Bradley (1786–92, p. 135). See discussion by Powell (1990, p. 94).
55 Carr & Carr (1981b, pp. 18–9) and Everist (1981b, pp. 224–5).
56 Lampert & Sanders (1973, p. 107).
57 Moore (1884, p. 117).
58 Whiting (1963, p. 276). See also Thieret (1958, p. 16).
59 Hooker (1830, p. 242).
60 Maiden (1899, p. 29). Maiden used an obsolete name, *Eugenia tierneyana*.
61 *Adelaide Observer*, 17 September 1853, p. 6.
62 Journal entry for 15 August 1864 (Taplin, 1859–79, p. 204).
63 Clarke (1985b, pp. 11, 13–4, 16).
64 F.M. Bailey, Early Adelaide. Peeps at Pioneering: Introduction of Plants, *The Register*, Adelaide, 6 June 1914, p. 17.
65 Low (1989, pp. 133–9).
66 Anonymous (1788 cited by Flannery 1999, p. 80 & Day, 2001, p. 35).
67 Maiden (1889, p. 9). See also Hardwick (2001, p. 33).
68 White (1790, p. 155). See discussion by Powell (1990, p. 93).
69 W.J. Hooker (cited Maiden, 1900, p. 119). Maiden used an obsolete plant name, *Nasturtium palustre*.
70 Bindon (1996, p. 249), Grey (1997, pp. 123–5), Gunn (1842, p. 42), Hardwick (2001, pp. 44–5), Powell (1990, p. 93), Ramson (1988, p. 712), Smith & Smith (1999, pp. 72–3) and Zola & Gott (1992, p. 21). In the context of this plant food, the use of 'warrigal' is possibly equivalent to 'Aboriginal' or simply 'wild'.
71 Cook (1893, p. 59). Use of greens to treat scurvy is discussed by Kodicek & Young (1969, p. 54) and Wharton (1893, pp. xxviii–xxix).
72 Maiden (1889, pp. 62–3). For an overview of New Zealand spinach use in colonial times, see editorial comments in *Australian Bushfoods*, issue 12, August-September 1999, pp. 20–1.
73 Tench (1788–92 [1996, p. 231]). This cabbage-tree would have been *Livistona australis*.
74 White (1790, p. 156).
75 Baker et al (1986, p. 90).
76 Hodgkinson (1845, p. 226).
77 Anonymous (1847 [1953, p. 175]). The plant species involved here would have been *Livistona australis*.
78 Von Mueller (cited Maiden, 1889, p. 40; Smyth, 1878, vol. 1, p. 213).
79 Crawford (1982, pp. 22, 28, 38–9), Hiddins (2001, pp. 19–21, 24–5), Levitt (1981, pp. 42, 147–8), McConnel (1930, p. 19), Puruntatameri (2001, pp.

35, 57, 61, 63, 142), Roth (1901b, pp. 9, 14), Smith (1991, pp. 10, 15, 18), Smith & Wightman (1990, pp. 7, 17), Specht (1958, p. 485), Sutton (1994, p. 37), Tindale (1925, p. 77), Wightman, Jackson et al (1991, p. 19) and Wightman, Roberts et al (1992, pp. 30, 32).

80 Backhouse (1843, p. 89). See also Walker (1888–99, p. 244).

81 Backhouse (1843, p. 121; Appendix D, p. xxxix–xl). Robinson (1830–32 [Plomley, 1966, pp. 287, 377, 631]), Plomley & Cameron (1993, pp. 2–3, 12, 19, 22) and Roth (1899, pp. 96–7) discussed edible tree-ferns in Tasmania.

82 Coutts (1970, Table 1, pp. 6–7), Gott & Conran (1991, p. 40), Hardwick (2001, pp. 112–13), Hope & Coutts (1971, Table 1, pp. 107–9), Low (1989, p. 130) and Zola & Gott (1992, p. 36). See also Bunce (1859a, pp. 69–70, 72).

83 Bunce (1859a, p. 8) and Maiden (1889, p. 33).

84 Richards (cited Tate, 1882, p. 136) and Maiden (1889, p. 17; 1900, p. 121). Maiden used an earlier name, *Claytonia balonensis*. See also Low (1989, p. 138).

85 Basedow (1925, pp. 91, 149), Cleland (1936, p. 9), Goddard & Kalotas (1988, pp. 120–1), Latz (1995, pp. 56, 134), Lawrence (1968, p. 61) and Stirling (1896, p. 65).

86 Bolam (1930, p. 51).

87 Maiden (1889, p. 55).

88 Woolls (cited Maiden, 1900, p. 122).

89 Douglas (1982, chapter 4). See also Clarke (2007, chapter 11) and Pyke (1968, chapter 4).

90 Farb & Armelagos (1980, p. 14).

91 Maiden (1889, pp. 38–9).

92 Herbert (1966, p. 99). and Maiden (1889, p. 66).

93 Barwick (1985, p. 185), Clarke (1996, 1998a) and Plomley & Henley (1990).

94 *Adelaide Observer*, 28 September 1844, p. 6.

95 Clarke (1996).

96 Clarke (1996, pp. 59–65) and Hagger (1979, pp. 17–18).

97 Collins (1798–1802, vol. 1, pp. 459; see also pp. 407, 424).

98 Collins (1798–1802, vol. 2, pp. 59–60, 87–9, 98–9).

99 *Sydney Gazette*, 2 December 1815, p. 2.

100 Bellingshausen (1819–21 [1945, pp. 328, 350]).

101 Collins (1798–1802, vol. 1, pp. 458–9) and Troy (1990, p. 52).

102 Threlkeld (1824–1859 [1974, pp. 120, 146, 223]).

103 Bunce (1859a, pp. 10, 33) and Ryan (1996, pp. 77–8).

104 Boyce (1970) and Mulvaney (1994).

105 Mitchell (1838, vol. 1, p. 1).

106 S. Leigh (1821, cited Threlkeld, 1824–1859 [1974, p. 333]) described body-scarring practices in the Sydney district.

107 Favenc (1888, pp. 102–4), Mitchell (1838, vol. 1, pp. 2, 351–2) and Salier (1931, pp. 8–10).

108 Tolmer (1882, vol. 1, pp. 319–323).

109 Clarke (1998a, p. 43).

110 Threlkeld (1824–1859 [1974, p. 57]).

111 Bellingshausen (1819–21 [1945, pp. 329–30]). Threlkeld (1824–1859 [1974, pp. 57, 99–100, 159, 167]) gave accounts of Aboriginal people tracking bushrangers around Sydney.

112 For example, refer to Lines (1994, p. 177).

113 Morgan (1852) provided an account of Buckley's life in the Australian bush. See also Barrett (1948, chapter 1).

114 Ramson (1988, p. 101).

115 Barrett (1948, chapter 3) and Laurie (1966).

116 Barrett (1948, chapter 2), Morrill (1864 [2006]) and Welch (2006). Morrill's surname has also been written as 'Morrell' and 'Murrells'.

117 Morrill (1864 [2006, pp. 55–6]). Thozet (1866) investigated the plant foods eaten by Morrill.

118 Bellchambers (1931, p. 167). For references to Aboriginal people eating grasstree cores and leaf bases see Angas (1847, pp. 84,203), Backhouse (1843, Appendix D, p. xxxviii), Bindon (1996, p. 265), Bunce (1859a, p. 30), Cleland (1966, pp. 134–5), Coutts (1970, Table 1, pp. 6–7), Gott & Conran (1991, pp. 64–5), Hardwick (2001, pp. 56–7), Hiddins (2001, p. 80), Hope & Coutts (1971, Table 1, pp. 107–9), Robinson (1830–31 [Plomley, 1966, pp. 188, 383]), Roth (1899, p. 96), Stewart & Percival (1997, p. 49) and von Mueller (cited Smyth, 1878, vol. 1, p. 213),

119 Bindon (1996, p. 138). See also Low (1988, p. 153).

120 F.M. Bailey, Early Adelaide. Peeps at Pioneering: Introduction of Plants, *The Register*, Adelaide, 6 June 1914, p. 17.

121 Curtis (1974, p. 338) and Low (1988, p. 180).

122 Maiden (1900, p. 125).

123 Symons & Symons (1994, p. 42).

124 Baker (1999, pp. 164–5).

125 Anonymous (1788 cited by Flannery 1999, p. 80 & Day, 2001, p. 35).

126 Bailey (1880, p. 26), Campbell (1932, pp. 80–1), Cribb & Cribb (1981, pp. 76, 86–7), Finney (1984, p. 45), Grimshaw et al (1994, p. 23), Hardwick (2001, p. 17), Lassak & McCarthy (1983, pp. 16, 76–7, 91–2, 182), Powell (1990, p. 94), Smith & Smith (1999, pp. 14–5) and Tench (1788–92 [1996, p. 231]).

127 Campbell (1932, p. 81).

128 Campbell (1932, pp. 81, 83).

129 Bailey (1880, p. 8). Bailey used an earlier name for this species, *Hardenbergia monophylla*. Reverend Woolls (in Wyndham, 1890, p. 118) identified 'native sarsaparilla' as *Kennedya monophylla,* which is an obsolete name

130 Clarke (1987, p. 8)

131 Campbell (1932).

132 Maiden (1889, p. 49; 1900, p. 128).

133 Bunce (1859a, p. 47), Robinson (1831, cited Plomley & Cameron, 1993, p. 11) and Smith & Smith (1999, pp. 110–1). Bunce incorrectly referred to this species as *Eucalyptus resinifera.*

134 Breton (1846, pp. 140–1).

135 Bunce (1859, p. 47). See also Carr & Carr (1981b, p. 17) and Robinson (1831 [Plomley, 1966, pp. 534, 538–9, 542, 556–7, 569, 580]), who referred to the use of the 'cider tree' and 'cider gum'. Bunce incorrectly referred to this species as *Eucalyptus resinifera.*

136 R.C. Gunn correspondence to W.J. Hooker, 9 May 1844 (cited Burns & Skemp, 1961, p. 99).

137 For the native pennyroyal refer to Bailey (1880, p. 19). See Cribb & Cribb (1981, pp. 78–9) and Lassak & McCarthy (1983, pp. 15, 19, 77, 88, 175) and Low (1989, p. 177) for descriptions of the use of other *Mentha* species, such as the Australian river mint. Brooker et al (1987, pp. 148–9) noted European settlers in both Australia and New Zealand using indigenous species of mint as a tea substitute.

138 In the case of Tasmania, refer to Backhouse (1843, Appendix D, p. xxxiii) and Bunce (1859a, pp. 23–4). For Kangaroo Island, see Gill (1909, p. 117) and Clarke (1996, p. 62).

139 Backhouse (1843, Appendix D, p. xxxiii). See also Gunn (1842, p. 38).

140 Gunn (1842, p. 38) and Maiden (1889, p. 4).

141 Backhouse (1843, Appendix D, p. xxxii). See also Gunn (1842, p. 37) and Smith & Smith (1999, pp. 11–12).

142 Brennan (1986, p. 34), Campbell (1932, pp. 78–80), Ramson (1966, p. 87; 1988, p. 667), Smith & Smith (1999, pp. 13, 16), Webb (1948, p. 7) and Zola & Gott (1992, p. 50).

143 Delbridge et al (1997, p. 2174) and Ramson (1966, p. 87; 1988, p. 667). Also written as 'tea-tree'.

144 Campbell (1932, pp. 78–9, 86) and Delbridge et al (1997, pp. 2174, 2209). The ti-palm, *Cordyline fruticosa*, was formerly known as *C. terminalis.*

145 P. Clarke (1986, p. 89).

146 Bailey (cited Maiden, 1900, p. 126).

147 Maiden (1900, p. 127).

148 N. Holtze (cited Maiden, 1900, p. 123).

149 or Australian settler folk remedies refer to Hagger (1979). Pearn & O'Carrigan (1983) provided an overview of the development of a colonial health system in Australia.

150 Cawte (1974, 1996), Clarke (2007, pp. 96–105), Maher (1999), Ngaanyatjarra et al (2003), Reid (1979, 1982, 1983) and Wiminydji & Peile (1978) have discussed Aboriginal notions of sickness and medicine. Hagger (1979) considered colonial health and medicine in Australia.

151 Crosby (2004, chapter 9).

152 For examples of Aboriginal pharmacopoeias, see Barr et al (1988), Clarke (2007, chapter 8), Henshall et al (1980), Kyriazis (1995), Levitt (1981, chapter 9), D. Bird Rose (1987), Watson (1994) and Webb (1960).

153 Hagger (1979, pp. 46–7).

154 Tench (1788–92 [1996, pp. 196–7]). The term *caradyee* appears to be a rendering of *garraaji*, meaning 'clever man' in the Dharug language of Sydney (Delbridge et al, 1997, p. 1189; Dixon et al, 1992, pp. 155–6, 213; Troy, 1994, p. 66, Ramson, 1988, p. 354). Troy wrote this as *garadyigan.*

155 Gason (1879, p. 79). Gason spelled Diyari as 'Dieyerie'.

156 Taplin (1879, p. 50). Taplin spelled Ngarrindjeri as 'Narrinyeri'.

157 Taplin (1874 [1879, p. 46]).

158 Bellingshausen (1819–21 [1945, pp. 332–3]).

159 Jorgenson (1837 [1991, pp. 93, 122]).

160 Dawson (1881, p. 56).

161 Campbell (1932, pp. 77–80). Cribb & Cribb (1981, chapter 3) has

provided an account of other species used by European settlers.
162 D. Considen (cited Campbell, 1932, p. 80).
163 Lassak & McCarthy (1983, pp. 55–6).
164 Cobcroft (1983, pp. 18, 27, 29–30, 32), D. Collins (cited Campbell, 1932, p. 83) and Powell (1990, p. 94). Rennie (1880, p. 119) provided a chemical analysis of the acids involved.
165 Clarke (1987, p. 6) and Lassak & McCarthy (1983, p. 56).
166 *Adelaide Observer*, 28 September 1844, p. 6. The plant concerned was presumably a species of *Leptospermum*.
167 Clarke (1996, p. 62) and Leigh (1839, pp. 160–1).
168 Bindon (1996, p. 206), Hiddins (2000b, p. 32; 2001, p. 17) and Low (1988, p. 122).
169 Gilmore (1935, p. 232). MacPherson (1925, 1939) provided overviews for the Aboriginal uses of acacia gum and eucalypt kino.
170 Gilmore (1935, p. 226).
172 Clarke (2007, pp. 99–105).
172 Clarke (2003a, pp. 85–6; 2007, pp. 92, 99).
173 Lassak & McCarthy (1983, p. 15), Powell (1990, p. 94) and Webb (1948, p. 7; 1969, p. 84).
174 White (1790, pp. 226–8). See also Baker & Smith (1902, pp. 166–8). Froggatt (1932, p. 103) described White as a botanical collector.
175 White (1790, pp. 178, 233, 235–6). See discussion by Bancroft (1886, pp. 4–5) and Cribb & Cribb (1981, pp. 71–2).
176 Cribb & Cribb (1981, pp. 71–4), Finney (1984, pp. 42, 45) and Hagger (1979, pp. 22–3).
177 Bailey (1880, p. 10).
178 Collins (1798–1802, vol. 1, p. 57).
179 White (1790, pp. 194–5). See also Bradley (1786–92, p. 136), Lampert & Sanders (1973, p. 106) and Tench (1788–92 [1996, p. 231]). Rennie (1886) gave a chemical analysis of the sweet principle of sweet tea.
180 Finney (1984, p. 35). Maiden (1910, pp. 126–9) gave an account of La Billardière on the 1791–94 expedition under the command of D'Entrecasteaux.
181 Bunce (1859a, p. 11).
182 Gunn (1842, p. 43).
183 Hercus (1986, p. 235).
184 Bunce (1859a, pp. 27–8). Bunce writes the generic name as '*Zierria*'.
185 Low (1988, p. 125).
186 Koch (1898, p. 113).
187 Latz (1995, pp. 253–4). In the Kimberley, Aboriginal people use this plant as a substitute for chewing tobacco (Clarke, 2007, p. 109).
188 Koch (1898, p. 114). Refer to Crawford (1982, p. 68) for references to Aboriginal use of this plant.
189 Webb (1948, p. 10).
190 Lassak & McCarthy (1983, pp. 89–90), Low (1989, p. 178) and Smith & Smith (1999, pp. 9–10). Lassak & McCarthy used the earlier name *Ocimum sanctum* var. *angustifolium*.
191 Hagger (1979, pp. 22–3). *Corymbia citriodora* was formerly known as *Eucalyptus citriodora*.
192 Roth (1903a, p. 39). Roth used the obsolete name *Ficus scabra*.
193 Bailey (1880, p. 21). Cited by Maiden (1889, p. 182). Brooker et al (1987, pp. 131–3) outlined use of the milkweeds in Australia and the Pacific.
194 Rowley (1972b) outlined European use of Aboriginal labour. Clarke (2001, 2003b) discussed Aboriginal labour in the southern marine industry. May (1994) described Aboriginal participation in the Queensland cattle industry. Refer to Wilkinson (1848, pp. 210, 352) and Hawker (1841–45 [1981, entry for 17 February 1845]) for records of European employment of Aboriginal collectors of gum in South Australia.
195 Clarke (2003b).
196 For a history of maize and its spread across the globe, see McCann (2005). For the potato, refer to Diamond (1998, pp. 118, 127–8, 132, 185, 187), McIntosh (1927) and Pollan (2002, chapter 4).

Chapter 3 Making Plant Names

1 Hunter (1787–92, cited McLaren, 1996, p. 22).
2 Dawes (1790–91). Flannery (1999, pp. 111–15) discussed William Dawes and Patyegarang, giving some examples of the recorded Eora words.
3 Jorgenson (1837, pp. 55–6).
4 Gunn (1842, pp. 44–5). See also Plomley & Cameron (1993, pp. 9,10, 24).
5 Gunn (1842, p. 36). See also Plomley & Cameron (1993, pp. 9–10). Gunn used an earlier name, *Geranium parviflorum*.
6 Dixon et al (1992, pp. 111–49) and Mills (1999–2000).
7 Aboriginal foragers used all these plants (Cherikoff & Isaacs, 1989; Clarke, 2007; Cribb & Cribb, 1982; Low, 1989, 1991a, 1991b; Maiden, 1889).
8 Aboriginal toolmakers in tropical Australia used the wood from the narrow trunk of the 'nasturtium tree' to make the butt end of spears (Kamminga, 1988, p. 49; Specht, 1958, p. 492; Roth, 1909, p. 191). Maiden (1889, p. 567) considered the species as having potential as a commercial timber for European use. According to Maiden, in some Queensland Aboriginal languages the nasturtium tree was known as 'tumkullum'. In the case of the 'peanut tree', Aboriginal people gathered the inner bark to make string (Clarke, 2007, p. 120).
9 Clarke (2007, p. 106).
10 These plants are all edible, given the appropriate treatment (Cherikoff & Isaacs, 1989; Clarke, 2007; Cribb & Cribb, 1982; Low, 1989, 1991a, 1991b; Maiden, 1889).
11 Australian Plant Name Index (http://www.anbg.gov.au) and Maiden (1900, p. 119). Sharr (1996, pp. 13, 164) gave the derivation of the scientific name. Clarke (2007, pp. 23, 28) described the Aboriginal significance of the species.
12 Mitchell (1838, vol. 1, p. 314).
13 Mitchell (1848, p. 77).
14 Dixon et al (1992, p. 113) and Ramson (1988, p. 107).
15 See Ramson (1988, p. 60) for 'blackboy'. The term 'gin' refers to an Aboriginal woman (Ramson, 1988, pp. 273–4).
16 For Aboriginal uses of the plant, refer to Clarke (2007, pp. 17, 19, 32–3, 60–1, 79, 105, 114, 120–2).
17 Brokensha (1975, pp. 44–5), Clarke (2007, p. 114) and Urban (2001, p. 176). An earlier scientific name for the Central Australian spearwood was *Tecoma doratoxylon*.
18 Ramson (1988, p. 705) and Urban (2001, p. 71).
19 Specht (1958, p. 495) and Kamminga (1988, p. 50).
20 Clarke (2007, p. 124).
21 Sharr (1996, p. 162) gave the name derivation. Baker (1919, pp. 110–11) listed the commercial wood uses.
22 Breton (1846, p. 139), Maiden (1889, pp. 311, 359) and Robinson (1832 [Plomley, 1966, pp. 633, 645]).
23 Baker (1919) provided weights of all Australian timbers used commercially.
24 *Eucalyptus leucoxylon* entry of Brown (1882).
25 Berlin (1992), Brown (1986), Duranti (2001), Foley (1997) and Lakoff (1987). For Aboriginal Australia, see Crawford (1982, pp. 16–7), Heath (1978), Kean (1991), Laramba Community Women (2003, pp. vii-ix), McKnight (1999), Nash (1997), Rudder (1978–79), Smith & Kalotas (1985, p. 326), Waddy (1979, 1982, 1988) and Walsh (1993).
26 Blake (1981), Henderson & Nash (2002), Schmidt (1993), Thieberger & McGregor (1994) and Yallop (1982).
27 Heath (1978, pp. 52–3). See also Levitt (1981, pp. 111–12) and Wightman, Roberts & Williams (1992, p. 36).
28 Palmer (1883, p. 101).
29 Warner (1958, pp. 278–9). Refer to Capell (1962), Mathews (1903) and O'Grady (1956) for descriptions of Aboriginal 'secret' languages.
30 Clarke (2007, p. 15).
31 Mühlhäusler & Fill (2001) and Mühlhäusler (2003) provided overviews of the study of language and environment. Harkins (1994) and Schmidt (1993) gave an overview of the status of Australian indigenous languages.
32 Lowe (2002) and Magarey (1899, pp. 120–1).
33 Wilhelmi (1861, p. 172).
34 Clarke (2007, pp. 12–13), Goddard & Kalotas (1988, pp. 14–5), Johnston & Cleland (1942, pp. 95, 102), Thomson (1962, pp. 272–3), Tindale (1978, pp. 160–2) and Wiminydji & Peile (1978, p. 506).
35 Clarke (2003a, pp. 43, 46–9, 57, 59, 69–70, 113, 153, 169; 2007, p. 12, 14, 21, 38, 57, 66–7, 73–4, 76–7, 80, 85–8). For discussion of female gathering activities in Aboriginal Australia, refer to Kaberry (1939) and Dahlberg (1981).
36 For instance, refer to Akerman (1978), Davis (1989, pp. 34–5), Heath (1978, pp. 44), Mathews (1904, pp. 227, 237), O'Connell et al (1983, p. 83), Sutton (1995, p. 154) and Wilhelmi (1861, p. 172).
37 Refer to the description of 'rubbish tucker' in Arthur (1996, p. 171)
38 Heath (1978, p. 46).
39 McEntee et al (1986). J. McEntee (pers. com.) provided spellings of these Adnyamathanha words in normal fonts.
40 McEntee et al (1986, p. 13 & pers. com). These authors used the names 'puffball' and *Podaxon* species, which probably refer to the stalked puffball fungus (*Podaxis pistillaris*).

41 McEntee et al (1986, pp. 11–12 & pers. com).
42 McEntee et al (1986, p. 14).
43 For a discussion of Aboriginal placenames, refer to Clark & Heydon (2002), Clarke (2003a, pp. 114–6, 220), Hercus et al (2002) and Tunbridge (1985b, 1987).
44 Attenbrow (2003, p. 10) and Troy (1994, p. 62).45 Blake (1981), Clarke (2003a, pp. 39–40), Thieberger & McGregor (1994) and Yallop (1982).
46 Dixon et al (1992, pp. 67, 72, 76, 78–80, 84–5) and Ramson (1988, pp. 341, 353, 440, 514, 662, 708, 742).
47 Spencer (1918, p. 9).
48 Donaldson (1985).
49 The fish names have been published in a natural history volume by Eckert & Robinson (1990, pp. 18–19). See Clarke (2003b, pp. 88–9) for a listing of contemporary Aboriginal language terms for the biota of the Lower Lakes.
50 Mühlhäusler (1996a & b). Using the words 'nardoo' and 'munyeru' as examples, Spencer (1918, vol. 15, p. 9) has a counter argument where Europeans have spread the use of these indigenous words.
51 Collins (1798–1802, vol. 1, p. 544; see also p. 209). Troy (1993a) and Turner (1972, chapter 10) discussed language contact in early New South Wales.
52 Lines (1994, p. 151) described this bias in language learning at the opposite end of Australia, for the Nyungar people on the southwest of Western Australia frontier.
53 Maslin (2001).
54 Maiden (1889, p. 435). See also Maiden (1903, pp. 990, 996), who wrote the name as 'bengaly'.
55 Ramson (1966, p. 102).
56 Ramson (1966, pp. 102–3).
57 Attenbrow (2003, p. 78), Dixon et al (1992, pp. 122–3), Maiden (1889, p. 633–4), Ramson (1966, pp. 113, 115; 1988, p. 711) and Troy (1994, p. 72). Other spelling variations in Australian English are 'warata', 'warratah', 'warrataw', 'warratau' and 'warrettah'.
58 Collins (1798–1802, vol. 2, p. 66).
59 Maiden (1889, p. 62).
60 Breton (1846, p. 134) wrote the name of this Tasmanian plant as 'warratah'.
61 Dixon et al (1992, pp. 117), Ramson (1966, pp. 113–5; 1988, p. 269) and Smith & Smith (1999, pp. 36–40). Other spelling variations in Australian English are 'geebong', 'jibbong', 'jibong' and 'tyibung'.
62 Moore (1884, pp. 17, 19) considered 'burrawang' or 'wild pineapple' to be *Macrozamia spiralis*, although more recently scholars (Dixon et al, 1992, p. 115; Ramson, 1988, p. 111) have identified it as *M. communis*. Also written as 'buddawong', burrawong', 'buruwang' and 'burwan'.
63 Maiden (1889, p. 41).
64 Dixon et al (1992, p. 115), Ramson (1966, pp. 113–5; 1988, p. 111), Sharpe (1994, p. 14) and Whiting (1963, p. 294). Aboriginal people in the Sydney Basin also appear to have used this term (Attenbrow, 2003, p. 78; Dawes, 1790–91).
65 Stewart & Percival (1997, p. 37). Symons & Symons (1994, p. 77) confirmed that the 'burrawang' (*Lepidozamia peroffskyana*) from around Brisbane also required processing.
66 Cawte (1996, p. 104). The poisonous properties of the cycads are summarised by Everist (1981a, pp. 226–48) and Webb (1969, p. 87).
67 Dixon et al (1992, p. 125) and Ramson (1988, p. 186). Other spellings include 'conjeboi' and 'cunjiboy'. Cunjevoi is also a term for a sea squirt.
68 Hiddins (2001, p. 31), Hodgkinson (1845, pp. 225–6) and Low (1989, pp. 9, 111, 216).
69 Attenbrow (2003, p. 116), Dixon et al (1992, p. 119) and Ramson (1966, pp. 104–5, 113–4, 116; 1988, p. 355). Troy (1994, p. 67) recorded this Dharug term as *garradjun*, while Attenbrow wrote it as *carrejun/ carrahjun*. Other spelling variations in Australian English are 'currajong' and 'currijong'.
70 Dixon et al (1992, p. 119) and Maiden (1889, p. 633). Maiden used the name *Sterculia diversifolia*.
71 Cunningham (1818, cited Lee, 1925, pp. 413–4).
72 The genera of 'kurrajongs' are *Brachychiton*, *Commersonia* and *Sterculia*. They are part of the Sterculiaceae family.
73 Maiden (1889, pp. 630–1). Maiden uses an earlier name, *Plagianthus sidoides*.
74 Maiden (1889, pp. 617–35).
75 Dixon et al (1992, p. 144) and Ramson (1966, pp. 113, 115, 117; 1988, pp. 415–6).
76 Dixon et al (1992, p. 134) and Ramson (1966, p. 113–4; 1988, p. 94). Brigalow is also written as 'bricklow'.
77 Bates (1918, p. 156). Also confirmed by the present author's fieldwork.
78 Dixon et al (1992, p. 132) and Ramson (1988, p. 54). Bindi-eye is also written as 'bindy-eye', 'bindei', 'bindii' and 'bindiyi'.
79 Jones (1995, pp. 126–7).
80 Dixon et al (1992, pp. 126–7, 130) and Ramson (1966, p. 114–5; 1988, pp. 34, 474). Note that Europeans have also used the term 'bangalow' when referring to the palm *Ptychosperma elegans* (Maiden, 1889, pp. 592, 631). Cunningham recorded the Aboriginal name as *bangla* (Cunningham, 1818 [cited Lee, 1925, p. 415]; see Chapter 5 of this volume).
81 Dixon et al (1992, pp. 113–5). Ramson (1966, pp. 114–5; 1988, p. 109) listed it as possibly a Wiradhuri word.
82 Clarke (2007, pp. 78–9, 83, 142), Smyth (1878, vol. 1, pp. 218–20, 244) and Steele (1984, pp. 212–3, 239–40).
83 Mills (1999–2000, p. 24).
84 Dixon et al (1992, p. 113), Maiden (1889, pp. 4, 359), Ramson (1966, pp. 114, 117; 1988, p. 79) and Smith & Smith (1999, p. 98). Another spelling variation in Australian English is 'boobyalla'. Robinson (1831 [Plomley, 1966, pp. 289, 364, 368, 389, 463]) mentions a watercourse called 'Boobyalla River' in northeast Tasmania. In the 1840s, there was also a pastoral property known as 'Boobyalla' located near the coast of northeast Tasmania (Burns & Skemp, 1961, pp. 116–7). In contemporary Australia, there are several places with this name in northeast Tasmania (see http://www.ga.gov.au/map/names/).
85 Gunn (1842, p. 37).
86 Clarke (2003b, p. 89; 2007, pp. 77–8).
87 Urban (2001, pp. 176–7).
88 Bunce (1859b, p. 71; see also p. 74).
89 Frankel (1982), Gott (1983), Gott & Conran (1991, pp. 6–7), Low (1989, pp. 104, 108) and Zola & Gott (1992, pp. 7–9).
90 Batey (1909–10, cited Frankel, 1982) and Smyth (1878, vol. 1, pp. 209, 238–44). For Aboriginal earth mound construction see also Etheridge (1893), Gott (1982a, p. 65; 1983, pp. 11–12) and Hallam (1975, pp. 12–13, 72, 74).
91 Smyth (1878, vol. 1, p. 209).
92 Dixon et al (1992, pp. 117, 223), Ramson (1988, p. 201) and Zola & Gott (1992, p. 26). Stone (1911, p. 445) recorded the word in the languages around Lake Boga in central northern Victoria as 'dillunge'. Across Victoria, Hercus (1986, pp. 179, 202, 220, 251) recorded it as *dilanj* for Wemba Wemba and the Djadjala dialect of Wergaia, and *dilanggi* in Muthimuthi (Madi Madi).
93 Gott & Conran (1991, p. 34), commenting upon the 'giant saltbushes' (nitre bushes) mentioned by Beveridge (1884, p. 39).
94 Cleland (1957, p. 154), Eyre (1845, vol. 2, p. 271), Low (1989, pp. 53–4, 69–70), von Mueller (cited Smyth, 1878, vol. 1, p. 215), Wilhelmi (1861, p. 173) and Zola & Gott (1992, p. 26). Name written as '*N. billardieri*' by earlier sources.
95 Low (1989, p. 217) and White (1994, p. 167).
96 Beveridge (1884, p. 36), Dixon et al (1992, pp. 127–8), Gott & Conran (1991, p. 8), Ramson (1988, p. 185) and Zola & Gott (1992, p. 8). Beveridge wrote the Aboriginal word as 'kumpung', while Stone (1911, p. 444) similarly recorded from Lake Boga in central northern Victoria that roots from rushes were 'gumbung'.
97 Clarke (1988, pp. 66, 69–70, 72–3) and Gott (1999).
98 Dixon et al (1992, pp. 139–40) and Ramson (1966, pp. 114–5, 117; 1988, p. 383). Stone (1911, p. 445) recorded the word in the languages around Lake Boga in central northern Victoria as 'malle'.
99 Smyth (1878, vol. 1, p. 211).
100 Dixon et al (1992, pp. 107–8) and Ramson (1988, p. 366).
101 Beveridge (1884, pp. 64–5).
102 Dixon et al (1992, p. 112), Gott & Conran (1991, p. 31), Maiden (1889,p. 30), Ramson (1988, p. 32) and Smith & Smith (1999, pp. 28–30). Other spelling variations in Australian English are 'ballat' and 'ballot', the former of which Maiden used for *Exocarpos latifolius* (Maiden's *Exocarpos latifolia*).
103 Smyth (1878, vol. 1, p. 210). Smyth's sources were Rev. J. Bulmer for Gippsland and Mr Hogan for Lake Condah.
104 Bindon (1996, pp. 137–8) and Maiden (1889, p. 30).
105 Dixon et al (1992, p. 118), Maiden (1889, pp. 57–8) and Ramson (1966, pp. 114–5; 1988; p. 299). The Aboriginal form of the word was probably *gunyang*, as it is generally spelled today. This species is possibly applied more specifically to *Solanum vescum*, which was confined to eastern Victoria (Gott & Conran, 1991, p. 37). Another written variation is 'koonyang'.

106 Cleland (1966, p. 122), Dixon et al (1992, p. 148), Hercus (1992, p. 32), Ramson (1988, pp. 753–4) and Teichelmann & Schürmann (1840, pt 2, p. 59).
107 Clarke (2007, pp. 120–1), Cleland (1939, p. 11; 1966, p. 122), Hassell (1936, p. 694), Meagher & Ride (1980, pp. 76–7), Willsteed et al (2006, p. 4) and Wreck Bay Community & Renwick (2000, p. 27).
108 Clarke (1985b, p. 13; 2003b, p. 89), Dixon et al (1992, p. 120), Gott & Conran (1991, p. 32), Ramson (1988, p. 410), Smith & Smith (1999, pp. 57–9) and Zola & Gott (1992, p. 21). Other spelling variations in Australian English for muntry are 'muntry, 'montry', 'muntree', 'muntri', 'muntari', 'muntaberry' and 'munterberry'.
109 Black (1943–65, p. 604) and Jessop & Toelken (1986, pt 2, p. 932).
110 Bonney (1994, p. 13) and Lang (cited Smyth 1878 vol. 1, p. 219).
111 Delbridge et al (1997, p. 2436) and Dixon et al (1990, p. 147). Written various as 'wurrulde', 'wurruldi', 'wirilti', 'wirrildar', 'weerilda' (Berndt & Berndt, 1993, p. 309; Black, 1943–65, p. 411; Meyer, 1843, p. 108; Taplin, 1879, p. 131).
112 Clarke (1994, p. 58, 261) and Jenkin (1979, p. 210).
113 Bindon (1996, p. 173), Dixon et al (1992, pp. 115–6), Meagher (1974, pp. 54–5, 65), Ramson (1988, p. 126), Smith (1982, p. 119) and Whiting (1963, p. 294). Other recordings of the name are 'baio', 'bayio', 'boyar', 'boyoo' and 'byyu'.
114 Bindon (1996, p. 173), Hassell (1936, p. 705), Meagher (1974, pp. 25, 34–6, 54–5), Smith (1982, pp. 119–20) and von Mueller (cited Smyth, 1878, vol. 1, p. 215).
115 Smith (1982, p. 117).
116 Dixon et al (1992, p. 126), Hallam (1975, pp. 12–14), Meagher (1974, pp. 24, 26, 34, 40, 54, 61–2, 64) and Ramson (1988, p. 712). Other spellings include 'warrain', 'warrang', 'warrine', 'woorine', 'warryn', 'worrain', 'wyrang' and 'warrein'.
117 Drummond (1862, p. 25).
118 Grey (1841, vol. 2, pp. 11–12).
119 Dixon et al (1992, pp. 137–8), Maiden (1889, p. 480) and Ramson (1966, pp. 114, 117–8; 1988, p. 332).
120 Baker (1919, pp. 155–8).
121 Bindon (1996, pp. 124, 134), Dench (1994, pp. 186–7), Dixon et al (1992, pp. 139, 146–7), Maiden (1889, pp. 444, 459, 508) and Ramson (1966, pp. 114–5, 117–8; 1988, pp. 346, 690, 711). Other spelling variations in Australian English for tuart are 'tewart', 'tooart', 'tooat' and 'touart'. Maiden refers to wandoo as *Eucalyptus redunca*.
122 J. Drummond (cited Maiden, 1918, p. 501). Drummond recorded the Aboriginal word for tuart as 'doatta'.
123 Maiden (1918, p. 501).
124 Dixon et al (1992, pp. 121–2), Maiden (1889, p. 31), McNicol & Hosking (1994, p. 92) and Ramson (1966, pp. 113, 115; 1988, pp. 511–2). Sometimes written as 'quondong', 'quandang' and 'quantong'. Maiden used the earlier name *Fusanus acuminatus*.
125 Matthews (1997). There is an Australian Quandong Industry Association that has occasionally held conferences.
126 Dixon et al (1992, p. 147) and Ramson (1988, p. 737).
127 Dixon et al (1992, pp. 136–7), Maiden (1889, p. 357) and Ramson (1966, pp. 114, 117; 1988, pp. 271–2). Other spelling variations in Australian English are 'gidgea', 'gidga', 'gidia', 'gidgee' and 'gidyea'. Maiden referred to this species as *Acacia homalophylla*.
128 Dixon et al (1992, pp. 142–3, 135–6), Hercus (1994, p. 52), Maiden (1889, pp. 349, 495), Ramson (1966, pp. 113–7; 1988, pp. 168, 407–8) and Smith & Smith (1999, p. 99). Other spelling variations in Australian English for coolibah are 'coolabah', 'coolibar', 'coolobar' and 'coolybah'. For mulga, other variations are 'malga', 'mulgah' and 'mulgar'.
129 Zola & Gott (1992, p. 29). Jessop & Toelken (1986, p. 162) and Stone (1911, p. 445) gave bitter quandong the alternative common name of ming. Zola & Gott used the name Madi Madi for Muthimuthi.
130 Clarke (2007, p. 105) and Stone (1911, p. 445).
131 Bancroft (1877–80), Basedow (1925, pp. 155–7), Carr & Carr (1981b, pp. 24–7), Cleland (1936, pp. 7–8; 1939, p. 13; 1957, pp. 158–9; 1966, p. 119), Cleland & Johnston (1933, p. 116; 1937–38, p. 338; 1939b, p. 177), Dobkin de Rios & Stachalek (1999), Hardy (1969, p. 14), Helms (1896, pp. 248–9), Hicks (1963), Johnston (1939), Johnston & Cleland (1933–34), Latz (1974; 1995), Maiden (1889, pp. 168–72), Peterson (1977), Reid (1977, pp. 159–61), Roth (1901b, p. 31), Smith (1991, p. 22), Smyth (1878, vol. 2, pp. 222–3), Tolcher (2003, pp. 23–4, 83–4, 110), Watson (1983) and Webb (1973, pp. 293–4). Jessop (1981, pp. 324–5) described the distribution of pituri.
132 Dixon et al (1992, pp. 130–1) and Ramson (1966, pp. 115–6; 1988, p. 483). The Aboriginal plant name has also been rendered as 'piturie', 'pitjuri', 'pitcheri', 'pitcherry', 'pitchery', 'pitchury', 'peturr', 'bedgery' and 'pedgery'.
133 Von Mueller (cited Smyth, 1878, vol. 1, p. 223) provided the information on pituri being used to give courage in fighting, while Latz (1995, p. 68) described pituri use for initiatory revelation.
134 Cleland & Johnston (1937–38, p. 341), Dixon et al (1992, p. 130), Finlayson (1952, p. 64), Peterson (1977), Symon (2005) and Watson (1983, p. 7).
135 Brokensha (1975, p. 29), Cleland (1936, p. 7; 1939, p. 14; 1957, p. 159), Cleland & Johnston (1933, pp. 116, 123; 1937–38, pp. 211, 341), Cleland & Tindale (1954, p. 85; 1959, p. 138), Latz (1995, pp. 62–4, 233–5) and O'Connell et al (1983, pp. 97–8).
136 Basedow (1904, p. 18; 1914, pp. 74–5, 79; 1925, p. 149) appears to have made the initial mistake (Cleland & Johnston, 1937–38, p. 336).
137 Dixon et al (1992, pp. 128–9), Koch (1898, p. 104), Maiden (1889, pp. 17, 123–4), Ramson (1988, p. 410). Maiden used the name *Claytonia balonensis*. Other spelling variations include 'monyeroo' and 'munyeru'.
138 Spencer (1918, vol. 15, p. 9). Spencer used *Claytonia* species, which is an obsolete name for this species.
139 Dixon et al (1992, p. 129), Koch (1898, p. 104) and Ramson (1988, p. 462). Other spelling variations in Australian English are 'periculia', 'parrakeelya', 'parakeelia', 'parakelia', 'parakilya' and 'parakylia'.
140 Maiden (1889, pp. 17, 53, 123). Maiden used an earlier name, *Claytonia balonensis*.
141 Dixon et al (1992, pp. 108–9) and Ramson (1988, p. 462). Witchetty is also spelled as 'widgery', 'witchety' and 'witjuti'. Note that the Adnyamathanha name for the witchetty bush is *nulpuru* (McEntee et al, 1986, p. 11). Tindale (1952, 1966) outlined Aboriginal use of 'witchetty grubs'.
142 Dixon et al (1992, p. 127) and Ramson (1988, p. 164). Conkerberry has also been written as 'coongaberry', 'konkleberry', 'kunkerberry' and 'koonkerberry'.
143 Palmer (1883, p. 96). Palmer used the obsolete scientific name *Carissa brownii*, which is now *C. spinarum*.
144 Dixon et al (1992, p. 124) and Ramson (1988, p. 743). Wongay is also spelled as 'wongai' and 'wongi'.
145 Dixon et al (1992, p. 138) and Ramson (1988, p. 271). Jitta is also written as 'ghittoe', 'jhito', 'jitter', 'jhitu', 'jidu' and 'jitto'.
146 Baker (1913, p. 168). The term 'kauri' is derived from a Maori language in New Zealand, where it originally concerned a botanically related species of tree (Ramson, 1988, p. 346).
147 For histories of the origin and development of plant taxonomy and the role of herbaria refer to Linnaeus (1753), Maiden (1912, pp. 49–73), Short (2003, pp. 326–8), Stearn (1957) and Webb (2003, chapter 2). For the establishment history of Australian museums, see Finney (1986, pp. 94–101), Green (1990), Jones (1996, part 1) and Kohlstedt (1983). Hudson (1975) provided a social history of museums across the world.
148 Sharr (1996, introduction), Short (2003, pp. 323–5) and Sivarajan (1991). Pearn (2001, pp. 9–12) and Stearn (1983) gave outlines of botanical naming practices. Carr & Carr (1981a & c) and Short (1990) provided a history of Australian systematic botany.
149 Finney (1984, chapter 10) and Moyal (1986, chapter 5).
150 Brown (1801–05 [2001, pp. 445–7]) used the names *Avicennia resinifera* (now *A. marina* subsp. *australasica*) and *Banksia pyriformis* (now *Xylomelum pyriforme*).

Chapter 4 George Caley in New South Wales

1 Pollan (2002, chapter 2).
2 Aitken (2006, pp. 75, 152, 156, 158, 187, 216). Bean (1908, part 4) discussed the growing of tropical plants in palm and stove houses at Kew.
3 Aitken (2006, pp. 212–4).
4 Aitken (2006, p. 196).
5 Joseph Pitton de Tournefort (cited Aitken, 2006, p. 124).
6 In 1735, C. Linnaeus published *Systema naturae* in Leiden, Holland. In 1753 he published *Species plantorum* in Stockholm, Sweden.
7 Aitken (2006, p. 57).
8 Aitken (2006, p. 75).
9 Jewett & McCausland (1958), MacKenzie (1991) and Sampson (1935).
10 Whittle (1970). Quote is the title of Part 3 of his book.
11 Aitken (2006, chapter 24), Barker & Barker (1990, pp. 52–3) and Finney (1984, chapter 3).
12 From 1787 to 1800 William Curtis published *The Botanical Magazine*

(vols 1–14). From 1801 to 1807 John Sims published *Curtis's Botanical Magazine* (vols 15–26).

13 Aitken (2006, pp. 156, 212), Barker & Barker (1990, pp. 42, 44–5, 53), Cavanagh (1990), Finney (1984, pp. 14, 47–8, 57, 60) and Webb (2003, pp. 22–4).

14 Finney (1984, p. 3) and Nelson (1990a, p. 285).

15 Aitken (2006, p. 161).

16 Finney (1984, chapter 3).

17 Collins (1798–1802, vol. 1, pp. 30–1), Jones (2007, chapter 1), Tench (1788–92 [1996, p. 91]).

18 Arnold (cited Whitley, 1933, p. 304). See also Finney (1984, pp. 131–2, 134, 139).

19 Arnold (cited Whitley, 1933, p. 305).

20 Arnold (cited Whitley, 1933, p. 305) and Finney (1984, p. 134).

21 Aitken (2006, pp. 115, 123, 136, 158, 163), Desmond (1989a, p. 2), Field (1993, p. 142), Lemmon (1968, chapter 3), Lyte (1983, p. 173) and Whittle (1970, pp. 63–4, 138).

22 Day (2001, pp. 47–58), Denoon & Mein-Smith (2000, pp. 86–94), Gilbert (1981), Macintyre (1999, pp. 17–35) and McLaren (1996, p. 19). Refer to MacKenzie (1991) for a broader view of imperialism and plant collecting.

23 Desmond (1995, chapter 8), Froggatt (1932), Hepper (1989) and Short (2003).

24 MacKenzie (1991, p. 9). Blunt (1978) provided a history of the Royal Gardens at Kew

25 Finney (1984, p. 105).

26 Finney (1984, p. 141) and Lyte (1983, p. 173).

27 Lyte (1983, p. xiii).

28 Aitken (2006, pp. 158, 180).

29 For examples, refer to Aitken (2006, pp. 139, 163, 174).

30 Aitken (2006, p. 163).

31 MacKenzie (1991, p. 9).

32 Sparrman (1772–76 [1975–76, vol. 2, pp. 248–51]).

33 Hastings (1989, p. 174), Ponsonby (1998, p. 19) and Short (2003, pp. 285–94). See Spruce (1908) for botanical notes.

34 Ponsonby (1998, p. 19).

35 Finney (1984, p. 32).

36 Aitken (2006, p. 138) and Lyte (1980, pp. 175–6).

37 Aitken (2006, pp. 140, 174, 177, 211, 214–5, 219), Desmond (1989b, pp. 16, 18–19), Field (1993, p. 146), Short (2003, pp. 329–34) and Whittle (1970, pp. 109–12).

38 Carter (1988, pp. 358, 487–8), Desmond (1995, pp. 99, 114, 363, 433), Else-Mitchell (1939, 1966), Hall (1978, p. 32), Lyte (1980, p. 176), McLaren (1996, p. 23), Maiden (1903, pp. 992–6), Moyal (1986, p. 24), Serle (1949, vol. 1, p. 140) and Webb (2003, pp. 5, 20, 70, 102–4). Please note that some correspondents, such as Governor King, referred to the plant hunter as 'Cayley', which is incorrect. Ritchie (1989, pp. 45, 49) also wrote his name as 'Cayley'.

39 Barker & Barker (1990, pp. 53–4), Bean (1908, p. 21), Moyal (1976, pp. 11–12) and Vallance et al (2001, p. 202).

40 Letter from Banks to John King, 13 December 1798 (cited Desmond, 1995, p. 123). See also Field (1993, p. 142).

41 Stearn (1974, pp. 21–2).

42 Maiden (1921a, p. 154).

43 Maiden (1921b, p. xxxi), Pearn (2001, pp. 104–5) and Stearn (1974, pp. 20–2).

44 Desmond (1989a, pp. 5–6).

45 Lyte (1980, p. 176). Refer also to McLaren (1996, p. 26) and Moyal (1986, p. 24).

46 J. Banks correspondence to P. G. King, 29 August 1804 (Banks papers, Mitchell Library, Sydney).

47 Caley cited Webb (1995, p. 32).

48 Caley cited Webb (1995, p. 55).

49 Hunter (1793, pp. 520–1). Jacques Arago (1822, cited Rolls, 2002, pp. 69–70), Collins (1798–1802, vol. 1, p. 550; vol. 2, p. 169) and Tench (1788–92, p. 197) described tree-climbing around Sydney. See also paintings by Joseph Lycett (1820–22 [1990, plates 1 & 2]).

50 C. Fraser (cited Hooker, 1830, p. 240).

51 G. Caley correspondence to J. Banks (12 March 1804, Banks papers, Mitchell Library, Sydney). See also Threlkeld (1824–1859 [1974, p. 316]).

52 Collins (1798–1802, vol. 2, p. 301).

53 Collins (1798–1802, vol. 1, p. 550).

54 G. Caley correspondence to J. Banks (1809, cited Currey, 1966, p. 178).

55 Collins (1798–1802, vol. 1, p. 599). Smith (2001) produced a biography of Bennelong. For a description of 'tribal boundaries' around Sydney, refer to Attenbrow (2003), Tindale (1974, Tribal Boundaries Map, S.E. sheet & p. 193) and Troy (1994).

56 Attenbrow (2003, chapter 7), Buku-larrnggay Mulka Centre (1999), Clarke (2003a, pp. 116, 134), McKnight (1999, p. 125), Peterson & Rigsby (1998), Sharp (2002) and Smyth (1993).

57 Powell (1990, p. 95) gave an account of vegetation clearance around Sydney by 1800.

58 R. Brown correspondence to J. Banks (30 May 1802, cited Moyal, 1976, p. 25).

59 Carter (1988, pp. 418, 473–4) and McLaren (1996, chapter 2).

60 McLaren (1996, p. 32).

61 Else-Mitchell (1939, p. 458).

62 Else-Mitchell (1939, 1966).

63 G. Caley correspondence to J. Banks (28 April 1803, Banks papers, Mitchell Library, Sydney).

64 Else-Mitchell (1939, p. 459) and Whitley (1933, p. 302).

65 G. Caley correspondence to J. Banks (12 October 1800, Banks papers, Mitchell Library, Sydney).

66 R. Brown correspondence to J. Banks (December 1804, Banks papers, Mitchell Library, Sydney).

67 Currey (1966, p. 140).

68 Else-Mitchell (1939, pp. 482–3).

69 G. Caley correspondence to J. Banks (1803–4, cited Else-Mitchell, 1939, pp. 468–9 and Finney, 1984, p. 107). For other Aboriginal views of monotremes, see Moyal (1986, p. 142; 2001, pp. 15–6, 84–6, 152).

70 J. Banks correspondence to G. Caley (8 April 1803, Banks papers, Mitchell Library, Sydney.

71 G. Caley correspondence to J. Banks (12 March 1804, Banks papers, Mitchell Library, Sydney). See also Moyal (2001, p. 16).

72 Moyal (2001, pp. 83–5, 150–2, 157, 188). Smyth (1878, vol. 1, pp. 248–9, 251–2) provided accounts of the platypus derived from Aboriginal sources.

73 G. Caley correspondence to J. Banks (1809, cited Currey, 1966, p. 177).

74 Finney (1984, p. 152).

75 Yarwood (1967).

76 Else-Mitchell (1939, pp. 528–9, 532).

77 Carter (1988, pp. 436–8), K.V. Smith (2005b), Threlkeld (1824–1859 [1974, p. 316]) and Webb (1995, pp. 55, 78–9, 87, 91, 125).

78 K.V. Smith (2005b) and Webb (1995, Appendix D, p. 175).

79 Tindale (1974, Tribal Boundaries Map, S.E. sheet & p. 193) and Troy (1994).

80 G. Caley correspondence to J. Banks (3 November 1808, Banks papers, Mitchell Library, Sydney). See also Currey (1966, p. 174) and Threlkeld (1824–1859 [1974, p. 316]).

81 G. Caley correspondence to J. Banks (25 September 1807, Banks papers, Mitchell Library, Sydney, 1807). See also Else-Mitchell (1939, p. 511) and Threlkeld (1824–1859 [1974, p. 316]).

82 Flinders (1814a, plate 1)

83 Currey (1966, p. 191), Carter (1988, pp. 436, 438, 443, 447) and Webb (1995, pp. 55, 60, 94, 96–99).

84 Lyte (1980, pp. 187–95, 199) and K.V. Smith (1992, pp. 45–6, 83, 175, 202).

85 Brook (2001), Dark (1966), Kenny (1973), Smith (2001) and Tench (1788–92 [1996, pp. 7–8, 117–9, 125–6, 134–150, 160–5]).

86 K.V. Smith (2005b).

87 K.V. Smith (2005b) and Threlkeld (1824–1859 [1974, p. 316]).

88 G. Suttor correspondence to J. Banks (12 November 1812, Banks papers, Mitchell Library, Sydney).

89 Webb (1995, pp. 98–9). An account of the trial was given on 28 September 1816, *Supplement to the Sydney Gazette*. Moowattin's execution on 1 November 1816 was listed in the *Sydney Gazette*, 2 November 1816, p. 2. Also see an account by 'A Friend to Missions', dated 9 February 1831, in the *Sydney Gazette* (published 15 February 1831, p. 3).

90 S. Marsden (2 December 1826, cited Threlkeld, 1824–1859 [1974, p. 348; see also p. 11]).

91 Else-Mitchell (1939, p. 459) and Maiden (1903).

92 Maiden (1903, p. 988).

93 Maiden (1903, p. 989). See Hastings (1989, p. 174) for an account of London's Great International Exhibition of 1862, and Wickens (1993, p. 88) for this exhibition and the one in Paris in 1855.

94 Maiden (1903, pp. 989–90).

95 Maiden (1903, p. 990).

96 Maiden (1903, p. 990). 'Barilgora' (inland grey box, *Eucalyptus*

microcarpa) was previously described as *E. hemiphloia*.

97 George Caley's notes on eucalypts (cited Webb, 1995, Appendix D, p. 176). See also Maiden (1903, pp. 988, 991, 996).

98 George Caley's notes on eucalypts (cited Webb, 1995, Appendix D, p. 175). Maiden (1903, pp. 990, 996) listed the 'mogargro ironbark' as *Eucalyptus crebra*, which is the narrow-leaved red ironbark.

99 Lyte (1980, p. 176).

100 Else-Mitchell (1966, pp. 194–5).

101 Whitley (1933, p. 302).

102 A. Cunningham (cited Maiden, 1903, p. 996).

103 Else-Mitchell (1939, p. 466) and Hall (1978, p. 32).

104 Currey (1966, p. 221), Hall (1978, p. 32), Jessop & Toelken (1986, p. 2075), Maiden (1903, p. 988), Sharr (1996, pp. 12, 97) and Webb (1995, Appendix C, p. 174).

Chapter 5 Allan Cunningham and the Mapping of Australia

1 Carter (1988, pp. 476–8, 482), Curry et al (2002, p. 2), Desmond (1995, pp. 102, 114–7, 433, 364–5), Finney (1984, pp. 140, 152), Froggatt (1932, pp. 113–9), Hall (1978, pp. 42–43), Lee (1925, pp. 167–68), Lyte (1980, pp. 176, 240), McMinn (1970, p. 41; 1971, p. 2), Maiden (1921b, pp. xxix–xxx), Perry (1966), Serle (1949, vol. 1, pp. 204–6) and Webb (2003, chapter 5).

2 Desmond (1989a, pp. 7–9), Moyal (1976, p. 13; 1986, p. 28) and Webb (2003, p. 87).

3 Barker & Barker (1990, pp. 54–5) and Bean (1908, p. 21).

4 McLaren (1996, pp. 18, 32, 35, 40, 51).

5 Carter (1988, pp. 474–5, 479–80), McLaren (1996, chapter 3), McMinn (1971, p. 3), Maiden & Cambage (1909), Oxley (1820) and Webb (2003, pp. 93–104). See Curry et al (2002) for maps of Cunningham's plant collecting localities.

6 Barker & Barker (1990, p. 55), Froggatt (1932, pp. 103–8), Moyal (1974, p. 39) and Oxley (1820, p. 375).

7 Oxley (1820, p. xiii).

8 Cunningham (1817, cited Webb, 2003, p. 98).

9 Cunningham (1817, cited Webb, 2003, p. 98).

10 Sharr (1996, p. 117).

11 Maiden (1889, pp. 115, 354).

12 Cunningham (1817, cited Lee, 1925, p. 276).

13 Jones (1969). For other descriptions of Aboriginal vegetation firing practices refer to Gill et al (1991), Hallam (1975), Latz (1995), Nicholson (1981), Pyne (1991) and Rose (1995).

14 Cunningham (1817, cited Lee, 1925, p. 227).

15 Cunningham (1817, cited Lee, 1925, pp. 275–6).

16 A. Cunningham, journal entries dated 29–30 July 1817 (cited Lee, 1925, pp. 271–2).

17 Oxley (1820, p. 138).

18 Oxley (1820, p. 139).

19 Oxley (1820, pp. 139–41).

20 Oxley (1820, pp. 302–3).

21 Oxley (1820, p. 302).

22 Oxley (1820, p. 310).

23 Cunningham (1818, cited Lee, 1925, pp. 413–4). The term 'currajong' normally applies to *Brachychiton* species. On this occasion, it was used for the native cottonwood (*Hibiscus heterophyllus*).

24 Cunningham (1818, cited Lee, 1925, p. 415).

25 Cunningham (1818, cited Lee, 1925, p. 413).

26 Cunningham (1818, cited Lee, 1925, p. 416).

27 Cunningham (1818, cited Lee, 1925, p. 418).

28 King (1827) and Whitley (1933, pp. 308–10). Curry & Maslin (1990) listed Cunningham's collecting localities during the 1817–22 trips with King.

29 King (1827).

30 Cunningham (1818, cited Lee, 1925, p. 378).

31 Cunningham (1818, cited Lee, 1925, p. 382).

32 King (1827, vol. 1, pp. 111–2). The 'sago palm' may have been prepared from cycad nuts.

33 Puruntatameri (2001, pp. 40–1). The species of zamia palm would have been *Cycas armstrongii*.

34 Cunningham (1817, cited Lee, 1925, p. 352).

35 Carter (1988, pp. 480–2). See King (1827, vol. 1, pp. 161–2).

36 King (1827, vol. 2, pp. 631–7).

37 King (1827, p. 636).

38 King (1827, vol. 1, p. 265). Possibly refers to the white sand lily (*Crinum angustifolium*), which according to Bindon (1996, p. 91) is poisonous, but can be eaten after leaching and baking.

39 King (1827, vol. 1, pp. 372–3). The 'cabbage palm' was the solitaire palm, but also used for 'cabbage' were probably the Alexandra, Cairns fan, Carpentarian and northern Kentia palms.

40 King (1827, vol. 2, p. 156). *Atriplex cinerea*, earlier known as *A. halimus*.

41 King (1827, vol. 2, p. 134).

42 King (1827, vol. 2, pp. 497–565).

43 McMinn (1971, p. 3).

44 Ritchie (1989, p. 49).

45 Allan Cunningham, 'Report to Darling, 134' (cited McMinn, 1970, p. 93).

46 Cunningham (1827, cited Lee, 1925, p. 569). The species of 'box' tree involved is unclear.

47 Cunningham (1827, cited Lee, 1925, p. 569).

48 Cunningham (1827, cited Lee, 1925, pp. 548, 552, 556, 569, 579).

49 This change in materials is reflected in the Aboriginal artefact collections of the South Australian Museum. See also Clarke (2003a, pp. 213–4).

50 Cunningham (1828, cited Lee, 1925, p. 587). See also C. Fraser (30 July 1828, cited Hooker, 1830, p. 259; 1 August 1828, cited Hooker, 1830, pp. 260–1).

51 Anderson (1996, p. 74), Backhouse (1843, p. 362), Hiddins (2000b, p. 6; 2001, p. 55) and Roberts et al (1995, pp. 8–9).

52 C. Fraser (4 July 1828, cited Hooker, 1830, p. 243; see also p. 259).

53 Cunningham (1828, cited Lee, 1925, p. 598).

54 C. Fraser (26 July 1828, cited Hooker, 1830, pp. 253–4).

55 C. Fraser (27 July 1828, cited Hooker, 1830, p. 256). Steele (1983, pp. 138–9) claimed it was the skin of a man.

56 Clarke (2003a, pp. 46–7, 97–8, 182–3), Maddock (1982, pp. 152–6) and Morphy (1984) have provided an overview of burial practices in Aboriginal Australia.

57 C. Fraser (5 July 1828, cited Hooker, 1830, p. 243).

58 Griffiths (1996, chapter 2).

59 MacDonald (2005) and Museums Australia (2005) gave an overview of indigenous human remains issues in Australia.

60 A. Cunningham correspondence to J. Banks (8 November 1819, cited Moyal, 1974, pp. 27–8).

61 King (1827, vol. 1, pp. 47–8).

62 McMinn (1970, p. 94).

63 Cunningham (1818, cited Lee, 1925, p. 318).

64 Aitken (2006, pp. 180, 185), Short (2003, p. 237) and Whittle (1970, p. 93).

65 Barker & Barker (1990, p. 62), Bunce (1859a, pp. 82–3), Desmond (1995, p. 140), Lee (1925, pp. 615–6), Moyal (1986, p. 65) and Salier (1931, p. 11).

66 Froggatt (1932, pp. 109–12) and Parsons (1966).

67 Mitchell (1838, vol. 1, pp. 177–200, 204, 331).

68 Favenc (1888, pp. 407–8).

69 Zouch (cited Favenc, 1888, pp. 407–8; Mitchell, 1838, vol. 1, pp. 353–5). Favenc and Mitchell gave his initial as 'W', although Parsons (1966) stated his first name was 'Henry'.

70 Threlkeld (1824–1859 [1974, pp. 132–3]).

71 Threlkeld (1824–1859 [1974, p. 133]).

72 Governor Sir Richard Bourke correspondence to Secretary of State for the colonies Baron Glenelg, 22 December 1835 (*Historical Records of Australia, Series 1*, vol. 18, pp. 235–7).

73 A. Cunningham correspondence to W.J. Hooker, 5 March 1836 (cited Webb, 2003, p. 117).

74 MacKenzie (1991, p. 8).

75 Sampson (1935, pp. 406–7).

76 Sampson (1935, pp. 407–8).

77 Cunningham (cited Lee, 1925, p. 318). See King (1827, vol. 1, pp. 18, 38, 222).

78 Cunningham (1823, cited Favenc, 1888, pp. 403–4).

79 Oxley (1817 [cited Favenc, 1888, p. 63]; 1820, pp. 50, 59). See also Cunningham (cited Webb, 2003, pp. 98–9). The acorns planted by Cunningham were presumably from English oak.

80 Cunningham (1818, cited Lee, 1925, p. 318). Other references to this are Cunningham (1817, cited Lee, 1925, pp. 194, 210, 292) and King (1827, vol. 2, pp. 123–4).

81 Cunningham (1818, cited Lee, 1925, p. 361). See also King (1825, vol. 1, pp. 72–3).

82 Bunce (1859a, p. 4), Cooper (1954), Finney (1984, p. 26) and Low (1999, chapter 3).

83 Pollan (2002, chapter 1).
84 For a description of the Austronesian migration out of Asia, refer to Diamond (1998, chapter 17), Flannery (1994, chapters 14–8) and Thorne & Raymond (1989). See Clarke (2007, chapter 10) for a discussion of the human movement of plants into northern Australia during the pre-European period.
85 Finney (1984, pp. 146–7) and Webb (2003, pp. 131–3).
86 Telford (1990).
87 McMinn (1971, pp. 1–2).
88 Aitken (2006, p. 156), Froggatt (1932, p. 117), McMinn (1971, p. 2), Perry (1966) and Webb (2003, pp. 123–6).
89 Sharr (1996, pp. 109–10).
90 Clarke (2007, pp. 119–20).
91 Smith & Kalotas (1985). An earlier name was *Lysiphyllum cunninghamii*.
92 McMinn (1971, p. 3).
93 Blandowski (1858, p. 127).

Chapter 6 Resident Plant Collectors and Aboriginal People

1 Moyal (1986, pp. 46–7), Serle (1949, vol. 1, p. 253), Sharr (1996, pp. 23, 117). and Webb (2003, p. 136). The year of James Drummond's birth is variously given as 1783–84 and 1786–87.
2 Barker & Barker (1990, pp. 55–9, 69) and Webb (2003, pp. 136, 140).
3 Nelson (1990b), Sharr (1996, p. 23) and Short (2003, pp. 217–28).
4 Erickson (1969, p. 9),
5 Fforde (2002) and Lowe (1995, chapter 3).
6 Robert Lyon (cited Lowe, 1995, p. 19).
7 J. Drummond (cited Erickson, 1969, p. 9).
8 Fforde (2002, p. 230), Hasluck (1967) and Lowe (1995, pp. 58–63). A variation of the spelling of Middgegooroo's name is 'Midgigoroo'.
9 Erickson (1969, p. 9).
10 Fforde (2002).
11 Portman (1989).
12 Webb (2003, pp. 140–1).
13 Erickson (1966, p. 325).
14 Erickson (1966, p. 102).
15 Erickson (1966, chapter 3).
16 James Drummond (cited Erickson, 1969, p. 31).
17 Webb (2003, p. 140).
18 Erickson (1969, pp. 38–9).
19 Erickson (1969, pp. 39–40).
20 Erickson (1969, pp. 40–1).
21 Erickson (1969, p. 41). See also Tilbrook (1983, pp. 6–7).
22 J. Drummond correspondence to W.J. Hooker, 30 October 1844 (cited Erikson, 1969, pp. 100–1).
23 Moyal (1976, pp. 39–41; 1986, pp. 49–50) and Webb (2003, p. 142). Drummond's notes appeared in Hooker's *Journal of Botany*, vol. II (1840), vols I and III (1850), vol. IV (1852), vol. V (1853), and vol. VI (1854).
24 Erickson (1969, pp. 89, 104, 110), Moyal (1986, p. 54) and Webb (2003, p. 144). Refer to Chisholm (1942) for a list of Gould's publications. Barrett (1938) and Chisholm (1966b) provided biographies of Gould.
25 Webb (2003, pp. 144, 146).
26 J. Gilbert correspondence to J. Gould (3 September 1839, National Library of Australia).
27 Chisholm (1973, p. 52).
28 W. Harvey, 1854 (cited Short, 2003, p. 135).
29 Drummond (1862, p. 25).
30 James Drummond correspondence to W.J. Hooker, 1839 (cited Short, 2003, pp. 124–5).
31 Drummond (1862, pp. 26–7). See also Backhouse (1843, p. 526), Bird & Beeck (1988, p. 116), Meagher (1974, pp. 25, 27, 58–9, 65), Meagher & Ride (1980, p. 75), Reid (1977, p. 15) and von Brandenstein (1988, pp. 19, 77, 167). Von Brandenstein recorded the Nyungar name as 'miern'.
32 Drummond (1862, pp. 26–7).
33 Clarke (2007, p. 89).
34 Barr et al (1988, pp. 214–9), Clarke (2007, p. 89), Levitt (1981, pp. 61, 64), Rowland (2002), Tindale (1981, p. 1862) and Worsnop (1897, p. 84).
35 Drummond (1862, p. 27). Von Brandenstein (1988, pp. 115, 165) recorded the Nyungar name as 'tyuubaq'.
36 Drummond (1862, p. 27).
37 Clarke (2003a, pp. 48–9).
38 Drummond (1862, pp. 27–8). The author gave the italic emphasis within the quote.
39 Drummond (1862, p. 28).
40 Clarke (2003a, pp. 120, 167, 173; 2007, pp. 21, 73–5, 77, 84–6, 88–9).
41 James Drummond correspondence to J.D. Hooker, dated late 1830s, (cited Maiden, 1918, p. 501). For a further account of 'doatta' root bark use, see Drummond (1862, p. 26).
42 Drummond (1862, p. 26).
43 Drummond (1862, p. 26).
44 J. Drummond correspondence to W.J. Hooker, 1841 (cited Erickson, 1969, p. 74; Short, 2003, p. 129). The term 'chinga' maybe a rendering of *djanga* (or *janga*), meaning the spirits of the dead (Bates 1912 [1992, 16]).
45 Drummond (1862, p. 25).
46 J. Drummond (cited Erickson, 1969, pp. 99–101).
47 J. Drummond (cited Erickson, 1969, p. 85).
48 Erickson (1969, p. 68).
49 Erickson (1969, pp. 75–9). See Erickson (1969, pp. 107–8) for later references to Kabinger.
50 J. Drummond (cited Erickson, 1969, p. 99).
51 J. Drummond correspondence to W.J. Hooker, 30 October 1844 (cited Erickson, 1969, p. 100).
52 James Drummond (cited Erickson, 1969, p. 99).
53 J. Drummond correspondence to W.J. Hooker, 21 February 1844 (cited Erickson, 1969, p. 92).
54 Chisholm (1973, p. 52) and Erickson (1969, pp. 105–6).
55 See chapter 7 of this volume.
56 Erickson (1969, pp. 107–9).
57 Erickson (1969, p. 113) and Webb (2003, p. 144).
58 J. Drummond in *The Inquirer*, 7 April & 14 April 1847 (cited in Barker, 1996, pp. 4–5). Barker (1996, pp. 2–3) discussed the relationship between Drummond's writings in *The Inquirer* and his letters published by W.J. Hooker, from 1847 to 1852.
59 J. Drummond correspondence to W.J. Hooker, 1851 (cited W.J. Hooker, *Journal of Botany and Kew Garden Misc.*, Vol. 4, 1952, p. 189). See Erickson (1969, p. 134) and Webb (2003, p. 149).
60 Webb (2003, p. 144).
61 Correspondence from C. Darwin to J. Drummond, 20 December 1860 (Letter 3026). Refer to **http://www.darwinproject.ac.uk/darwinletters/calendar/entry–3026.html** (accessed 26 September 2007).
62 Webb (2003, p. 144).
63 Bailey (1899–1902, vol. 6, p. 1929) and Sharr (1996, pp. 23, 117).
64 See chapter 9 of this volume.
65 De Vries (2005), Lines (1994), Moyal (1986, pp. 107–9), Pickering (1929, pp. 30–6) and Webb (2003, pp. 136, 153–4).
66 G. Molloy correspondence to J. Mangles, 21 March 1837 & 25 January 1838 (cited Pickering, 1929, pp. 54–62). See also Bassett (1981), De Vries (2005, p. 48), Jordan (2005, pp. 144–5), Lines (1994, pp. 218–2, 236, 257–9, 280, 283, 294) and Webb (2003, p. 142).
67 Lines (1994, pp. 258, 278, 295), Moyal (1986, pp. 48, 109) and Webb (2003, p. 145).
68 De Vries (2005, pp. 66–7), Lindley (1840) and Lines (1994, pp. 261–2).
69 Lines (1994, chapter 12).
70 G. Molloy correspondence to E. Besley, 20 November 1833 (cited Pickering, 1929, p. 45).
71 G. Molloy correspondence to E. Besley, 13 November 1833 (cited Lines, 1994, p. 183).
72 Fforde (2002, pp. 229–30) and Lines (1994, pp. 183–4). Fforde spelled the Aboriginal man's name as 'Gallypert'.
73 G. Molloy correspondence to E. Besley, 20 November 1833 (cited Pickering, 1929, pp. 44–5).
74 G. Molloy correspondence to Mrs Storey, 8 December 1834 (cited Pickering, 1929, pp. 48–50). She also Lines (1994, pp. 184–5).
75 Lines (1994, pp. 165–6).
76 G. Molloy correspondence to J. Mangles, 14 March 1840, 15 April 1840, June 1840 & 11 April 1842 (cited Pickering, 1929, pp. 72–3, 77–9, 83–4). See also Lines (1994, pp. 284–5, 308).
77 Bindon (1996, p. 186), Lines (1994, p. 285) and Von Brandenstein (1988, p. 22).
78 G. Molloy correspondence to J. Mangles, 20 January 1841 (cited Hasluck 1955, p. 225 and Pickering, 1929, p. 79; also cited in part by Lines, 1994, p. 309 and De Vries, 2005, p. 64). For similar references, see Hasluck (1955, pp. 192, 206).
79 De Vries (2005, p. 54; see also p. 59), Lines (1994, pp. 297, 301) and Pickering (1929, p. 78).
80 G. Molloy correspondence to J. Mangles, 14 June 1841 (cited Pickering, 1929, p. 82).
81 Lines (1994, p. 278).

82 G. Molloy correspondence to J. Mangles, 31 January 1840 (cited Lines, 1994, pp. 278–9).
83 De Vries (2005, pp. 27–8, 49).
84 De Vries (2005, pp. 52, 55).
85 Lines (1994, pp. 244–5) and Webb (2003, p. 136).
86 G. Molloy correspondence to J. Mangles, 8 September 1838 (cited Pickering, 1929, p. 63).
87 Lines (1994, pp. 326–9, 332) and Webb (2003, p. 159).
88 Barker (1996, p. 1) and Erickson (1969, pp. 86–7).
89 Baulch (1961), Buchanan (1990), Burns & Skemp (1966), Hall (1978, p. 64), Moyal (1986, pp. 50–1) and Webb (2003, pp. 136, 161).
90 Barker & Barker (1990, p. 56), Buchanan (1990, pp. 180, 190) and Webb (2003, p. 161).
91 Reynolds (1995, pp. 117–9, 127–8) and Ryan (1996, pp. 110–12, 120, 143, 145–6, 149–51, 155).
92 R.W. Lawrence journal, 11 November 1830 (cited Burns & Skemp, 1961, p. 12).
93 R.C. Gunn correspondence to W.J. Hooker, 12 September 1834 (cited Burns & Skemp, 1961, pp. 34–5).
94 R.C. Gunn correspondence to W.J. Hooker, 12 September 1834 (cited Burns & Skemp, 1961, pp. 34–5).
95 Burns & Skemp (1961, p. 19), Moyal (1976, pp. 80–1) and Webb (2003, pp. 161–3, 166–7).
96 Barker & Barker (1990, p. 56), Buchanan (1990, p. 188), Moyal (1976, pp. 40–3) and Webb (2003, p. 181).
97 Baulch (1961, p. xvi), Holden & Holden (2001, p. 92) and Rolls (2002, p. 133).
98 Buchanan (1990, p. 188) and Webb (2003, p. 181). See Slater (1978, pp. 9, 14) for an account of the Tasmanian emu.
99 Gunn (1847). See discussion by Holden & Holden (2001, pp. 91–3).
100 For overviews of Tasmanian indigenous history, refer to Crowley (1996), Reynolds (1995) and Ryan (1996).
101 Baulch (1961, p. xvii), Buchanan (1990, p. 180) and Webb (2003, p. 169).
102 R.C. Gunn correspondence to W.J. Hooker, 15 February 1838 (cited Burns & Skemp, 1961, p. 71).
103 R.C. Gunn correspondence to W.J. Hooker, 15 February 1838 (cited Burns & Skemp, 1961, p. 71).
104 Gunn (1842, p. 36).
105 Gunn (1842).
106 Backhouse (1841). See comment by Webb (2003, p. 175). Buchanan (1990, p. 190) outlined the relationship between Gunn and Backhouse.
107 Gunn (1842, p. 35).
108 Gunn (1842, p. 44).
109 Gunn (1842, p. 36).
110 Gunn (1842, p. 48).
111 Jorgenson (1837) and Robinson (cited Plomley, 1966, 1987). For an overview of Tasmanian Aboriginal plant use, refer to Plomley & Cameron (1993).
112 Gunn (1842, p. 49).
113 Backhouse (1843, Appendix D, p. xxxi).
114 Backhouse (1843, Appendix D, p. xl).
115 Baulch (1961, pp. xvi, xix), Buchanan (1990, pp. 181, 191) and Moyal (1976, pp. 41–2; 1986, pp. 50–1).
116 Turrill (1963) provided a biography of J.D. Hooker. For J.D. Hooker's involvement with Charles Darwin, see Moyal (1986, p. 144) and Tyler (1980, p. 53).
117 Hall (1978, pp. 10–11), Johnston (1916), Marks (1969), Serle (1949, vol. 1, pp. 34–5) and White (1950).
118 Johnston (1916, p. 5) and White (1950, pp. 108–10) discussed Bailey's trips.
119 Bailey (1889, p. 3). See Meston (1889) for an official account of the Bellenden-Ker Ranges Expedition, 1899.
120 Bailey (1880, 1889, 1899–1902, 1913). Johnston (1916, pp. 7–10) and White (1950, pp. 110–12) provided bibliographies of Bailey's main botanical work.
121 Refer to chapter 10.
122 F.M. Bailey, *The Register*, Adelaide, 6 June 1914, p. 17.
123 Bailey (cited White, 1950, p. 106).
124 Brown (1993, p. 3734), Delbridge et al (1997, p. 2456), Dixon et al (1992, pp. 202, 221) and Ramson (1988, p. 751).
125 F.M. Bailey, *The Register*, Adelaide, 6 June 1914, p. 17.
126 Berndt (1981) and Clarke (2003a, pp. 47–9, 69–70; 2007, pp. 21, 38, 57, 67, 73–4, 76–7, 80, 85–8) gave summaries of foraging activities specific to women.
127 Clarke (1985a, 1986b, 1988, 1998b) and Gott (1982a, 1983, 1999) provided overviews of temperate Aboriginal use of roots as plant foods.
128 F.M. Bailey cited Maiden (1889, p. 37).
129 F.M. Bailey cited Gott (1983, p. 8).
130 Clarke (2007, pp. 16, 20, 38, 40, 73).
131 F.M. Bailey, *The Register*, Adelaide, 6 June 1914, p. 17.
132 Bailey referred to the ming as *Fusanus persicarius*, which is an obsolete name.
133 F.M. Bailey, *The Registerr*, Adelaide, 6 June 1914, p. 17. See also F.M. Bailey in the *Adelaide Observer*, 13 June 1914, p. 41.
134 Stephens (1890, p. 487). See also Worsnop (1897, pp. 4, 15).
135 F.M. Bailey, Early Adelaide. Peeps at Pioneering: Introduction of Plants, *The Register*, Adelaide, 6 June 1914, p.17.
136 Clarke (2003a, pp. 190, 215).

7 Leichhardt and the Riddle of Inland Australia

1 Favenc (1888, pp. 465–74) has provided a chronological summary of major expeditions of exploration in Australia.
2 Sturt (1833). See Hardy (1969, chapter 2).
3 Eyre (1845).
4 Lewis (2006, pp. 1–2).
5 Chisholm (1941), Erdos (1967), Hall (1978, p. 84), Hiddins (2000a), Moyal (1986, pp. 63–5), Pearn (1990, p. 83; 2001, pp. 224–6), Roderick (1988, part 1), Serle (1949, vol. 2, p. 28–30) and Short (2003, pp. 150–60).
6 Roderick (1988, pp. 106, 144, 146–7) described the connection between Leichhardt and Humboldt.
7 Barker & Barker (1990, pp. 59–61), Hoare (1969, p. 29) and Moyal (1986, pp. 63, 130).
8 P.B. Webb correspondence to W.J. Hooker, 1841 (cited Webb, 2003, p. 43). See also Roderick (1988, pp. 149, 151–2).
9 Barker & Barker (1990, p. 60).
10 Hoare (1969, p. 28) and Short (2003, p. 150).
11 Leichhardt correspondence, 3 April 1848 (cited Favenc, 1888, p. 417).
12 Moyal (1986, p. 63).
13 Leichhardt gave an account of Aboriginal use of the bunya pines in correspondence he sent to his mother, 27 August 1843 (Leichhardt, 1842–48 [1944, p. 31]). See also Roderick (1988, pp. 211, 216, 220–1).
14 F.D.L. Leichhardt correspondence to his mother, 27 January 1843 (Leichhardt, 1842–48 [1944, p. 26]).
15 Roderick (1988, pp. 215–6).
16 Roderick (1988, p. 216). Nicky received his Christian name when the German missionaries baptised him.
17 Roderick (1988, p. 216).
18 Roderick (1988, p. 216).
19 Smith (1793–95). See Aitken (2006, p. 152).
20 Brown (1810–30, 1814). See Barker & Barker (1990, p. 60) for a discussion of the botanical literature available to Leichhardt.
21 Hooker (1834–42).
22 Aitken (2006, p. 152, 155), Finney (1984, chapter 10), Moyal (1986, chapters 1–2) and Sharr (1996, pp. xxi-xxiii).
23 Maiden (1921a, p. 165).
24 Barker & Barker (1990, pp. 60–1), Froggatt (1932, p. 122) and Hoare (1969, p. 29).
25 Barker & Barker (1990, p. 60).
26 Leichhardt (1842–48 [1968, vol. 3, pp. 865–6]) for correspondence to D. Archer, dated 14 May 1846.
27 Leichhardt (1842–48 [1968, vol. 2, p. 683]) for letter to D. Archer, dated 24 November 1843. Leichhardt (1847a, p. 305) also compared other plant species to the 'bread-fruit,' referring to 'Another little tree, belonging to the Hamelieae D.C.' that he considered was related to the 'little bread fruit of the upper Lynd.'
28 Leichhardt (1842–48 [1968, vol. 2, p. 678]) for letter to D. Archer, dated 19 October 1843. Symons & Symons (1994, p. 16) gave another account of the bread-fruit ('buerwi'), although it is not identified.
29 Leichhardt (1847a, 1813–48). For John Gilbert's account, refer to Chisholm (1973).
30 Sturt (1849).
31 Leichhardt (1847a, p. xviii).
32 Chisholm (1973, p. 75).
33 Threlkeld (1824–1859 [1974, p. 46]). Elsewhere, his birth year was said to be 1820 (Threlkeld, 1824–1859 [1974, p. 315]).
34 Threlkeld (1824–1859 [1974, p. 315]). See Threlkeld (1824–1859 [1974, pp.

360–4]) for the lists of Aboriginal people living around Lake Macquarie in 1828 and 1833.
35 See Clarke (2003a, pp. 49–50) for an overview of Aboriginal name avoidance customs.
36 Threlkeld (1824–1859 [1974, p. 70]).
37 Threlkeld (1824–1859 [1974, p. 70]).
38 Threlkeld (1824–1859 [1974, p. 70]).
39 Hiddins (2000a, p. ix).
40 Leichhardt (1847a, p. 158).
41 Leichhardt (1847a, p. 12). See also Chisholm (1973, pp. 88–9).
42 Leichhardt (1847a, pp. 12, 52). See also Chisholm (1973, pp. 90, 101, 113).
43 Leichhardt (1842–48 [1968, vol. 3, pp. 915–6]) for letter dated 19 November 1846.
44 Leichhardt (1847a, pp. 103–4). See also Chisholm (1973, pp. 145–7).
45 Gilbert (cited Chisholm, 1973, p. 146).
46 Leichhardt (1842–48 [1968, vol. 3, p. 906]) for letter dated 27 September 1846.
47 Chisholm (1941, 1966a), Hall (1978, p. 59), Leichhardt (1847a, pp. 201–2) and Moyal (1986, pp. 54–63). Also Leichhardt (1842–48 [1968, vol. 3, pp. 841, 844–5, 880, 906, 965–6, 1038, 1041, 1046]) for letters dated 24 January 1846, 18 June & 27 September 1846, 21 October 1847. Barker & Barker (1990, pp. 71–2) discussed the relationship between Leichhardt and Gilbert. See Leichhardt (1847b) for a description of the landscape on the trip.
48 Erickson (1969, pp. 43, 45–7, 67, 75–9, 83–5, 89, 101, 111) and Moyal (1976, pp. 61–3; 1986, p. 54).
49 Chisholm (1966a) and Moyal (1986, pp. 55, 61).
50 J. Gilbert correspondence to J. Gould (19 September 1840, National Library of Australia).
51 Clarke (2007, p. 128) and Earl (1846, p. 240).
52 Moyal (1974, p. 62).
53 Leichhardt (1847a, p. 36). Similar references are Leichhardt (1847a, pp. 69, 151, 215, 218, 220, 226, 235–6, 240, 271, 288, 296–7, 301, 319, 336).
54 Leichhardt (1847a, p. 43).
55 Leichhardt (1847a, p. 144).
56 Leichhardt (1847a, p. 32).
57 Chisholm (1944).
58 McLaren (1996, pp. 205–23) and Symons & Symons (1997). Rose (1994) provided a contrary view, with which I disagree, claiming that explorers such as Leichhardt and Mitchell had made their remarks on Aboriginal burning practices from the generalisations gained from British farmers, rather than from their own experience. Jackes (1990) discussed the botany of the 1844–45 expedition.
59 Leichhardt (1847a, pp. 219–20, 338–44).
60 Leichhardt (1847a, p. 6).
61 Leichhardt (1847a, p. 22).
62 Leichhardt (1847a, pp. 83, 200, 205, 218, 259, 336).
63 Leichhardt (1847a, p. 287).
64 Leichhardt (1847a, pp. 27, 145, 162, 178, 190–1, 194, 231, 267, 278, 305, 309–10, 312, 315, 322, 334, 344–5). See Cooper & McLaren (1997).
65 Leichhardt (1847a, p. 60). Similar references are Leichhardt (1847a, pp. 76, 98–9, 106, 122, 128, 176, 182, 185–7, 189–90, 199, 205, 213, 223, 234, 259, 265, 304, 333).
66 For accounts of the Aboriginal use of the long yam, refer to Bindon (1996, p. 102), Hiddins (2001, p. 134), Levitt (1981, pp. 41, 136–7), Puruntatameri (2001, pp. 8, 44, 143–4) and Stewart & Percival (1997, p. 19). The long yam is sometimes referred to as the 'pencil yam'.
67 Leichhardt (1847a, p. 71). See also Leichhardt (1847a, pp. 271–2, 292, 297). Brennan (1986, p. 101) provided an identification of plant.
68 Leichhardt (1847a, p. 215). See also Maiden (1900, p. 118).
69 Jackes (1990, p. 166) provided a list of the fruits tried by Leichhardt.
70 Gilbert, November 1844 (cited Chisholm, 1973, p. 103).
71 Leichhardt (1847a, p. 27).
72 Leichhardt (1847a, pp. 325–6). See also Leichhardt (1842–48 [1968, vol. 3, pp. 884, 890]) for letters dated 24 January & 12 August 1846. The eugenia was probably *Eugenia reinwardtiana*. The identification of 'Allamurr'/'Murnatt' as the water chestnut was provided by K. Akerman (pers. comm.). Earl (1846, p. 245) described the importance of these rush roots, 'marwait', as an Aboriginal source of root vegetable.
73 Leichhardt (1847a, p. 243). The *Terminalia* species was probably *T. hadleyana* subsp. *carpentariae*. See Bindon (1996, p. 245).
74 Leichhardt (1847a, pp. 161, 168). See also Leichhardt (1842–48 [1968, vol. 3, pp. 871, 890]) for letters dated 20 May & 12 August 1846. Maiden (1900, p. 118) referred to Leichhardt's account.
75 Brennan (1986, p. 90), Crawford (1982, pp. 2, 17, 21–2, 48–9), Hiddins (2000b, p. 25), Levitt (1981, pp. 41, 111–12), McConnel (1930, pp. 8, 20; 1953, p. 7), Puruntatameri (2001, pp. 8, 69), Rae et al (1982, pp. 45, 47), Roth (1901b, p. 14), Specht (1958, p. 490), Sutton (1994, p. 37) and Wightman, Roberts et al (1992, p. 36).
76 Leichhardt (1847a, p. 194).
77 Blake et al (1998, pp. 100–1), Hiddins (2001, pp. 1, 6), Levitt (1981, pp. 36, 111–12), McConnel (1930, pp. 8, 20; 1953, pp. 7, 41), Puruntatameri (2001, p. 69), Roth (1901b, p. 14), Sutton (1994, p. 37), Symons & Symons (1994, p. 83), Wightman, Roberts et al (1992, p. 36) and Yunupingu (1995, p. 59).
78 Leichhardt (1847a, pp. 263–4). Also see Bunce (1859a, pp. 128–9).
79 Leichhardt (1847a, pp. 259–60, 263–4). See Bindon (1996, p. 191).
80 For example, refer to J. Gilbert (cited Chisholm, 1973, p. 85) and Leichhardt (1847a, pp. 41, 121, 163).
81 Leichhardt (1847a, p. 152).
82 Leichhardt (1847a, pp. 140, 170–1, 195, 202, 316).
83 Leichhardt (1847a, p. 160).
84 Leichhardt (1847a, pp. 182, 194) and Leichhardt (1842–48 [1968, vol. 3, p. 843]) for letter dated 24 January 1846.
85 Leichhardt (1847a, p. 30).
86 Chisholm (1973, p. 104).
87 Leichhardt (1847a, pp. 214, 219, 229, 268, 289, 302, 314, 326, 334–5).
88 For examples of Mitchell's party providing metal objects to Aboriginal people during his River Darling expedition of 1835, refer to Mitchell (1838, vol. 1, pp. 193, 204, 212, 214, 216, 219, 221, 225, 235, 247–51, 258, 260, 264, 266–7, 287–8, 292–3, 295–6, 304–5).
89 Clarke (2003a, p. 181).
90 Leichhardt (1847a, p. 214).
91 Berndt & Berndt (1954), Clarke (2003a, pp. 177–183, 186), Macknight (1976) and Urry & Walsh (1981).
92 Leichhardt (1847a, p. 210). Leichhardt stated that 'Mareka' would have been equivalent to 'Marega' recorded by King (1827, vol. 1, p. 135).
93 Flinders (1814b [2000, pp. 203–8]).
94 Attenbrow (2003, pp. 102–3, 124–5) discussed the Aboriginal use and trade of materials derived from European sources in the Sydney region. Clarke (2003a, chapter 7) gave a general overview of Aboriginal trading practices.
95 Akerman (1980), Clarke (2003a, pp. 107–11), McBryde (1986; 1987), McCarthy (1938–40) and Specht & White (1978).
96 Clarke (2003a, pp. 100–1).
97 Leichhardt (1847a, p. 314).
98 Chisholm (1966a; 1973, pp. 203–8), Erikson (1969, p. 111), Leichhardt (1847a, pp. 201–3) and Moyal (1986, pp. 55, 63). According to Jackes (1990, p. 168), Gilbert was struck in the neck, rather than the chest.
99 B.J. Dalton (1988, cited Jackes, 1990, p. 168).
100 Roderick (1988, pp. 328–9).
101 Leichhardt (1847a, p. 202). Jackes (1990, p. 168) provided a photograph of Gilbert's gravesite.
102 Leichhardt (1847a, p. 212). Jackes (1990, p. 169) gave the plant identification for the raspberry jam tree, using the generic name *Canthormium* (now called *Cathormion*).
103 Leichhardt (1847a, p. 215).
104 Leichhardt (1847a, pp. 345–7).
105 Froggatt (1932, p. 124) and Moyal (1986, pp. 63, 65).
106 L. Leichhardt correspondence with G. Durando (20 May 1846, Mitchell Library, Sydney).
107 Favenc (1888, pp. 164–5)
108 Leichhardt (1842–48 [1968, vol. 3, pp. 933, 937]) for letters dated 1 & 2 August 1847.
109 Leichhardt (1842–48 [1968, vol. 3, pp. 995–6]) for letter dated 22 February 1848. See also Bunce (1859a, pp. 86, 92).
110 Leichhardt (1842–48 [1968, vol. 3, pp. 914–5, 946]) for letters dated 9 & 19 November 1846 & 20 October 1847.
111 Leichhardt (1842–48 [1968, vol. 3, p. 937]) for letter dated 2 August 1847.
112 Leichhardt (1842–48 [1968, vol. 3, p. 912]) for letter dated 9 November 1846.
113 Leichhardt (1842–48 [1968, vol. 3, pp. 995–6; see also pp. 1026–8]) for letter dated 22 February 1848.
114 Bunce (1859a, pp. 121–2).
115 Howitt (1904, p. 313). For other accounts of Aboriginal body scarring practices, see Collins (1798–1802, vol. 1, p. 552), Elkin (1964, pp. 195–9,

202, 204–5), Foelsche (1895, p. 194), Howitt (1904, pp. 743–6) and Worsnop (1897, pp. 2, 163–4).
116 Favenc (1888, p. 165).
117 Bunce (1859a, pp. 102, 121). See Holden (1966) for an account of Bunce's relationship with Leichhardt.
118 Connell (1980), Favenc (1888, pp. 165–9) and Leichhardt (1842–48 [1968, vol. 3, pp. 979–1013, 1029–30]).
119 Connell (1980, pp. 20–1).
120 Leichhardt correspondence, 3 April 1848 (cited Favenc, 1888, p. 417).
121 Threlkeld (1824–1859 [1974, p. 70]).
122 Connell (1980, chapter 9). Roderick (1988, pp. 499–505) outlined the various theories about Leichhardt's disappearance.
123 Roderick (1988, p. 452) described the Leichhardt family tree. Some sources give Classen's Christian name as 'Adolf' rather than 'August'. The surname 'Classen' is sometimes spelled 'Klausen'.
124 Favenc (1888, p. 166).
125 Connell (1980, chapter 2).
126 Lewis (2006) has summarised the various Leichhardt search expeditions.
127 Lewis (2006, pp. 15–17).
128 Everist (1981, pp. 31–3, 638–9), Foley (2006, pp. 35–6, 48–57) and Ohlendorf (1996).
129 Bindon (1996, p. 183), Cribb & Cribb (1981, p. 79) and Lassak & McCarthy (1983, p. 89).
130 M. Aurousseau (Introduction of Leichhardt, 1842–48 [1968]) and Hoare (1969, p. 27).

Chapter 8 Von Mueller and Australian Botany

1 Pearn (1990, p. 98; 2001, pp. 272–4) and Ross (1996).
2 Daley (1924, pp. 5–6), Hall (1978, pp. 97–8), Maroske (1996), Morley & Toelken (1983, pp. 15–6), Morris (1974), Moyal (1976, pp. 172–3; 1986, pp. 149–53), Serle (1949, vol. 2, pp. 167–70), Short (2003, pp. 161–66), Webb (2003, chapter 10) and Willis (1949, chapters 1–4). Veitch (1981) wrote a novel based on von Mueller's later life.
3 Ritchie (1989, pp. 51–2).
4 Webb (2003, chapter 10).
5 Home et al in von Mueller (1825–96 [1998–2006, vol. 2, p. 31]).
6 F. von Mueller correspondence to R. Wagner, 25 January 1862 (von Mueller, 1825–96 [1998–2006, vol. 2, pp. 31–2, 128–9]).
7 F. von Mueller correspondence to W.J. Hooker, March 1855 (von Mueller, 1825–96 [1998–2006, vol. 1, pp. 201–2]). Plant identified by Zola & Gott (1992, p. 35).
8 Von Mueller (1856, p. 336). The gunyang was later mentioned in von Mueller's list of Victorian Aboriginal plant foods (Von Mueller [cited Smyth, 1878, vol. 1, p. 213]).
9 See chapter 3 of this volume.
10 Mueller (cited in Smyth, 1878, vol. 1, pp. 212–4). Orchard (1997) discussed von Mueller's plant collecting network in the Murray Basin.
11 Backhouse (1841), Dawson (1881), Gunn (1842).
12 Clarke (1985a, 1985b, 1986a, 1986b, 1987, 1988, 1998b), Gott (1982a, 1982b, 1984, 1985a, 1985b, 1999), Gott & Conran (1991), Hope & Coutts (1971), Morris (1943) and Zola & Gott (1992).
13 Archer & Maroske (1996, p. 193).
14 Archer & Maroske (1996, p. 191) and Sharr (1996, p. 95). See also Darragh (1996) and Gillbank (1996a&b).
15 Archer & Maroske (1996, p. 192).
16 *The West Australian*, 1883 (cited Archer & Maroske, 1996, p. 190).
17 Archer & Maroske (1996, p. 192).
18 Hall (1978, pp. 53–4), Maiden (1921, p. 154) and Noye (1972).
19 Jones (2007, pp. 160–1). See Sayers (1996, pp. 81–2, 115) for a brief biography of Biliamuk Gapal (Billiamook).
20 Maiden (1921, p. 159).
21 Refer to Jones (1996, pp. 216–24) for a history of Foelsche and his connections with the South Australian Museum.
22 Foelsche (1881, p. 9). Record appears to be associated with South Australian Museum ethnographic specimens, A39001–4.
23 Clarke (2007, pp. 26, 101, 123).
24 Foelsche (1895, pp. 194–5).
25 Foelsche (1895, p. 197).
26 E.C. Stirling (cited by the editor [Foelsche, 1895, p. 190]). The editor gave the italic emphasis within the quote.
27 Hall (1978, p. 54) and Sharr (1996, p. 127).
28 Jordan (2005, p. 144).
29 For a discussion of gender and nineteenth century botany, refer to Shteir (1996).
30 Moyal (1986, p. 114).
31 Collins (1988). The Ellis Rowan collection is held at the National Library of Australia, Canberra. Samuel (1961) wrote a fictionalised account of her life.
32 Hazzard (1982, 1988) and Moyal (1986, pp. 115–6).
33 See drawing in Samuel (1961, p. 127).
34 Rowan (1898, p. 89).
35 Rowan (1898, p. 95).
36 Rowan (1898, p. 122).
37 Rowan (1898, pp. 90–1).
38 Rowan (1898, pp. 123–4). Rowan wrote the island's name as 'Marbiag'.
39 Hewson (1982, p. 17).
40 Hazzard (1982, p. 8) and Lounsberry (1899, 1900, 1901).
41 Hazzard (1988).
42 Kempe (1880–82). See Daley (1924, p. 52), Gillbank (1996a, p. 225) and Gillbank & Maroske (1996, pp. 216–7).
43 H. Kempe correspondence to F. von Mueller, 20 January 1881 (von Mueller, 1825–96 [1998–2006, vol. 3, pp. 202–4]). Species identification via Eastern and Central Arrernte name (Henderson & Dobson, 1994, p. 365).
44 Kempe (1880–82, p. 133). Kempe used a now obsolete botanical name, *Canthium latifolium*.
45 Kempe (1880–82, p. 22). Kempe used a now obsolete botanical term, *Marsdenia Leichhardtiana*.
46 Kempe (1891, p. 1).
47 Beard (1981, p. xv), Gillbank & Maroske (1996), Morley & Toelken (1983, pp. 15–6), Short (2003, pp. 167) and Tietkens (1891, pp. 14, 27–31).
48 Gillbank (1996a, pp. 219, 223–5).
49 Gillbank (1996a, p. 224) and Short (2003, p. 180).
50 Short (2003, pp. 172–3, 178–80). Everist (1981a, p. 368) and Giles (1889, vol. 2, pp. 142–4, 310–13) specifically discussed the camel poison.
51 See chapter 9 of this volume.
52 Cohn (1996), Daley (1924, pp. 15–16) and von Mueller (1858b).
53 Willis (1949, p. 35).
54 Parkin (1996a, p. 170).
55 Von Mueller correspondence to W.J. Hooker, dated 14 January 1857 (cited Short, 2003, p. 166). See also Cohn (1996, p. 165).
56 Von Mueller (1858b, p. 143).
57 Von Mueller (1858b, pp. 143–4).
58 Von Mueller (1858b, pp. 142–4).
59 Von Mueller (1858b, p. 142).
60 Von Mueller (1858b, p. 138). For references to Leichhardt's use of water lily tubers, see Leichhardt (1847a, pp. 161, 168, 194).
61 Daley (1924, p. 14).
62 Webb (2003, pp. 240–5).
63 Sharr (1996, p. 165).
64 Daley (1924, p. 33) and Sharr (1996, pp. 164–5).
65 Labilliere (cited Daley, 1924, p. 41).
66 Daley (1924, p. 46).
67 Von Mueller (1858a, 1876, 1888). McPhee (1996) and Webb (2003, pp. 247–8) discussed von Mueller naturalisation projects.
68 Low (1989, p. 73). See also Wakefield (1959, 1961).
69 Arbury (1998).
70 Cribb & Cribb (1975, pp. 53–5), Gott & Conran (1991, p. 35), Isaacs (1987, pp. 76, 228, 238), Low (1989, pp. 41, 64, 72–3, 220–1) and Zola & Gott (1990, pp. 37, 49. Plants commonly known as 'native raspberry' include *Rubus moluccanus*, *R. moorei* and *R. parvifolius*.
71 Von Mueller (cited Smyth, 1878, vol. 1, p. 213).
72 Hogan (cited Smyth, 1878, vol. 1, p. 210) and Isaacs (1987, pp. 76, 228).
73 Isaacs (1987, p. 238).
74 Von Mueller (cited Maiden, 1889, p. 55).
75 Maiden (1889, p. 55). See also Gunn (1842, pp. 37–8).
76 Moyal (1986, p. 104) and Rix (1978, chapter 1).
77 Moorehead (1963, p. 39, see also p. 36).
78 Moorehead (1963, p. 165).
79 Giles (1889, vol. 1, pp. 29, 84, 157, 262; vol. 2, pp. 202–3, 226–7) and Jessop (1981, pp. xv–xvi).
80 Paddle (1996).
81 Parkin (1996b, p. 216; 1996c).
82 Von Mueller (1888, p. 1).
83 Sharr (1996, pp. xxiv–xxv), Webb (2003, p. 248) and Willis (1990, p. 2). See also von Mueller (1888, pp. iv–v).

84 Von Mueller (1888, p. 135). He referred to the warran yam as *Dioscorea hastifolia*.
85 Backhouse (1843, p. 540).
86 Grey (1841, vol. 2, pp. 11–12) for a description of the warran.
87 Hallam (1975, p. 12; see also pp. 13–14, 72–4).
88 Grey (1841, vol. 2, p. 124). See also Bates (1901–14, [1985, p. 261]), Bindon (1996, p. 101), Carr & Carr (1981b, p. 14), Hallam (1975, pp. 72, 74), Low (1989, p. 106), Roth (1903b, p. 48) and von Brandenstein (1988, p. 131).
89 Tietkens (1891, pp. 1, 8, 11) remarked on the perceived need for date palms being established around permanent waterways in northern and Central Australia.
90 Von Mueller (1888, p. 290).
91 Von Mueller (1888, p. 291). See also F. von Mueller correspondence to W. Gill, 19 April 1894 (von Mueller, 1825–96 [1998–2006, vol. 3, p. 674]).
92 Jessop (1981, p. 495), Jessop & Toelken (1986, vol. 4, p. 1995) and Low (1999, pp. 23, 329). See *Across the Outback* newsletter (January 2007, no. 29, p. 1) for an article concerning a Aboriginal community project for removing date palms from parts of the Dalhousie Mound Springs complex in the Witjira National Park of northern South Australia.
93 Moyal (1976, pp. 174, 187, 193; 1986, pp. 145–6, 153) and Webb (2003, pp. 245–6).
94 Darwin (1878, p. 337).
95 Maroske (2006, p. 154).
96 Sharr (1996, pp. 125, 165–6) and Webb (2003, p. 247).
97 Brand Miller et al (1993, pp. 17, 186–8), Brennan (1986, p. 56), Hiddins (2000b, p. 26; 2001, pp. viii, 87), Lands (1987, pp. 11, 18), Puruntatameri (2001, pp. 8, 85, 142), Rae et al (1982, p. 46), Smith & Kalotas (1985, p. 346), Smith & Wightman (1990, p. 25), Sutton (1994, p. 38), Wightman, Roberts et al (1992, p. 46) and Wightman & Smith (1989, p. 20).
98 Smith & Kalotas (1985, p. 337).
99 Bentham (1863–78). See discussion by Barker & Barker (1990, pp. 66–7) and Webb (2003, chapter 9 & pp. 236, 256).
100 Thozet (1866, p. 3).
101 Sharr (1996, p. 165).

Chapter 9 Inland Explorers and Aboriginal Knowledge

1 Swayne (1868).
2 Quote comes from a Moorehead (1963) chapter title, 'The Ghastly Blank'.
3 Thozet (1866, p. 5). Froggatt (1932, pp. 129–30) provided a biography of Thozet.
4 Beckler (1859–62), Blainey (1976, pp. 218–22), Bonyhady (1991), Favenc (1888, chapter 9 & 10), Moorehead (1963), Murgatroyd (2002) and Wills (1863).
5 Clark (1987, pp. 138–9) and Day (2001, pp. 108–9).
6 Refer to the *Progress Reports and Final Report of the Exploration Committee of the Royal Society of Victoria, 1863*. Mason & Firth Printers, Melbourne.
7 Moyal (1976, p. 172 & chapter 11; 1986, pp. 149, 161).
8 John Macadam transcript of evidence (Burke & Wills Commission, 1862, p. 13).
9 Moorehead (1963, chapter 4).
10 Moorehead (1963, p. 56).
11 Moorehead (1963, chapter 5).
12 Moorehead (1963, pp. 56–8).
13 Moorehead (1963, p. 103). McPherson's name was sometimes written as Macpherson and MacPherson. The standard adopted in this book is the spelling used by the Burke & Wills Commission (1862).
14 Moorehead (1963, p. 103) and Willis (1962, p. 252).
15 Edward Wecker transcript of evidence (Burke & Wills Commission, 1862, p. 62) and Dr Hermann Beckler transcript of evidence (Burke & Wills Commission, 1862, pp. 160, 163, 184).
16 Herman Beckler, 27 December 1860 (cited Spencer, 1918, p. 13). *Marsilea* was sometimes written as *Marsilia*.
17 Moorehead (1963, p. 105)
18 Newland (1889, p. 4).
19 Newland (1889, p. 8).
20 Moorehead (1963, pp. 65–8 & chapter 6).
21 William Brahé transcript of evidence (Burke & Wills Commission, 1862, p. 34).
22 William Wills, on a Sunday in February 1861 (cited by Burke & Wills Commission, 1862, p. 229).
23 For a description of the Aboriginal use of pencil yams, refer to Bindon (1996, p. 262), Clarke (2007, pp. 15, 28, 82) and Latz (1995, pp. 296–7).
24 William Wills, 21 April 1861 (cited by Burke & Wills Commission, 1862, p. 238). See also Moorehead (1963, pp. 88, 99, 144). For other references to the use of portulaca, see William Wills, 25 March, 8 April, 31 May 1861 (cited by Burke & Wills Commission, 1862, pp. 235–6, 238, 246). See also John King transcript of evidence (Burke & Wills Commission, 1862, pp. 84–5).
25 For Cook's use of portulaca, see Maiden (1900, p. 122). Von Mueller (1858b, p. 143) described its use on Gregory's expedition.
26 Clarke (2007, pp. 13, 83, 92), Maiden (1900, pp. 121–2) and Smyth (1878, vol. 1, pp. 213–4).
27 Smyth (1878, vol. 1, p. 214).
28 William Wills, 17 April 1861 (cited by Burke & Wills Commission, 1862, p. 237).
29 William Brahé transcript of evidence (Burke & Wills Commission, 1862, pp. 35–7). See also Moorehead (1963, pp. 86–7).
30 Earl & McCleary (1994, p. 684), Hagger (1979, pp. 93–4) and Ramson (1988, p. 37).
31 Thomas McDonough transcript of evidence (Burke & Wills Commission, 1862, p. 51; see also p. 42). William Wright's dairy (Burke & Wills Commission, 1862, p. 188) does contain a reference to 'preserved vegetables'.
32 John King transcript of evidence (Burke & Wills Commission, 1862, p. 75).
33 John King transcript of evidence (Burke & Wills Commission, 1862, p. 94) and William Brahé transcript of evidence (Burke & Wills Commission, 1862, p. 214).
34 John King transcript of evidence and his narrative (Burke & Wills Commission, 1862, pp. 92–4, 252). See also Moorehead (1963, chapter 10).
35 William Wills, 21 April 1861 (cited by Burke & Wills Commission, 1862, p. 238).
36 William Wills, 7 May 1861 (cited by Burke & Wills Commission, 1862, p. 242). See also Moorehead (1963, pp. 118–9). For a description of the use of pituri, see Clarke (2007, pp. 106, 108–9, 125).
37 John King transcript of evidence (Burke & Wills Commission, 1862, p. 253). See also Moorehead (1963, p. 140).
38 William Wills, 7 May 1861 (Burke & Wills Commission, 1862, pp. 242–3). See also Murgatroyd (2002, pp. 255, 257).
39 William Wills, 10 May 1861 (Burke & Wills Commission, 1862, p. 243).
40 Maiden (1889, p. 119) described *Atriplex holocarpa* and *A. spongiosa* (pop saltbush) as forage plants.
41 Reuther (1981, entry nos 2756a63 & 387). Koch (1898, p. 110) interpreted this term, written in the form of *paldroo*, as a species of 'native spinach', although this plant does not match the description of either the Wills or Reuther species. See also Gray (1997, p. 125).
42 William Wills, 11, 13, 17 May 1861 (Burke & Wills Commission, 1862, pp. 243–4).
43 William Wills, 24 May–26 June 1861 [sic: 28 June 1861] (Burke & Wills Commission, 1862, pp. 245–51).
44 John King narrative (Burke & Wills Commission, 1862, p. 253).
45 Clarke (2007, pp. 58, 84–6, 88) and Thomson (1962, p. 13).
46 William Wills, 26 June 1861 [sic: 28 June 1861] (Burke & Wills Commission, 1862, p. 253). Also cited by Moorehead (1963, p. 143, also see pp. 119–20, 141–2, 145).
47 John King narrative (Burke & Wills Commission, 1862, pp. 254–5). See also Moorehead (1963, p. 145).
48 John King narrative (Burke & Wills Commission, 1862, p. 255). See also Moorehead (1963, p. 146).
49 Allen (1974), Clarke (2007, pp. 75–6), Horne & Aiston (1924, p. 7), Lawrence (1968, p. 57), Tindale (1977, p. 346) and Tunbridge (1985a, p. 14).
50 Wills (1863). See also Bonyhady (1991), Moorehead (1963) and Murgatroyd (2002).
51 E.J. Welch (cited Favenc, 1888, pp. 419–24) and John King's narrative (Burke & Wills Commission, 1862, pp. 251–6). See also Moorehead (1963, pp. 147–8).
52 John King's narrative (Burke & Wills Commission, 1862, pp. 255–6).
53 Alfred W. Howitt transcript of evidence (Burke & Wills Commission, 1862, pp. 101–6). Sir William Foster Stawell transcript of evidence (Burke & Wills Commission, 1862, p. 142).

54 E.J. Welch (cited Favenc, 1888, p. 421). See also Moorehead (1963, pp. 138–9).
55 Earl & McCleary (1994, p. 684).
56 Moorehead (1974).
57 Moorehead (1963, chapters 11 & 12) and Murgatroyd (2002, pp. 329–30). Refer to Stanner (1972) for a biography of Howitt.
58 Alfred W. Howitt transcript of evidence (Burke & Wills Commission, 1862, p. 104). The 'wild spinach' is possibly either Chenopodium auricomum (see description of this species by Maiden, 1889, pp. 15–6) or a species of Tetragonia.
59 Willis (1962, pp. 254–5).
60 Bonyhady (1991, p. 27).
61 Howitt (in Smyth, 1878, vol. 2, pp. 302–3). Note that a bushel is a dry measure in the imperial system, equivalent to 8 gallons or 36.4 litres.
62 Moorehead (1963, p. 137).
63 Moorehead (1963, pp. 152–3).
64 Burke & Wills Commission (1862, pp. 4–5).
65 Bailey (1880, p. 8). Cited by Maiden (1889, p. 57). There is no mention of Bailey's theory in the Sesbania and Marsilea entries of The Queensland Flora (Bailey, 1899–1902, part 2, pp. 398–9 & part 6, p. 1929).
66 Zarkawi et al (2005).
67 T.A. Gulliver (cited Bailey, 1880, p. 8).
68 Bancroft (1884, p. 107).
69 Maiden (1889, pp. 42, 57).
70 Bancroft (1894, p. 215).
71 Koch (1898, p. 117).
72 Howitt & Siebert (1904, p. 121). See also the Mura-Mura mythology of Ngardu-etya (Ngarduetya) (Howitt, 1904, pp. 794–5; Howitt & Siebert, 1904, pp. 121–3). Kimber (1984, p. 17) presented an overview of other Nardoo Dreaming accounts from Central Australia.
73 Refer to 'Notes' in The Victorian Naturalist, vol. 27, no. 1, p. 16 [5 May 1910].
74 Bonyhady (1991).
75 Lees (1915, p. 134).
76 Lees (1915, p. 135).
77 Lees (1915, p. 135).
78 Howitt & Siebert (1905, p. 122). This fact was confirmed later by Johnston & Cleland (1943, pp. 153, 155).
79 Spencer (1918). Identification also later confirmed by Newland (1922, pp. 12–4).
80 Spencer (1918, pp. 10–11).
81 Spencer (1918, p. 12).
82 Bonyhady (1991, pp. 137–9, 411, 148, 177, 195–6, 206, 257, 290, 297), Daley (1931, p. 29), Kynaston (1981, pp. 193–5), Moorehead (1963, pp. 19, 61, 82, 103, 117–23, 140–2, 145–8, 152, 165), Murgatroyd (2002, pp. 151, 253, 255), Smyth (1878, vol. 1, pp. 216–7), Spencer (1918) and Thomas (1906, p. 114).
83 Dixon et al (1992, p. 120), Maiden (1889, pp. 42–4, 135), Ramson (1966, pp. 115–6; 1988, p. 418) and Spencer (1918, vol. 15, p. 9). Other spelling variations in Australian English, which as a form of speaking tends to not recognise the initial 'ng', are 'nardu', 'gnardu', 'ardo' and 'ardoo'.
84 Bates (1918, p. 160), Cleland (1957, pp. 153–4), Cleland & Johnston (1937–38, p. 335), Cleland et al (1925, p. 108), Favenc (1888, p. 418), Gason (1879, p. 76), Johnston & Cleland (1943, p. 151–2), Kemsley (1951, pp. 339–40), Kimber (1984, pp. 19–20), Riches (1964), Roth (1897, p. 92), Spencer (1918) and Worsnop (1897, pp. 81–2). There are two species: common nardoo and short-fruited nardoo.
85 Howitt (in Smyth, 1878, vol. 2, p. 302).
86 Basedow (1925, p. 150), Bonyhady (1991, pp. 138, 141), Cherikoff & Isaacs (1989, pp. 131, 184, 190, 195), Earl & McCleary (1994), Horne & Aiston (1924, pp. 52–7, 176), Isaacs (1987, pp. 115, 225), Low (1989, pp. 85, 87, 212) and Murgatroyd (2002, pp. 257–8, 261–3).
87 Earl & McCleary (1994), Everist (1981a, pp. 772–5), McCleary & Chick (1977) and Moran (2004).
88 Low (1989, p. 87). Brand Miller et al (1993) omitted nardoo from their survey of the nutritional composition of wild Australian foods, probably because of its poor reputation.
89 Horne & Aiston (1924, p. 52–3).
90 Howitt, 1861 (cited Spencer, 1918, p. 14).
91 S.T. Gason (cited Smyth, 1878, vol. 1, p. 216).
92 Bedford (1887, pp. 104–5).
93 Bancroft (1894, p. 216).
94 Earl & McCleary (1994, p. 684).
95 Clarke (2007, pp. 58, 84–6, 88) and Thomson (1962, p. 13).
96 Clarke (2003a, pp. 83–4, 110, 120, 144–8, 153), McBryde (1987, pp. 271–3), McCourt (1975, pp. 135–40), Newland (1889, p. 5), Tindale (1977) and Worsnop (1897, p. 98).
97 Howitt (cited Maiden, 1889, p. 44).
98 Lees (1915, p. 133).
99 Reuther (1981, entry no. 2767b).
100 Newland (1922, pp. 12–14).
101 Smyth (1878, vol. 1, p. 209). This statement appears inconsistent with data that Smyth (1878, vol. 1, p. 214) provided from Dr Gummow.
102 Von Mueller (cited Kynaston, 1981, p. 193). See also Beckler (1859–62 [1993, pp. 64–5, 74–5]), Bonyhady (1991, pp. 147–9) and Moorehead (1963, p. 103).
103 Moorehead (1963, pp. 31, 56).
104 Gillbank (1996a, pp. 220–1), Moorehead (1963, p. 105), Pearn (2001, pp. 68–70) and Willis (1962, pp. 251–3, 255–6).
105 Von Mueller (1862, cited Smyth [1878, vol. 1, p. 216]).
106 Bancroft (1884, p. 105).
107 Maiden (1889, p. 135; see also p. 42).
108 Thomas (1906, p. 114).
109 Daley (1931, p. 29).
110 F. Chapman (cited Spencer, 1918, vol. 14, p. 171; exhibits listed on vol. 14, p. 172 & vol. 15, p. 12).
111 Ken Barratt (cited Bonyhady, 1991, p. 141).
112 Kemsley (1951, p. 440).
113 Moorehead (1963, p. 1).
114 For examples refer to Cleland (1966), Hyam (1943) and Morris (1943).
115 Wilhelmi (1861, p. 172).
116 Published form of the paper is Crawfurd (1868).
117 Crawfurd (1868, p. 112) and Thozet (1866). Smyth (1878, vol. 1, pp. 227–34) reproduced Thozet's list.
118 George Bentham, cited Crawfurd (1868, p. 116).
119 Crawfurd (1868, p. 116).
120 Crawfurd (1868, p. 113).
121 East (1889, pp. 1–2).
122 For examples refer to Cleland (1966), Hyam (1943) and Morris (1943).
123 Murgatroyd (2002, pp. 261–2).
124 Coleman (1938, pp. 37–8).
125 Earl & McCleary (1994, p. 683) and McCleary & Chick (1977).
126 William Wills, 31 May 1861 (Burke & Wills Commission, 1862, p. 245). See also Earl & McCleary (1994, p. 683) and Murgatroyd (2002, p. 263). For thiaminase in mussels, refer to McCleary & Chick (1977).
127 E.J. Welch (cited in 'Notes', The Victorian Naturalist, vol. 27, no. 1, p. 16 [5 May 1910]).
128 Crowley (1981), Forrest (1875) and McLaren (1996, pp. 247, 250).
129 Wood (1981, chapter 3).
130 Crowley (1976) and Wood (1981, chapter 1). Njaki-Njaki has also been written as 'Njaggi Njaggi' and Kalaamaya is sometimes spelled as 'Kalarmai'.
131 Roberts (1972) and Wood (1981, chapter 2).
132 Forrest (1875, p. 17).
133 Forrest (1875, pp. 22–4, 45, 52, 55–6, 60–1, 63, 65–6, 71–2), Reynolds (1990, p. 25) and Wood (1981, chapter 3).
134 Forrest (1875, pp. 26–30).
135 Forrest (1875, p. 71) and Hall (1978, p. 54).
136 Forrest (1875, pp. 77, 80, 85, 92–3, 110, 145) and Wood (1981, chapter 4).
137 Forrest (1875, pp. 157, 185, 195, 203, 206, 228, 232, 236, 241, 244, 248, 251, 255–6, 266–7, 274) and Wood (1981, chapter 6).
138 Favenc (1888, p. 262), Forrest (1875, map facing p. 148) and Wood (1981, pp. 86–9).
139 Forrest (1875, p. 180).
140 Wood (1981, p. 115).
141 Wood (1981, pp. 113–7).
142 Portman (1989).
143 Forrest (1875, pp. 263, 266, 325–7).
144 Forrest (1875, pp. 227, 231).
145 Albrecht (1959), Bindon (1996, p. 143), Bindon & Gough (1993, p. 14), Clarke (2007, pp. 75, 78), Cleland (1957, p. 154; 1966, p. 149), Cleland & Johnston (1933, p. 118; 1937–38, pp. 209, 214, 336, 339), Cleland & Tindale (1954, pp. 82–3; 1959, p. 131), Finlayson (1952, p. 64), Goddard (1992, p. 16), Hiddins (2000, p. 32; 2001, p. 42), Latz (1995, pp. 196–7), Mutitjulu Community & Baker (1996, p. 38), O'Connell et al (1983, pp. 93, 95), Tindale (1941a, p. 9) and Turner (1994, p. 14).
146 Forrest (1875, p. 226).
147 McLaren (1996, pp. 247–8). Clarke (2003a, pp. 140–4; 2007, pp. 79–84), Gara (1985) and Magarey (1894–95) provided overviews of Aboriginal methods of finding water.
148 Carnegie (1898, pp. 188–9).
149 Wells (1902, p. 59).

Chapter 10 The Study of Aboriginal Plant Use

1 Hiatt (1996) provided an overview of the history of Australian anthropology.
2 For a history of Australian botany, refer to Orchard (1999) and Short (1990, 2003).
3 Refer to chapter 2 of this volume.
4 Hagger (1979, chapter 13).
5 Mackerras (1969) and Pearn (1990, p. 12; 2001, pp. 54–7).
6 Bancroft (1872; 1877–80).
7 Bancroft (1877–80, pp. 4–8, 11–2, 14–19), Basedow (1925, pp. 155–7), Bedford (1887, pp. 110–12), Cleland (1936, pp. 7–8; 1939, p. 13; 1957, pp. 158–9; 1966, p. 119), Cleland & Johnston (1933, p. 116; 1937–38, p. 338; 1939b, p. 177), Helms (1896, pp. 248–9), Hicks (1963), Johnston (1939), Johnston & Cleland (1933–34), Maiden (1889, pp. 168–72), Peterson (1977), Roth (1901b, p. 31), Smyth (1878, vol. 2, pp. 222–3) and Watson (1983). See chapter 3 of this volume for a word derivation.
8 Jessop (1981, pp. 324–5).
9 Bancroft (1877–80, pp. 2, 5–6).
10 Bonyhady (1991, p. 137), Moorehead (1963, pp. 118, 123, 137) and Murgatroyd (2002, p. 253).
11 Moorehead (1963, p. 118).
12 Bailey (1880, p. 17) and Foley (2006).
13 Clarke (2007, p. 105) and Woolls (cited Maiden, 1889, p. 172).

14 Clarke (2007, p. 124 & endnote 97 on p. 160) and Woolls (cited Maiden, 1889, p. 150).
15 Correspondence from F. von Mueller to F.M. Bailey (cited Bancroft, 1877–80, pp. 1–2). See also Foley (2006, p. 33).
16 Bancroft (1886, p. 13) and Foley (2006).
17 Ohlendorf (1996).
18 Liversidge (1880, p. 123).
19 Liversidge (1880, p. 124).
20 For details of this exhibition refer to Cundall (1886).
21 Hall (1978, pp. 12–13) and Pearn (1990, p. 12; 2001, pp. 58–60).
22 Bancroft (1888 [cited Webb, 1948, p. 9]).
23 Webb (1948, pp. 10–11).
24 Rennie (1880, 1886).
25 Baker & Smith (1902, 1905, 1910). See Hall (1978, pp. 118–9) for H.G. Smith's biography.
26 Webb (1948, p. 9). For twentieth century pharmacological surveys of Australian plants, refer to Barr et al (1988), Collins et al (1990), Everist (1974), Flower (2000), Hicks & Le Messurier (1935), Lassak & McCarthy (1983), Reid (1977), Semple (1998), Semple et al (1998), Stack (1989) and Webb (1948, 1960, 1969, 1973, 1977). For an overview, see Clarke (2003c, pp. 24–5).
27 Palmer (1883, p. 93).
28 Howitt & Siebert (1904, p. 100).
29 Palmer (1883).
30 Palmer (1883, p. 93).
31 Palmer (1883, p. 93).
32 For example, Dawson (1881) and Smith (1880). Woods (1879) republished work by authors such as Schürmann (1846) and Taplin (1874).
33 For example, Howitt (1904), Ling Roth (1899) and Walker (1889–99).
34 For a discussion of 'memory culture', see Berndt (1974, pp. 22, 25), Clarke (1995, p. 152) and Tonkinson (in Berndt & Berndt, 1993, p. xix).
35 Smyth (1878, vol. 1, pp. 183–252).
36 Smyth (1878, vol. 1, p. 210).
37 See Smyth (1878, vol. 1, pp. 212–4) for von Mueller's account.
38 Smyth (1878, vol. 2, pp. 169–74).
39 Smyth (1878, vol. 2, p. 170).
40 Curr (1886–87).
41 Curr (1883), Forster (1969) and Serle (1949, vol. 1, p. 206).
42 Curr (1883, p. 88).
43 Taplin (1879).
44 Andrew (1986, pp. 5–8), Andrew & Clissold (1986, pp. 9–11), Clissold (1986), Hall (1978, p. 19) and Robertson (1979). See Black (1922–29) for his published flora of South Australia.
45 Black (1917, 1920). Black also assisted in preparation of the vocabulary and ethnographic material published by Bates (1918).
46 Black (1875–86 [1986, p. 118]). See Black (1920, p. 77) for further discussion of the 'retroflexed fricative (Somersetshire) r'.
47 Black (1917, pp. 8–12).
48 Black (1917, pp. 3–8).
49 Black (1920, p. 86).
50 Black (1943–65, p. 604). Refer to chapter 3 of this volume for an account of the 'monterry'.
51 Black (1922, p. 570) and Sharr (1996, p. 72).
52 For a biography of Daisy Bates, refer to Salter (1971). For a personal account of her life as a missionary, see Bates (1947).
53 Bates (1918).
54 For example Albrecht (1959), Bates (1901–14, [1985, p. 263]; 1918; 1947]), Meyer (1843, 1846), Robinson (in Plomley, 1976), Schürmann (1844, 1846), Stone (1911), Strehlow (1907–1920), Taplin (1859–79, 1874, 1879), Teichelmann & Schürmann (1840) and Threlkeld (1824–1859 [1974]).
55 Hill (2002, pp. 2–3, 24, 48–50, 56, 65, 489, 510, 526–7, 625), Jones & Sutton (1986) and Reuther (1981). The Bethesda Mission site is on the shore of Lake Killalpaninna, near the bed of the Cooper Creek.
56 Reuther collection, South Australian Museum Archives and the Aboriginal artefact collection, South Australian Museum. Refer to Jones & Sutton (1986) for a history of the collection.
57 Darwin (1859 [1878], 1871 [1888]).
58 For example, see Frazer (1890, 1910), Lang (1905) and Lubbock (1870). Hiatt (1996) and Mulvaney (1985) provided overviews.
59 Frazer (1890, vol. 2, p. 133).
60 For example, at the turn of the nineteenth century the South Australian Museum Director Edward C. Stirling began actively acquiring Aboriginal artefacts from around Australia (Jones, 1996).
61 Moyal (1986, pp. 109–111) and Webb (2003, pp. 208–15).
62 Ritchie (1989, p. 53).
63 Ritchie (1989, pp. 50–1).
64 Lumholtz (1889).
65 Bailey (1889, p. 3). See Meston (1889) for an official account of the Bellenden-Ker Range Expedition, 1889.
66 Ritchie (1989, p. 51) and Stephens (1974). The South Australian Museum Archives holds a photographic album of Meston's photographs from the 1905 trip to Bellenden-Ker Range.
67 Branagan (1996, p. 45) and Moyal (1986, pp. 65–8).
68 Jessop (1981, pp. xvi–xvii), McLaren (1996, chapter 8), Morton & Mulvaney (1996), Moyal (1986, pp. 68–9) and Spencer (1896). For aims of the expedition, see W.A. Horn in Spencer (1896, p. v).
69 Stirling (1896, pp. 37–40) and Baker & Nesbitt (1996, p. 115).
70 Stirling (1896, p. 37). Reference cited by Stirling is Schulze (1891, p. 216).
71 Griffiths (1996, pp. 67–8), Kuper (1983, pp. 2, 4–5, 19), Morphy (1988, pp. 52–61), Moyal (1986, pp. 147, 162), Mulvaney (1990; 1996, pp. 4–5) and Short (2003, pp. 184–9).
72 Preface written by Spencer (Spencer & Gillen, 1927, p. vii).
73 Spencer (1896) and Stirling (1896)
74 Maiden (1896, p. 197). Tate (1896) wrote the main botany report. Maiden (1890) had an earlier interest in resin derived from soft spinifex grass.
75 Stirling (1896, p. 59). This species formerly known as Marsdenia leichhardtiana or Leichardtia australis (renamed as Marsdenia australis).
76 Bindon (1996, p. 165), Cleland (1957, p. 152), Cleland & Johnston (1933, p. 122; 1937–38, p. 338; 1939b, p. 177), Cleland & Tindale (1954, pp. 82, 85; 1959, p. 136), Hiddins (2000b, p. 30; 2001, p. 120), Latz (1995, pp. 224–5), O'Connell et al (1983, p. 93), Reid (1977, pp. 29–30), Scott (1972, p. 95), Sweeney (1947, p. 290) and Turner (1994, pp. 19–20).
77 Mulvaney et al (1997, 2000) and Short (2003, pp. 190–1).
78 Mulvaney (1996, p. 11), Mulvaney et al (1997, pp. 54, 114, 181) and Short (2003, pp. 190–1).
79 Mulvaney et al (1997, chapter 6).
80 Strehlow (1907–1920).
81 Spencer & Gillen (1899, 1904, 1912, 1927).
82 Donaldson & Donaldson (1985).
83 Reynolds (1988) and Pearn (1990, pp. 112–3; 2001, pp. 321–2).
84 Roth (1897, p. v).
85 Roth (1897).
86 Roth (1897, p. 93).
87 Roth (1897, p. 100).
88 Roth (1901b, p. 9). Also relevant for plants as sources of material to make artefacts are Roth (1901a, 1902, 1904, 1909 & 1910).
89 Roth (1901b, p. 15).
90 Roth (1903a, p. 11).
91 Roth (1903a, p. 11).
92 Pearn (2001, p. 322).
93 Hastings (1989, p. 174), Hepper (1982, chapter 9) and Lyte (1983, p. 178).
94 Bean (1908, part 1, chapter 4), Field (1993, p. 144), Hepper (1982, chapter 10), Moyal (1976, pp. 40–1), Sampson (1935, p. 405) and Wickens (1993, pp. 85–7).
95 W.J. Hooker (cited Wickens, 1993, p. 136).
96 Hooker (1858) and Ponsonby (1998).
97 Maiden (1891, 1892). Webb (2003, chapter 11) provides a biography of Maiden.
98 Maiden (1928, 1931) gave a history of the Sydney Botanic Gardens.
99 Webb (2003, pp. 7, 27, 273–5).
100 Maiden (1889).
101 Maiden (1900, p. 117).
102 Maiden (1889, pp. 146–7).
103 Short (2003, p. 192).
104 M. Koch correspondence to W. Thiselton-Dyer, Kew, dated 6 March 1899 (cited Short, 2003, p. 192).
105 Koch (1898, p. 101).
106 For example, see Cleland & Johnston (1933, 1937–38, 1939a, 1939b), Cleland & Tindale (1954, 1959) and Johnston & Cleland (1942, 1943).
107 Koch (1898, p. 110). Johnston & Cleland (1943, p. 153) recorded the Aboriginal plant name as 'kuluwa' or 'kulua'.
108 Koch (1898, p. 116), who used the 'not satisfactorily identified' botanical name of Cyperus subulatus. See discussion by Clarke (2007, pp. 24, 39, 41, 74) and Latz (1995, p. 158).
109 Koch (1900, p. 83). Koch used a now obsolete botanical name for the grey mulga, which is currently called Acacia brachystachya.
110 Balick & Cox (1996, p. 3), Berlin (1992, pp. 4–5), Clément (1998), Ford (1978, p. 33) and Wickens (1990, p. 16). Cotton (1996), Cunningham (2001), Jain (1987) and Martin (1995) have provided the field of ethnobotany with method manuals.
111 Harshberger (1896).
112 Ford (1978, p. 44).
113 Wickens (1993, p. 84). See also Wickens (1990, p. 14).
114 Clarke (2003c).

References

Abbie, A.A. 1976. *The original Australians*. Revised edn. Seal Books, Adelaide.

Aitken, R. 2006. *Botanical riches. Stories of botanical exploration*. Melbourne University Press at the Miegunyah Press, Melbourne.

Akerman, K. 1978. Ngarla and mei: living on bush foods in the Central Kimberley. *Earth Gardener*. Vol. 2, no. 3, pp. 20–2.

Akerman, K. 1980. Material culture and trade in the Kimberleys today. Pp. 243–51 in R.M. Berndt & C.H. Berndt (eds) *Aborigines of the west. Their past and their present*. University of Western Australia Press, Perth.

Albrecht, F.W. 1959. *The natural food supply of the Australian Aborigines*. Aborigines' Friends' Association, Adelaide.

Allen, H.R. 1974. The Bagundji of the Darling Basin: cereal gatherers in an uncertain environment. *World Archaeology*. Vol. 5, pp. 309–22.

Anderson, C. 1996. Traditional material culture of the Kuku-Yalanji of Bloomfield River, north Queensland. *Records of the South Australian Museum*. Vol. 29, part 1, pp. 63–83.

Anderson, K.J. & Gale, F. (eds) 1992. *Inventing places. Studies in cultural geography*. Longman Cheshire, Melbourne.

Andrew, M. 1986. History of the diaries. Pp. 5–8 in M. Andrew & S. Clissold (ed.) *The diaries of John McConnell Black*. Vol. 1. Investigator Press, Adelaide.

Andrew, M. & Clissold, S. (ed.) 1986. *The diaries of John McConnell Black*. Vol. 1. Investigator Press, Adelaide.

Angas, G.F. 1847. *Savage life and scenes in Australia*. 2 volumes. Smith, Elder & Co., London.

Anonymous. 1847. *Settlers and convicts or recollections of sixteen years' labour in the Australian backwoods by an emigrant mechanic (Alexander Harris?)*. Republished in 1953 by Melbourne University Press, Melbourne.

Arbury, J. 1998. Breeding brambles. *The Garden. Journal of the Royal Horticultural Society*. Vol. 123, part 8, pp. 578–83.

Archer, B. & Maroske, S. 1996. Sarah Theresa Brooks – plant collector for Ferdinand Mueller. *Victorian Naturalist*. Vol. 113, pt. 4, pp. 188–94.

Arthur, J.M. 1996. *Aboriginal English. A cultural study*. Oxford University Press, Melbourne.

Attenbrow, V. 2003. *Sydney's Aboriginal past. Investigating the archaeological and historical records*. University of New South Wales Press, Sydney.

Bach, J. 1966. Dampier, William (1651–1715). Pp. 277–8 in D. Pike (ed.) *Australian dictionary of biography. Volume 1. 1788–1850, A–H*. Melbourne University Press, Melbourne.

Backhouse, J. 1841. *Extracts from the letters of James Backhouse. Whilst engaged in a religious visit to Van Diemen's Land, New South Wales, and South Africa, accompanied by George Washington Walker*. Harvey & Darton, London.

Backhouse, J. 1843. *A narrative of a visit to the Australian colonies*. Hamilton, Adams & Co., London.

Bailey, F.M. 1880. Medicinal plants of Queensland. *Proceedings of the Linnean Society of New South Wales*. Vol. 5, part 1, pp. 1–29.

Bailey, F.M. 1889. *Report on new plants, preliminary to general report on botanical results of Meston's expedition to the Bellenden-Ker Range*. Government Printer, Brisbane.

Bailey, F.M. 1899–1902. *The Queensland flora*. 6 volumes. Diddams, Brisbane.

Bailey, F.M. 1913. *Comprehensive catalogue of Queensland plants both indigenous and naturalised*. A.J. Cumming, Queensland Government Printer, Brisbane.

Baker, L. & Nesbitt, B.J. 1996. The role of Aboriginal ecological knowledge in the Horn Expedition and contemporary ecological research. Pp. 115–22 in S.R. Morton & D.J. Mulvaney (eds) *Exploring Central Australia. Society, the environment and the 1894 Horn Expedition*. Surrey Beatty & Sons, Sydney.

Baker, M.J., Corringham, R. & Dark, J. 1986. *Native plants of the Sydney region*. Three Sisters Productions, Winmalee, New South Wales.

Baker, R.M. 1999. *Land is life. From bush to town. The story of the Yanyuwa people*. Allen & Unwin, Sydney.

Baker, R.T. 1913. *Cabinet timbers of Australia*. Technical Education Series No. 18. William Applegate Gullick, New South Wales Government Printer, Sydney.

Baker, R.T. 1919. *The hardwoods of Australia and their economics*. Technical Education Series No. 23. William Applegate Gullick, New South Wales Government Printer, Sydney.

Baker, R.T. & Smith, H.G. 1902. *A research on the eucalypts, especially in regard to their essential oils*. Technical Education Series No. 13. William Applegate Gullick, New South Wales Government Printer, Sydney.

Baker, R.T. & Smith, H.G. 1905. *Some west Australian eucalypts and their essential oils*. Reprint from *The Pharmaceutical Journal* (9 & 16 September 1905 issues).

Baker, R.T. & Smith, H.G. 1902. *A research on the pines of Australia*. Technical Education Series No. 16. William Applegate Gullick, New South Wales Government Printer, Sydney.

Balick, M.J. & Cox, P. A. 1996. *Plants, people and culture. The science of ethnobotany*. Scientific American Library Series No. 60. New York.

Balint, R. 2005. *Troubled waters. Borders, boundaries and possession in the Timor Sea*. Allen & Unwin, Sydney.

Bancroft, J. 1872. *The pituri poison*. Paper read before the Queensland Philosophical Society. Government Printer, Brisbane.

Bancroft, J. 1877–80. *Papers on pituri and Duboisia*. Warwick & Sapsford, Brisbane.

Bancroft, J. 1884. Food of the Aborigines of Central Australia. *Proceedings of the Royal Society of Queensland*. Vol. 1, pp. 104–7.

Bancroft, J. 1886. *Contribution to pharmacy from Queensland*. Warwick & Sapsford, Brisbane.

Bancroft, T.L. 1894. On the habit and use of nardoo (*Marsilea drummondii*, A.Br.), together with some observations on the influence of water-plants in retarding evaporation. *Proceedings of the Linnean Society of New South Wales*. Vol. 8, part 2, pp. 215–7.

Banks, J. 1768–71. *The Endeavour journal of Joseph Banks 1768–1771*. Edited by J.C. Beaglehole. 1962. Angus & Robertson, Sydney.

Barker, R.M. 1996. James Drummond's newspaper accounts of his collecting activities, in particular his 4th Collection and *Hakea victoria* (Proteaceae). *Nuytsia*. Vol. 11, part 1, pp. 1–9.

Barker, R.M. & Barker, W.R. 1990. Botanical contributions overlooked: the role and recognition of collectors, horticulturalists, explorers and others in the early documentation of the Australian flora. Pp. 37–85 in P.S. Short (ed.) *History of systematic botany in Australasia. Proceedings of a symposium held at the University of Melbourne*. Australian Systematic Botany Society, Melbourne.

Barr, A., Chapman, J., Smith, N. & Beveridge, M. 1988. *Traditional bush medicines. An Aboriginal pharmacopoeia*. Greenhouse Publications, Melbourne.

Barrallier, F. 1802. *Journal of the expedition into the interior of New South Wales, 1802, by order of... Governor Philip Gidley King*. Reprinted from the *Historical Records of New South Wales*, volume 5, appendix A in 1975 by Marsh Walsh, Melbourne.

Barrett, C. 1938. *The birdman. A sketch of the life of John Gould*. Whitcombe & Tombs, Melbourne.

Barrett, C. 1948. *White blackfellows. The strange adventures of Europeans who lived among savages*. Hallcraft Publishing, Melbourne.

Barwick, D.E. 1985. This most resolute lady: a biographical puzzle. Pp. 185–239 in D.E. Barwick, J. Beckett & M. Reay (eds) *Metaphors of interpretation. Essays in honour of W.E.H. Stanner*. Australian National University Press, Canberra.

Basedow, H. 1904. Anthropological notes made on the South Australian Government North-west Prospecting Expedition, 1903. *Transactions of the Royal Society of South Australia*. Vol. 28, pp. 12–51.

Basedow, H. 1914. Journal of the Government North-west Expedition (1903). *Proceedings of the Royal Geographical Society of South Australia*. Vol. 15, pp. 57–242.

Basedow, H. 1925. *The Australian Aboriginal*. F.W. Preece & Sons, Adelaide.

Bassett, M. 1981. Augusta and Mrs Molloy. Pp. 357–73 in D.J. Carr & S.G.M. Carr (eds) *People and plants in Australia*. Academic Press, Sydney.

Bates, D.M. 1901–14. *The native tribes of Western Australia*. Edited by I. White, 1985. National Library of Australia, Canberra.

Bates, D.M. 1912 (1992). Haunted Places of the West. Pp. 14–9. in P.J. Bridge (ed.)

Aboriginal Perth and Bibbulmun biographies and legends. Hesperian Press, Carlisle, Western Australia.

Bates, D.M. 1918. Aborigines of the west coast of South Australia. Vocabularies and ethnographical notes. *Transactions & Proceedings of the Royal Society of South Australia*. Vol. 42, pp. 152–67.

Bates, D.M. 1947. *The passing of the Aborigines*. Australian edn. John Murray, London.

Bauer, F.H. 1969. Climate and man in north-western Queensland. Pp. 51–63 in F. Gale & G.H. Lawton (eds) *Settlement and encounter. Geographical studies presented to Sir Grenfell Price*. Oxford University Press, Melbourne.

Baulch, W. 1961. Ronald Campbell Gunn. F.R.S. F.L.S. A biographical note. *Records of the Queen Victoria Museum. New Series*. No. 12, pp. xiii–xix.

Bean, W.J. 1908. *The Royal Botanic Gardens, Kew. Historical and descriptive*. Cassell, London.

Beard, J.S. 1981. Vegetation of Central Australia. Pp. xxi–xxvi in J.P. Jessop (ed.) *Flora of Central Australia*. The Australian Systematic Botany Society. Reed Books, Sydney.

Beaton, J.M. 1982. Fire and water: aspects of Australian Aboriginal management of cycads. *Archaeology in Oceania*. Vol. 17, pp. 51–8.

Beckett, J.R. 1987. *Torres Strait Islanders. Custom and colonialism*. Cambridge University Press, Cambridge.

Beckler, H. 1859–62. *A journey to Cooper's Creek*. Translated by S. Jeffries & M. Kertesz. 1993. Melbourne University Press at the Miegunyah Press & State Library of Victoria, Melbourne.

Bedford, C.T. 1887. Reminiscences of a surveying trip from Boulia to the South Australian border. *Proceedings & Transactions of the Royal Society of Australasia. Queensland Branch*. Vol. 2, pp. 99–113.

Bellchambers, T.P. 1931. *A nature-lover's notebook*. Nature Lovers League, Adelaide.

Bellingshausen, T. 1819–21. *The voyage of Captain Bellingshausen to the Antarctic seas 1819–21*. Translated from the Russian, edited by F. Debenham. Published in 1945. 2 volumes. Second series, nos 91–92. Hakluyt Society, London.

Bentham, G. 1863–78. *Flora Australiensis. A description of the plants of the Australian territory*. 19 volumes. L. Reeve, London.

Berlin, B. 1992. *Ethnobiological classification. Principles of categorization of plants and animals in traditional societies*. Princeton University Press, Princeton, New Jersey.

Berndt, C.H. 1981. Interpretations and 'facts' in Aboriginal Australia. Pp. 153–203 in F. Dahlberg (ed.) *Woman the gatherer*. Yale University Press, New Haven.

Berndt, R.M. 1974. *Australian Aboriginal religion. Fascicle one – introduction; the south-eastern region*. E.J. Brill, Leiden.

Berndt, R.M. & Berndt, C.H. 1954. *Arnhem Land. Its history and its people*. F.W. Cheshire, Melbourne.

Berndt, R.M. & Berndt, C.H. 1989. *The speaking land. Myth and story in Aboriginal Australia*. Penguin Books, Melbourne.

Berndt, R.M., Berndt, C.H., with Stanton, J.E. 1993. *A world that was. The Yaraldi of the Murray River and the lakes, South Australia*. Melbourne University Press at the Miegunyah Press, Melbourne.

Beveridge, P. 1884. Of the Aborigines inhabiting the Great Lacustrine and Riverine Depression of the Lower Murray, Lower Murrumbidgee, Lower Lachlan, and Lower Darling. *Journal & Proceedings of the Royal Society of New South Wales*. Vol. 17, pp. 19–74.

Bindon, P. 1996. *Useful bush plants*. Western Australian Museum, Perth.

Bindon, P. & Gough, D. 1993. Digging sticks and desert dwellers. *Landscope*. Spring, pp. 11–16.

Bird, C. & Beeck, C. 1988. Traditional plant foods in the southwest of Western Australia: the evidence from salvage ethnography. Pp. 113–22 in B. Meehan & R. Jones (eds) *Archaeology with ethnography. An Australian perspective*. Prehistory Pacific Studies, Australian National University, Canberra.

Black, J.M. 1875–86. *The diaries of John McConnell Black. Volume 1 diaries one to four – 1875–1886*. Edited by M. Andrew & S. Clissold. 1986. Investigator Press, Adelaide.

Black, J.M. 1917. Vocabularies of three South Australian languages, Wirrung, Narrinyeri and Wongaidya. *Transactions of the Royal Society of South Australia*. Vol. 41, pp. 1–8.

Black, J.M. 1920. Vocabularies of four South Australian languages, Adelaide, Narrunga, Kukata, and Narrinyeri, with special references to speech sounds. *Transactions of the Royal Society of South Australia*. Vol. 44, pp. 76–93.

Black, J.M. 1922. Additions to the flora of South Australia. No. 20. *Transactions of the Royal Society of South Australia*. Vol. 46, pp. 565–71.

Black, J.M. 1922–29. *Flora of South Australia. Parts 1–4*. Handbook of the Flora & Fauna of South Australia. Government Printer, Adelaide.

Black, J.M. 1943–65. *Flora of South Australia. Parts 1–4*. Second edn. Handbook of the Flora & Fauna of South Australia. Government Printer, Adelaide.

Blackburn, J. 1979. *The white men*. Times, London.

Blainey, G. 1976. *Triumph of the nomads. A history of ancient Australia*. Sun Books, Melbourne.

Blainey, G. 1977. *The tyranny of distance. How distance shaped Australia's history*. Macmillan, London.

Blake, B.J. 1981. *Australian Aboriginal languages. A general introduction*. Angus & Robertson, Sydney.

Blake, N.M., Wightman, G. & Williams, L. 1998. *Iwaidja ethnobotany. Aboriginal plant knowledge from Gurig National Park, northern Australia*. Northern Territory Botanical Bulletin No. 23. Parks & Wildlife Commission of the Northern Territory, Darwin.

Blandowski, W. 1858. Recent discoveries in natural history on the Lower Murray. *Transactions of the Philosophical Society of Victoria*. Vol. 2, pp. 124–37.

Blunt, W. 1978. *In for a penny. A prospect of Kew Gardens: their flora, fauna and falballas*. Hamish Hamilton in association with the Tryon Gallery, London.

Bolam, A.G. 1930. *The trans-Australian wonderland*. Baker & Co., Melbourne.

Bonney, N. 1994. *Uses of native plants in the south east of South Australia by the indigenous peoples before 1839*. Celebration South East Volume 4. Southeast Book Promotions, Naracoorte, South Australia.

Bonyhady, T. 1991. *Burke and Wills. From Melbourne to myth*. David Ell Press, Sydney.

Boyce, D. 1970. *Clarke of the Kindur. Convict, bushranger, explorer*. Melbourne University Press, Melbourne.

Bradley, W. 1786–92. *A voyage to New South Wales. December 1786–May 1792*. ML Safe1/14. State Library of New South Wales, Sydney.

Branagan, D.F. 1996. John Alexander Watt: geologist on the Horn Expedition. Pp. 42–58 in S.R. Morton & D.J. Mulvaney (eds) *Exploring Central Australia. Society, the environment and the 1894 Horn Expedition*. Surrey Beatty & Sons, Sydney.

Brand Miller, J.C., James, K.W. & Maggiore, P. M.A. 1993. *Tables of composition of Australian Aboriginal foods*. Aboriginal Studies Press, Canberra.

Brennan, K. 1986. *Wildflowers of Kakadu*. Publisher by author, Jabiru, Northern Territory.

Breton, W.H. 1846. An excursion to the Western Range. *Tasmanian Journal of Natural Science*. Vol. 7, pp. 121–41.

Brokensha, P. 1975. *The Pitjantjatjara and their crafts*. Aboriginal Arts Board, Australia Council, Sydney.

Brook, J. 2001. The forlorn hope: Bennelong and Yemmerrawannie go to England. *Australian Aboriginal Studies*. No. 1, pp. 36–47.

Brooker, S.G., Cambie, R.C. & Cooper, R.C. 1987. *New Zealand medicinal plants*. Heinemann, Auckland.

Brown, C.H. 1986. The growth of ethnobiological nomenclature. *Current Anthropology*. Vol. 27, part 1, pp. 1–19.

Brown, J.E. 1882. *The forest flora of South Australia*. E. Spiller, Government Printer, Adelaide.

Brown, L. 1993. *The new shorter Oxford English dictionary on historical principles*. Reprinted. Clarendon Press, Oxford.

Brown, R. 1801–05. *Nature's Investigator. The diary of Robert Brown in Australia, 1801–1805*. Compiled by T.G. Vallance, D.T. Moore & E.W. Groves. 2001. Australian Biological Resources Study, Canberra.

Brown, R. 1810–30. *Prodromus florae Novae Hollandiae et insulae Van Diemen*. 1810. *Supplementum Primum*. 1830. With an introduction by W.T. Stearn. 1960. Engelmann, Weinheim.

Brown, R. 1814. *General remarks, geographical and systematical on the botany of Terra Australis*. G.&W. Nichol, London.

Bruneteau, J-P. 1996. *Tukka. Real Australian food*. Angus & Robertson, Sydney.

Buchanan, A.M. 1990. Ronald Campbell Gunn 1808–1881. Pp. 179–92 in P.S. Short (ed.) *History of systematic botany in Australasia. Proceedings of a symposium held at the University of Melbourne*. Australian Systematic Botany Society, Melbourne.

Buku-larrnggay Mulka Centre. 1999. *Saltwater. Yirrkala bark paintings of sea country. Recognising indigenous sea rights*. Jennifer Isaacs Publishing, New South Wales, in association with Buku-larrnggay Mulka Centre.

Bunce, D. 1859a. *Travels with Dr. Leichhardt in Australia*. Steam Press, Melbourne.

Bunce, D. 1859b. *Language of the Aborigines of the colony of Victoria* ... Thomas Brown, Geelong, Victoria.

Burke & Wills Commission. 1862. *Report of the commissioners appointed to enquire into and report upon the circumstances connected with the sufferings and death of Robert O'Hara Burke and William John Wills the Victoria explorers*. Parliamentary Paper No. 97. John Ferres, Government Printer, Melbourne. Digitised and available at http://www.burkeandwills.net.au

Burns, T.E. & Skemp, J.R. 1961. *Van Diemen's Land correspondents. 1827–49*. Queen Victoria Museum, Launceston.

Burns, T.E. & Skemp, J.R. 1966. Gunn, Ronald Campbell (1808–1881). Pp. 492–3. in D. Pike (ed.) *Australian dictionary of biography. Volume 1. 1788–1850, A–H*. Melbourne University Press, Melbourne.

Butlin, N.G. 1993. *Economics and the Dreamtime. A hypothetical history*. Cambridge University Press, Cambridge.

Campbell, W.S. 1932. The use and abuse of stimulants in the early days of settlement in New South Wales. *Journal & Proceedings of the Royal Australian Historical Society*. Vol. 18, pt 2, pp. 74–99.

Capell, A. 1962. Language and social distinction in Aboriginal Australia. *Mankind*. Vol. 5, no. 12, pp. 514–22.

Carnegie, D.W. 1898. *Spinifex and sand. A narrative of five years pioneering and exploration in Western Australia*. 1973 facsimile edn. Penguin Books, Melbourne.

Carr, D.J. & Carr, S.G.M. (eds) 1981a. *People and plants in Australia*. Academic Press, Sydney.

Carr, D.J. & Carr, S.G.M. (eds) 1981b. The botany of the first Australians. Pp. 3–44 in D.J. Carr & S.G.M. Carr (eds) *People and plants in Australia*. Academic Press, Sydney.

Carr, D.J. & Carr, S.G.M. (eds) 1981c. *Plants and man in Australia*. Academic Press, Sydney.

Carter, H.B. 1988. *Sir Joseph Banks 1743–1820*. British Museum (Natural History), London.

Cavanagh, T. 1990. Australian plants cultivated in England, 1771–1800. Pp. 273–83 in P.S. Short (ed.) *History of systematic botany in Australasia. Proceedings of a symposium held at the University of Melbourne*. Australian Systematic Botany Society, Melbourne.

Cawte, J. 1974. *Medicine is the law*. University Press of Hawaii, Honolulu.

Cawte, J. 1996. *Healers of Arnhem Land.* University of New South Wales Press, Sydney.
Charlesworth, M. (ed.) 1998. *Religious business. Essays on Australian Aboriginal spirituality.* Cambridge University Press, Cambridge.
Charlesworth, M., Morphy, H., Bell, D. & Maddock, K. (eds) 1984. *Religion in Aboriginal Australia. An anthology.* University of Queensland Press, St Lucia.
Cherikoff, V. & Isaacs, J. 1989. *The bush food handbook. How to gather, grow, process and cook Australian wild foods.* Ti Tree Press, Sydney.
Chisholm, A.H. 1941. *Strange new world. The adventures of John Gilbert and Ludwig Leichhardt.* Angus & Robertson, Sydney.
Chisholm, A.H. 1942. John Gould's Australian prospectus. *Emu.* Vol. 42, pp. 74–84.
Chisholm, A.H. 1944. Introduction. In F.W.L. Leichhardt, *Dr. Ludwig Leichhardt's letters from Australia during the years March 23, 1842, to April 3, 1848.* Pan, Melbourne.
Chisholm, A.H. 1966a. Gilbert, John (1810?–1845). Pp. 441–2 in D. Pike (ed.) *Australian dictionary of biography. Volume 1. 1788–1850, A–H.* Melbourne University Press, Melbourne.
Chisholm, A.H. 1966b. Gould, John (1804–1881). Pp. 465–7 in D. Pike (ed.) *Australian dictionary of biography. Volume 1. 1788–1850, A–H.* Melbourne University Press, Melbourne.
Chisholm, A.H. 1973. *Strange journey. The adventures of Ludwig Leichhardt and John Gilbert.* Rigby, Adelaide.
Clark, I.D. & Heydon, T. 2002. *Dictionary of Aboriginal placenames of Victoria.* Victorian Aboriginal Corporation for Languages, Melbourne.
Clark, M. 1957. *Sources of Australian history.* Oxford University Press, Oxford.
Clark, M. 1987. *A short history of Australia.* Third edn. NAL Penguin, New York.
Clarke, P. 1986. *Hell and paradise: The Norfolk, Bounty, Pitcairn saga.* Viking, Ringwood, Victoria.
Clarke, P.A. 1985a. The importance of roots and tubers as a food source for southern South Australian Aborigines. *Journal of the Anthropological Society of South Australia.* Vol. 23, part 6, pp. 2–12.
Clarke, P.A. 1985b. Fruits and seeds as food for southern South Australian Aborigines. *Journal of the Anthropological Society of South Australia.* Vol. 23, part 9, pp. 9–22.
Clarke, P.A. 1986a. Aboriginal use of plant exudates, foliage and fungi as food and water sources in southern South Australia. *Journal of the Anthropological Society of South Australia.* Vol. 24, part 3, pp. 3–18.
Clarke, P.A. 1986b. The study of ethnobotany in southern South Australia. *Australian Aboriginal Studies.* No. 2, pp. 40–7.
Clarke, P.A. 1987. Aboriginal uses of plants as medicines, narcotics and poisons in southern South Australia. *Journal of the Anthropological Society of South Australia.* Vol. 25, part 5, pp. 3–23.
Clarke, P.A. 1988. Aboriginal use of subterranean plant parts in southern South Australia. *Records of the South Australian Museum.* Vol. 22, pt 1, pp. 63–76.
Clarke, P.A. 1994. *Contact, conflict and regeneration. Aboriginal cultural geography of the Lower Murray, South Australia.* Postgraduate thesis. University of Adelaide, Adelaide.
Clarke, P.A. 1995. Myth as history: the Ngurunderi mythology of the Lower Murray, South Australia. *Records of the South Australian Museum.* Vol. 28, part 2, pp. 143–57.
Clarke, P.A. 1996. Early European interaction with Aboriginal hunters and gatherers on Kangaroo Island, South Australia. *Aboriginal History.* Vol. 20, part 1, pp. 51–81.
Clarke, P.A. 1998a. The Aboriginal presence on Kangaroo Island, South Australia. Pp. 14–48 in J. Simpson & L. Hercus (eds) *Aboriginal portraits of 19th century South Australia.* Aboriginal History Monograph. Australian National University, Canberra.
Clarke, P.A. 1998b. Early Aboriginal plant foods in southern South Australia. *Proceedings of the Nutrition Society of Australia.* Vol. 22, pp. 16–20.
Clarke, P.A. 2001. The significance of whales to the Aboriginal people of southern South Australia. *Records of the South Australian Museum.* Vol. 34, part 1, pp. 19–35.
Clarke, P.A. 2003a. *Where the ancestors walked. Australia as an Aboriginal landscape.* Allen & Unwin, Sydney.
Clarke, P.A. 2003b. Twentieth century Aboriginal hunting and gathering practices in the rural landscape of the Lower Murray, South Australia. *Records of the South Australian Museum.* Vol. 36, part 1, pp. 83–107.
Clarke, P.A. 2003c. Australian ethnobotany – an overview. *Australian Aboriginal Studies.* No. 2, pp. 21–38.
Clarke, P.A. 2007. *Aboriginal people and their plants.* Rosenberg Publishing, Sydney.
Cleary, T. 1993. *Poignant regalia. 19th century Aboriginal breastplates and images.* Historic Houses Trust of New South Wales, Glebe, New South Wales.
Cleland, J.B. 1936. Ethno-botany in relation to the Central Australian Aboriginal. *Mankind.* Vol. 2, no. 1, pp. 6–9.
Cleland, J.B. 1939. Some aspects of the ecology of the Aboriginal inhabitants of Tasmania and southern Australia. *Papers & Proceedings of the Royal Society of Tasmania.* Issued separately, 1940, pp. 1–18.
Cleland, J.B. 1957. Ethno-ecology. Our natives and the vegetation of southern Australia. *Mankind.* Vol. 5, no. 4, pp. 149–62.
Cleland, J.B. 1966. The ecology of the Aboriginal in South and Central Australia. Pp. 111–58 in B.C. Cotton (ed.) *Aboriginal man in South and Central Australia. Part 1.* South Australian Government Printer, Adelaide.
Cleland, J.B., Black, J.M. & Reese, L. 1925. The flora of the north-east corner of South Australia, north of Cooper's Creek. *Transactions of the Royal Society of South Australia.* Vol. 49, pp. 103–20.
Cleland, J.B. & Johnston, T.H. 1933. The ecology of the Aborigines of Central Australia; botanical notes. *Transactions of the Royal Society of South Australia.* Vol. 57, pp. 113–24.
Cleland, J.B. & Johnston, T.H. 1937–38. Notes on native names and uses of plants in the Musgrave Ranges region. *Oceania.* Vol. 8, no. 2, pp. 208–15; no. 3, pp. 328–42.
Cleland, J.B. & Johnston, T.H. 1939a. Aboriginal names and uses of plants at the Granites, Central Australia. *Transactions of the Royal Society of South Australia.* Vol. 63, no. 1, pp. 22–6.
Cleland, J.B. & Johnston, T.H. 1939b. Aboriginal names and uses of plants in the northern Flinders Ranges. *Transactions of the Royal Society of South Australia.* Vol. 63, no. 2, pp. 172–9.
Cleland, J.B. & Tindale, N.B. 1954. The ecological surroundings of the Ngalia natives in Central Australia and native names and uses of plants. *Transactions of the Royal Society of South Australia.* Vol. 77, pp. 81–6.
Cleland, J.B. & Tindale, N.B. 1959. The native names and uses of plants at Haasts Bluff, Central Australia. *Transactions of the Royal Society of South Australia.* Vol. 82, pp. 123–40.
Clément, D. 1998. The historical foundations of ethnobiology (1860–1899). *Journal of Ethnobiology.* Winter vol. 18, pt 2, pp. 161–87.
Clissold, S. 1986. Biography. Pp. 9–11 in M. Andrew & S. Clissold (ed.) *The diaries of John McConnell Black.* Investigator Press, Adelaide.
Cobcroft, M. 1983. Medical aspects of the Second Fleet. Pp. 13–33 in J.H. Pearn & C. O'Carrigan (eds) *Australia's quest for colonial health. Some influences on early health and medicine in Australia.* Department of Child Health, Royal Children's Hospital, Brisbane.
Cohn, H.M. 1996. Botanical researches in Intertropical Australia: Ferdinand Mueller and the North Australian Exploring Expedition. *Victorian Naturalist.* Vol. 113, pt.4, pp. 163–8.
Coleman, E. 1938. One man's meat. *Walkabout.* Vol. 4, pt 11 (1 September), pp. 36–8.
Collins, D. 1798–1802. *An account of the English colony in New South Wales: with remarks on the dispositions, customs, manners, &c. of the native inhabitants of that country.* Two volumes. T. Cadell & W. Davies, London.
Collins, D.J., Culvenor, C.C.J., Lamberton, J.A., Loder, J.W. & Price. J.R. 1990. *Plants for medicine. A chemical and pharmacological survey of plants in the Australian region.* Commonwealth & Scientific Industrial Research Organisation, East Melbourne.
Collins, K. 1988. *Ellis Rowan 1848–1922. A biographical sketch.* Famous Australian Series. Oz Publishing, Brisbane.
Connell, G. 1980. *The mystery of Ludwig Leichhardt.* Melbourne University Press, Melbourne.
Cook, J. 1768–71. *The journals of Captain James Cook on his voyages of discovery. The voyage of the Endeavour 1768–1771.* Edited by J.C. Beaglehole. 1968. Published for the Hakluyt Society by Cambridge University Press, London.
Cook, J. 1768–79. *The explorations of Captain James Cook in the Pacific as told by selections of his own journals 1768–1779.* Edited by A. Grenfell Price, 1957. Limited Editions Club, New York.
Cook, J. 1893. *Captain Cook's journal during his first voyage round the world made in H.M. Bark 'Endeavour' 1768–71.* Edited by W.J.L. Wharton, with notes and introduction. Elliot Stock, London.
Cooper, H.M. 1952. *French exploration in South Australia.* Published by the author, Adelaide.
Cooper, H.M. 1954. Kangaroo Island's wild pigs: their possible origin. *South Australian Naturalist.* Vol. 28, pt 5, pp. 57–61.
Cooper, W. & McLaren, G. 1997. The changing dietary habits of nineteenth-century Australian explorers. *Australian Geographer.* May 1997, vol. 28, pt 1, pp. 97–106.
Cotton, C.M. 1996. *Ethnobotany. Principles and applications.* John Wiley & Sons, Chichester.
Couper Black, E. 1966. Population and tribal distribution. Pp. 97–109 in B.C. Cotton (ed.) *Aboriginal man in South and Central Australia.* South Australian Government Printer, Adelaide.
Coutts, P.J.F. 1970. *The archaeology of Wilson's Promontory.* Australian Aboriginal Studies No. 28. Prehistory & Material Culture Series No. 7. Australian Institute of Aboriginal Studies, Canberra.
Crawford, I.M. 1982. Traditional Aboriginal plant resources in the Kalumburu area: aspects in ethno-economics. *Records of the Western Australian Museum Supplement No. 15.* Western Australian Museum, Perth.
Crawfurd, J. 1868. On the vegetable and animal food of the natives of Australia in reference to social position, with a comparison between the Australian and some other races of man. *Transactions of the Ethnological Society of London. New Series.* Vol. 6, pp. 112–22.
Cribb, A.B. & Cribb, J.W. 1981. *Wild medicine in Australia.* Fontana/Collins, Sydney.
Cribb, A.B. & Cribb, J.W. 1982. *Wild food in Australia.* Revised edn. Fontana/Collins, Sydney.
Crosby, A.W. 2004. *Ecological imperialism. The biological expansion of Europe, 900–1900.* Second edn. Cambridge University Press, Cambridge.
Crowley, F.K. 1976. Windich (Windiitj), Tommy (1840–1876). Pp. 422–3 in B. Nairn (ed.) *Australian dictionary of biography. Volume 6. 1851–1890, R–Z.* Melbourne University Press, Melbourne.
Crowley, F.K. 1981. Forrest, Sir John [Baron Forrest] (1847–1918). Pp. 544–51 in B. Nairn & G. Serle (eds) *Australian dictionary of biography. Volume 8. 1891–1939, Cl–Gib.* Melbourne University Press, Melbourne.
Crowley, T. 1996. Early language contact in Tasmania. Map 27 & pp. 25–32 in S.A. Wurm, P. Mühlhäusler, D.T. Tryon (eds), *Atlas of languages of intercultural communication in the Pacific, Asia, and the Americas.* 3 volumes. Mouton de Gruyter, Berlin.
Cumpston, J.S. 1970. *Kangaroo Island 1800–1836.* Roebuck, Canberra.
Cundall, F. (ed.) 1886. *Reminiscences of the Colonial and Indian Exhibition.* William

Clowes & Sons, London.

Cunningham, A.B. 2001. *Applied ethnobotany. People, wild plant use and conservation.* Earthscan Publications, London.

Curr, E.M. 1883. *Recollections of squatting in Victoria. Then called the Port Phillip District (From 1841 to 1851).* Second edn. Melbourne University Press, Melbourne.

Curr, E.M. 1886–87. *The Australian race. Its origin, languages, customs, place of landing in Australia, and the routes by which it spread itself over that continent.* 4 volumes. Trubner, London.

Currey, J.E.B. 1966. *Reflections on the colony of New South Wales. George Caley.* Lansdowne Press, Melbourne.

Curry, S. & Maslin, B. 1990. Cunningham's collecting localities while botanist on Lieutenant Phillip Parker King's survey of coastal Australia, December 1817 to April 1822. Pp. 137–48 in P.S. Short (ed.) *History of systematic botany in Australasia. Proceedings of a symposium held at the University of Melbourne.* Australian Systematic Botany Society, Melbourne.

Curry, S., Maslin, B. & Maslin, J. 2002. *Allan Cunningham. Australian collecting localities.* Australian Biological Resources Study, Canberra.

Curtis, B.V. 1974. Spare a spot for the *Santalum. Australian Plants.* June issue. Vol. 7, pp. 337–8.

Dahlberg, F. (ed.) 1981. *Woman the gatherer.* Yale University Press, New Haven.

Daley, C. 1924. *Baron Sir Ferdinand von Mueller. Botanist, explorer, and geographer.* Government Printer, Melbourne.

Daley, C. 1931. Food of the Australian Aborigines. *Victorian Naturalist.* Vol. 48, pp. 23–31.

Dampier, W. 1697. *A new voyage round the world. The journal of an English buccaneer.* Edited & revised 1998. Hummingbird Press, London.

Dampier, W. 1697–1703. *Dampier's voyages. Consisting of a new voyage round the world, a supplement to the voyage round the world, two voyages to Campeachy, a discourse of winds, a voyage to New Holland, and a vindication, in answer to the chimerical relation of William Furnell.* Edited by John Masefield, 1906. E. Grant Richards, London.

Dampier, W. 1703. *A voyage to New Holland.* J. Knapton, London.

Dan, N. 1983. The medical men of the First Fleet. Pp. 3–12 in J.H. Pearn & C. O'Carrigan (eds) *Australia's quest for colonial health. Some influences on early health and medicine in Australia.* Department of Child Health, Royal Children's Hospital, Brisbane.

Dark, E. 1966. Bennelong (1764?–1813). Pp. 84–5 in D. Pike (ed.) *Australian dictionary of biography. Volume 1. 1788–1850, A–H.* Melbourne University Press, Melbourne.

Darragh, T.A. 1996. Mueller and personal names in zoology and palaeontology. *Victorian Naturalist.* Vol. 113, pt.4, pp. 195–7.

Darwin, C. 1878. *On the origin of species by means of natural selection or the preservation of favoured races in the struggle for life.* Sixth edn. John Murray, London.

Darwin, C. 1888. *The descent of man and selection in relation to sex.* Second edn. John Murray, London.

David, B. 1998. Introduction: 'a mountain once seen never to be forgotten'. Pp. 1–26 in B. David (ed.) *Ngarrabullgan. Geographical investigations in Djungan country, Cape York Peninsula.* Monash Publications in Geography & Environmental Science No. 51. Monash University, Clayton, Victoria.

David, B. 2002. *Landscapes, rock-art and the Dreaming. An archaeology of preunderstanding.* Continuum, London.

Davidson, D.S. 1934. Australian spear-traits and their derivations. *Journal of the Polynesian Society.* Vol. 43, pp. 41–72, 143–62.

Davidson, D.S. 1936. Australian throwing-sticks, throwing clubs, and boomerangs. *American Anthropologist.* Vol. 38. Pp. 76–100.

Davis, S. 1989. *Man of all seasons. An Aboriginal perspective of the natural environment.* Angus & Robertson, Sydney.

Dawes, W. 1790–91. *Vocabulary of the language of New South Wales in the neighbourhood of Sydney.* Unpublished notebooks in the William Marsden collection [M686] in the library of the School of Oriental & African Studies, London. Copy of notebooks held by the Mitchell Library, Sydney.

Dawson, J. 1881. *Australian Aborigines.* Robertson, Melbourne.

Day, D. 2001. *Claiming a continent. A new history of Australia.* Harper Collins Publishers, Sydney.

De Vries, S. 2005. Farming fresh fields. Pp. 3–69 in *Great pioneer woman of the outback.* Harper Collins, Sydney.

Delbridge, A., Bernard, J.R.L., Blair, D., Butler, S., Peters, P. & Yallop, C. 1997. *The Macquarie dictionary.* Third edn. Macquarie University, Sydney.

Dench, A. 1994. Nyungar. Pp. 173–92 in N. Thieberger & W. McGregor (eds) *Macquarie Aboriginal words.* Macquarie Library, Macquarie University, New South Wales.

Denoon, D. & Mein-Smith, P. 2000. *A history of Australia, New Zealand, and the Pacific.* Blackwell Publishers, Oxford.

Desmond, R. 1989a. Strange and curious plants (1772–1820). Pp. 1–9 in F.N. Hepper (ed.) *Plant hunting for Kew.* Her Majesty's Stationery Office, London.

Desmond, R. 1989b. From rhododendrons to tropical herbs (1820–1939). Pp. 11–20 in F.N. Hepper (ed.) *Plant hunting for Kew.* Her Majesty's Stationery Office, London.

Desmond, R. 1995. *Kew. The history of the Royal Botanic Gardens.* The Harvill Press with the Royal Botanic Gardens, Kew.

Diamond, J. 1998. *Guns, germs and steel. A short history of everybody for the last 13,000 years.* Vintage, London.

Dixon, R.M.W., Ramson, W.S. & Thomas, M. 1992. *Australian Aboriginal words in English. Their origin and meaning.* Oxford University Press Australia, Melbourne.

Dobkin de Rios, M. & Stachalek, R. 1999. The *Duboisia* genus, Australian Aborigines and suggestibility. *Journal of Psychoactive Drugs.* Vol. 32, no. 2, pp. 155–61.

Donaldson, I. & Donaldson, T. (eds) 1985. *Seeing the first Australians.* Allen & Unwin, Sydney.

Donaldson, T. 1985. Hearing the First Australians. Pp. 76–91 in I. & T. Donaldson (eds) *Seeing the first Australians.* Allen & Unwin, Sydney.

Douglas, M.T. 1982. *In the active voice.* Routledge & Kegan Paul, London.

Drummond, H. [sic. J.] 1862. Useful products of Western Australia. *The Technologist.* Vol. 2, pp. 25–8.

Duranti, A. (ed.) 2001. *Linguistic anthropology. A reader.* Blackwell Publishers, Malden, MA.

Earl, G.W. 1846. On the Aboriginal tribes of the northern coast of Australia. *Royal Geographic Society of London.* Vol. 16, pp. 239–51.

Earl, J.W. & McCleary, B.V. 1994. Mystery of the poisoned expedition. *Nature.* Vol. 368, no. 6473, pp. 683–4.

East, J.J. 1889. The Aborigines of south and Central Australia. *Proceedings of the Field Naturalist Section of the Royal Society of South Australia.* Pamphlet.

Eckert, J. & Robinson, R.D. 1990. The fishes of the Coorong. *The South Australian Naturalist,* vol. 65, no. 1, pp. 4–30.

Elkin, A.P. 1964. *The Australian Aborigines. How to understand them.* Fourth edn. Angus & Robertson, Sydney.

Else-Mitchell, R. 1939. George Caley. His life and work. *Royal Australian Historical Society. Journal & Proceedings.* Vol. 25, pt 6, pp. 437–542.

Else-Mitchell, R. 1966. Caley, George (1770–1829). Pp. 194–5 in D. Pike (ed.) *Australian dictionary of biography. Volume 1. 1788–1850, A–H.* Melbourne University Press, Melbourne.

Erdos, R. 1967. Leichhardt, Friedrich Wilhelm Ludwig (1813–1848). Pp. 102–4 in D. Pike (ed.) *Australian dictionary of biography. Volume 2. 1788–1850, I–Z.* Melbourne University Press, Melbourne.

Erickson, R. 1966. Drummond, James (1784–1863). Pp. 325–7 in D. Pike (ed.) *Australian dictionary of biography. Volume 1. 1788–1850, A–H.* Melbourne University Press, Melbourne.

Erickson, R. 1969. *The Drummonds of Hawthornden.* Lamb Paterson, Perth.

Etheridge, R. 1893. The "mirrn-yong" heaps at the North-west Bend of the River Murray. *Transactions, Proceedings & Report of the Royal Society of South Australia.* Vol. 17, pp. 21–4.

Everist, S.L. 1981a. *Poisonous plants of Australia.* Revised edn. Australian Natural Science Library. Angus & Robertson, Sydney.

Everist, S.L. 1981b. The history of poisonous plants in Australia. Pp. 223–55 in D.J. Carr & S.G.M. Carr (eds) *Plants and man in Australia.* Academic Press, Sydney.

Ewing, T.J. 1846. Statistics of Tasmania, 1838–1841: Extracts from the introductory letter to his Excellency Sir John Franklin. *Tasmanian Journal of Natural Science.* Vol. 7, pp. 141–50.

Eyre, E.J. 1845. *Journals of expeditions of discovery.* 2 volumes. Boone, London.

Farb, P. & Armelagos, G. 1980. *Consuming passions. The anthropology of eating.* Houghton Mifflin Company, Boston.

Favenc, E. 1888. *The history of Australian exploration from 1788 to 1888.* Turner & Henderson, Sydney.

Fforde, C. 2002. Yagan. Pp. 229–41 in C. Fforde, J. Hubert & P. Turnbull (eds) *The dead and their possessions: repatriation in principle, policy and practice.* Routledge, London.

Field, D.V. 1993. In the wake of the *Endeavour.* Banks's botanical legacy. *Endeavour, New Series.* Vol. 17, no. 3, pp. 141–6.

Finlayson, H.H. 1952. *The red centre. Man and beast in the heart of Australia.* Angus & Robertson, Sydney.

Finney, C.M. 1984. *To sail beyond the sunset. Natural history in Australia 1699–1829.* Rigby, Adelaide.

Flannery, T.F. 1994. *The future eaters. An ecological history of the Australasian lands and people.* Reed Books, Sydney.

Flannery, T.F. 1997. The fate of empire in low- and high-energy ecosystems. Pp. 46–59 in T. Griffiths & L. Robin (eds) *Ecology and empire. Environmental history of settler societies.* Melbourne University Press, Melbourne.

Flannery, T.F. (ed.) 1998. *The explorers.* Introduced by T.F. Flannery. Text Publishing, Melbourne.

Flannery, T.F. (ed.) 1999. *The birth of Sydney.* Introduced by T.F. Flannery. Text Publishing, Melbourne.

Flannery, T.F. 2000. The indefatigable Matthew Flinders. Pp. vii–xxxiv as an introduction to M. Flinders, 1814, *Terra Australis.* Text Publishing, Melbourne.

Flannery, T.F. (ed.) 2002. *The birth of Melbourne.* Introduced by T.F. Flannery. Text Publishing, Melbourne.

Flinders, M. 1814a. *A voyage to Terra Australis. Undertaken for the purpose of completing the discovery of that vast country, and prosecuted in the years 1801, 1802, and 1803, in his majesty's ship the Investigator ...* 2 volumes, including folio of the *Atlas of Terra Australis.* G. & W. Nicol, London.

Flinders, M. 1814b. *Terra Australis.* Edited and introduced by T.F. Flannery. 2000. Text Publishing, Melbourne.

Flower, R. 2000. *Characterisation of anti-viral compounds in Australian bush medicines.* Research Paper No. 00/006. Rural Industries Research & Development Corporation, Canberra.

Foelsche, P. 1881. Notes on the Aborigines of North Australia. *Transactions of the Royal Society of South Australia.* Vol. 5, pp. 1–18.

Foelsche, P. 1895. On the manners, customs, &c., of some tribes of the Aborigines, in the neighbourhood of Port Darwin and the west coast of the Gulf of Carpentaria, north Australia. *Journal of the Anthropological Institute of Great Britain & Ireland.* Vol. 24, pp. 190–8.

Foley, P. 2006. *Duboisia myoporoides*: the medical career of a native Australian plant. *Historical Records of Australian Science*. Vol. 17, pt 1, pp. 31–69.

Foley, W.A. 1997. *Anthropological linguistics. An introduction*. Blackwell Publishers, Cambridge, MA.

Ford, R.I. 1978. Ethnobotany: historical diversity and synthesis. Pp. 33–49 in R.I. Ford (ed.) *The nature and status of ethnobotany*. Anthropological Papers. Museum of Anthropology, University of Michigan No. 67. Ann Arbour, Michigan.

Forrest, J. 1875. *Explorations in Australia*. Sampson Low, Marston, Low & Searle, London. Facsimile edn published by the Libraries Board of South Australia, Adelaide, 1969.

Forster, H.W. 1969. Curr, Edward Micklethwaite (1820–1889). P. 508 in D. Pike (ed.) *Australian dictionary of biography. Volume 3. 1851–1890, A–C*. Melbourne University Press, Melbourne.

Foster, R.G.K., Monaghan, P. & Mühlhäusler, P. 2003. Early forms of Aboriginal English in South Australia, 1840s–1920s. *Pacific Linguistics* No. 538. Research School of Pacific and Asian Studies, Australian National University, Canberra.

Frankel, D. 1982. An account of Aboriginal use of the yam-daisy. *The Artefact*. Vol. 7, pp. 43–5.

Frazer, J.G. 1890. *The golden bough. A study in comparative religion*. 2 volumes. Macmillan, London.

Frazer, J.G. 1910. *Totemism and exogamy. A treatise on certain early forms of superstition and society*. Macmillan, London.

Froggatt, W.W. 1932. The curators and botanists of the Botanic Gardens, Sydney. *Journal & Proceedings of the Royal Australian Historical Society*. Vol. 18, pt 3, pp. 101–33.

Gandevia, B. 1981. 'A-going for greens'. Pp. 256–65 in D.J. Carr & S.G.M. Carr (eds) *Plants and man in Australia*. Academic Press, Sydney.

Gara, T. 1985. Aboriginal techniques for obtaining water in South Australia. *Journal of the Anthropological Society of South Australia*. Vol. 23, pt 2, pp. 6–11.

Gason, S. 1879. The 'Dieyerie' Tribe. Pp. 66–86 in G. Taplin (ed.) *Folklore, manners, customs and languages of the South Australian Aborigines*. South Australian Government Printer, Adelaide.

Gilbert, L.A. 1966. Banks, Sir Joseph (1743–1820). Pp. 52–5 in D. Pike (ed.) *Australian dictionary of biography. Volume 1. 1788–1850, A–H*. Melbourne University Press, Melbourne.

Gilbert, L.A. 1981. Plants, politics and personalities in colonial New South Wales. Pp. 220–58 in D.J. Carr & S.G.M. Carr (eds) *People and plants in Australia*. Academic Press, Sydney.

Giles, E. 1889. *Australia twice traversed. The romance of exploration, being a narrative compiled from the journals of five exploring expeditions into and through central South Australia and Western Australia, from 1872 to 1876*. Sampson, Low, Marston, Searle & Rivington, London.

Gill, A.M., Moore, P.H.R. & Armstrong, J.P. 1991. *Bibliography of fire ecology in Australia*. Department of Bush Fire Services, Sydney.

Gill, T. 1909. A cruise in the S.S. 'Governor Musgrave'. *Proceedings of the Royal Geographical Society of Australasia. South Australian Branch*. Volume 10, pp. 90–184.

Gillbank, L. 1996a. Mueller's naming of places and plants in Central Australia – Victorian eponyms. *Victorian Naturalist*. Vol. 113, pt.4, pp. 219–26.

Gillbank, L. 1996b. Review of 'The natural Art of Louisa Atkinson' by Elizabeth Lawson. *Victorian Naturalist*. Vol. 113, pt.4, pp. 229–30.

Gillbank, L. & Maroske, S. 1996. Behind the botany of the Horn Expedition: Ferdinand Mueller's documentation of the larapintine flora. Pp. 209–24 in S.R. Morton & D.J. Mulvaney (eds) *Exploring Central Australia. Society, the environment and the 1894 Horn Expedition*. Surrey Beatty & Sons, Sydney.

Gilmore, M. 1935. *More recollections*. Angus & Robertson, Sydney.

Goddard, C. 1992. *Pitjantjatjara/Yankunytjatjara to English dictionary*. Second edn. Institute of Aboriginal Development, Alice Springs, Northern Territory.

Goddard, C. & Kalotas, A. (eds) 1988. *Punu. Yankunytjatjara plant use*. Angus & Robertson, Sydney.

Good, P. 1801–03. *The journal of Peter Good. Gardener on Matthew Flinders voyage to Terra Australis 1801–03*. Published in 1981. Introduced & edited by P.I. Edwards. Bulletin of the British Museum (Natural History). Historical Series Vol. 9, London.

Goodale, J.C. 1971. *Tiwi wives. A study of the women of Melville Island, north Australia*. University of Washington Press, Seattle.

Gott, B. 1982a. Ecology of root use by the Aborigines of Southern Australia. *Archaeology in Oceania*. Vol. 17, pp. 59–67.

Gott, B. 1982b. *Kunzea pomifera* - Dawson's 'Nurt'. *The Artefact*. Vol. 7, part 1, pp. 3–17.

Gott, B. 1983. Murnong – '*Microseris scapigera*'. A study of a staple food of Victorian Aborigines. *Australian Aboriginal Studies*. No. 2, pp. 2–17.

Gott, B. 1984. Victorian ethnobotanical records. *Australian Aboriginal Studies*. No. 1, p. 56.

Gott, B. 1985a. Plants mentioned in Dawson's *Australian Aborigines*. *The Artefact*. Vol. 10, pp. 3–14.

Gott, B. 1985b. The use of seeds by Victorian Aborigines. Pp. 25–30 in G.P. Jones (ed.) *The food potential of seeds from Australian native plants*. Proceedings of a colloquium held at Deakin University on 7 March 1984. Deakin University Press, Victoria.

Gott, B. 1999. Cumbungi, *Typha* species: a staple Aboriginal food in southern Australia. *Australian Aboriginal Studies*. No. 1 pp. 33–50.

Gott, B. & Conran, J. 1991. *Victorian Koorie plants. Some plants used by Victorian Koories for food, fibre, medicines and implements*. Yangernnanock Women's Group, Hamilton, Victoria.

Green, J.W. 1990. History of early Western Australian herbaria. Pp. 23–7 in P.S. Short (ed.) *History of systematic botany in Australasia. Proceedings of a symposium held at the University of Melbourne*. Australian Systematic Botany Society, Melbourne.

Grey, G. 1841. *Journals of two expeditions of discovery in north-west and Western Australia during the years 1837, 38 and 39* ... Boone, London.

Grey, M. 1997. A new species of *Tetragonia* (Aizoaceae) from arid Australia. *Telopea*. Vol. 7, part 2, pp. 119–27.

Griffin, T. & McCaskill, M. (eds). 1986. *Atlas of South Australia*. South Australian Government Printer, Adelaide.

Griffiths, T. 1996. *Hunters and collectors. The antiquarian imagination in Australia*. Cambridge University Press, Cambridge.

Grimshaw, P. , Lake, M., McGrath, A. & Quartly, M. 1994. *Creating a nation*. McPhee Gribble, Melbourne.

Gunn, R.C. 1842. Remarks on the Indigenous vegetable products of Tasmania available as food for man. *Tasmanian Journal of Natural Science*. Pp. 35–52.

Gunn, R.C. 1847. On the Bunyip of Australia Felix. *Tasmanian Journal of Natural Science*. Vol. 3, no. 2, pp. 147–9.

Haddon, A.C. (ed.) 1904–35. *Reports of the Cambridge Anthropological Expedition to Torres Straits*. Cambridge University Press, Cambridge.

Hagger, J. 1979. *Australian colonial medicine*. Rigby, Adelaide.

Hall, N. 1978. *Botanists of the eucalypts*. Commonwealth Scientific & Industrial Research Organisation, Melbourne.

Hallam, S.J. 1975. *Fire and hearth. A study of Aboriginal usage and European usurpation in South-western Australia*. Australian Institute of Aboriginal Studies, Canberra.

Hardwick, R.J. 2001. *Nature's larder. A field guide to the native food plants of the NSW South Coast*. Homosapien Books, Jerrabomberra, New South Wales.

Hardy, B. 1969. *West of the Darling*. Jacaranda Press, Milton, Queensland.

Harkins, J. 1994. *Bridging two worlds. Aboriginal English and crosscultural understanding*. University of Queensland Press, St Lucia.

Harshberger, J.W. 1896. The purposes of ethnobotany. *Botanical Gazette*. Vol. 21, pp. 146–54.

Hasluck, A. 1955. *Portrait with background. A life of Georgiana Molloy*. Oxford University Press, Melbourne.

Hasluck, A. 1967. Yagan (– 1833). P. 632 in in D. Pike (ed.) *Australian dictionary of biography. Volume 1. 1788–1850, A–H*. Melbourne University Press, Melbourne.

Hassell, E. 1936. Notes on the ethnology of the Wheelman tribe of southwestern Australia. *Anthropos*. Vol. 31, pp. 679–711.

Hastings, R.B. 1989. The Sir Joseph Banks Centre and the economic botany collections at Kew. *Endeavour, New Series*. Vol. 13, no. 4, pp. 174–8.

Hawker, J.C. 1841–45. *Journal of an expedition to the River Murray, against the natives, in order to recover sheep taken by them from Messrs Field and Inman on their overland journey from New South Wales to South Australia; also to protect another overland party expected almost immediately*. Transcribed by I. Palios, 1981. State Library of South Australia, Adelaide.

Hazzard, M. 1982. Forgotten heritage. The life-work of Ellis Rowan. Pp. 1–15 in *Flower paintings of Ellis Rowan. From the collection of the National Library of Australia*. National Library of Australia, Canberra.

Hazzard, M. 1988. Rowan, Marian Ellis (1848–1922). Pp. 465–6 in G. Searle (ed.) *Australian dictionary of biography. Volume 11. 1891–1939, Nes–Smi*. Melbourne University Press, Melbourne.

Head, L. 2000a. *Second nature. The history and implications of Australia as Aboriginal landscape*. Syracuse University Press, New York.

Head, L. 2000b. *Cultural landscapes and environmental change*. Arnold, London.

Heath, J. 1978. Linguistic approaches to Nunggubuyu ethnozoology and ethnobotany. Pp. 40–55 in L.R. Hiatt (ed.) *Australian Aboriginal concepts*. Australian Institute of Aboriginal Studies, Canberra.

Helms, R. 1896. Anthropology of the Elder Expedition. *Transactions, Proceedings & Report of the Royal Society of South Australia*. Vol. 16, part 3, pp. 238–332.

Henderson, J. & Dobson, V. 1994. *Eastern and Central Arrernte to English dictionary*. Arandic Languages Dictionaries Program, Language Centre, Institute for Aboriginal Development, Alice Springs.

Henderson, J. & Nash, D. 2002. *Language in native title*. Aboriginal Studies Press, Canberra.

Henshall, T., Jambijinpa, D., Spencer, J.N., Kelly, F.J., Bartlett, P., Mears, J., Coulshed, E., Robertson, G.J. & Granites, L.J. 1980. *Ngurrju maninja kurlangu. Yapa nyurnu kurlangu. Bush medicine*. Revised edn. Warlpiri Literature Production Centre, Yuendumu, Northern Territory.

Hepper, F.N. (ed.) 1982. *Kew Gardens for science and pleasure*. Her Majesty's Stationery Office, London.

Hepper, F.N. (ed.) 1989. *Plant hunting for Kew*. Her Majesty's Stationery Office, London.

Herbert, D.A. 1966. Bidwill, John Carne (1815–1853). Pp. 98–9 in D. Pike (ed.) *Australian dictionary of biography. Volume 1. 1788–1850, A–H*. Melbourne University Press, Melbourne.

Hercus, L.A. 1986. *The languages of Victoria. A late survey. Parts 1–2*. Australian Aboriginal Studies No. 17, Linguistic Series No. 6. Australian Institute of Aboriginal Studies, Canberra.

Hercus, L.A. 1992. *A Nukunu dictionary*. Australian Institute of Aboriginal & Torres Strait Islander Studies, Canberra.

Hercus, L.A. 1994. Paakantyi. Pp. 41–60 in N. Thieberger & W. McGregor (eds) *Macquarie Aboriginal words*. Macquarie Library, Macquarie University, New South Wales.

Hercus, L.A. & Simpson, J. 2002. Indigenous Placenames: an Introduction. Pp. 1–23 in L. Hercus, F. Hodges, & J. Simpson (eds) *The land is a map. Placenames of indigenous origin in Australia*. Pandanus Books, Australian National University, Canberra.

Hewson, H. 1982. Ellis Rowan. An appreciation. Pp. 17–20 in *Flower paintings of Ellis Rowan. From the collection of the National Library of Australia*. National Library of Australia, Canberra.

Hiatt, L.R. (ed.) 1975. *Australian Aboriginal mythology. Essays in honour of W.E.H. Stanner*. Australian Institute of Aboriginal Studies, Canberra.

Hiatt, L.R. 1996. *Arguments about Aborigines. Australia and the evolution of social anthropology*. Cambridge University Press, Cambridge.

Hicks, C.S. 1963. Climatic adaptation and drug habituation of the Central Australian Aborigine. *Perspectives in Biology & Medicine*. Vol. 7, no. 1, pp. 39–57.

Hicks, C.S. & Le Messurier, H. 1935. Preliminary observations on the chemistry and pharmacology of the alkaloids of *Duboisia hopwoodii*. *Australian Journal of Experimental Biology & Medical Science*. Vol. 13, pp. 175–88.

Hiddins, L.J. 2000a. Introduction. Pp. v–x in L. Leichhardt 1847. *Journal of an overland expedition in Australia, from Moreton Bay to Port Essington, a distance of upward of 3000 miles, during the years 1844–1845*. T. & W. Boone, London. Reprinted by Corkwood Press, Adelaide.

Hiddins, L.J. 2000b. *The bush tucker guide. 60 of the most common species in northern Australia illustrated region-by-region*. Viking Penguin Books, Melbourne.

Hiddins, L.J. 2001. *Bush tucker field guide*. Penguin Books, Melbourne.

Hill, B. 2002. *Broken song. T.G.H. Strehlow and Aboriginal possession*. Knopf, Sydney.

Hill, R.S. 2004. Origins of the southeastern Australian vegetation. *Philosophical Transactions: Biological Sciences. Royal Society*. Vol. 359, pp. 1537–49.

Hoare, M.E. 1969. The life and training of an explorer. *Records of the Australian Academy of Science*. Vol. 1, no. 4, pp. 27–31.

Hodgkinson, C. 1845. *Australia from Port Macquarie to Moreton Bay: with descriptions of the natives, their manners and customs; the geology, natural productions, fertility, and resources of the region ...* T. & W. Boone, London.

Holden, R. & Holden, N. 2001. *Bunyips. Australia's folklore of fear*. National Library of Australia, Canberra.

Holden, R.H. 1966. Bunce, Daniel (1813–1872). Pp. 176–7 in D. Pike (ed.) *Australian dictionary of biography. Volume 1. 1788–1850, A–H*. Melbourne University Press, Melbourne.

Hooker, W.J. 1830 A journal of a two months' residence on the banks of the rivers Brisbane and Logan, on the east coast of New Holland. *Hooker's Botanical Miscellany*. Vol. 1, pp. 237–356.

Hooker, W.J. 1834–42. *Journal of botany*. 4 volumes. Longman, Rees, Orme, Brown, Green & Longman, London.

Hooker, W.J. 1858. *Museum of Economic Botany. A popular guide to the useful and remarkable vegetable products in the two museum buildings of the Royal Gardens of Kew*. Longman, London.

Hope, G.S & Coutts, P. J.F. 1971. Past and present Aboriginal food resources at Wilsons Promontory, Victoria. *Mankind*. Vol. 8, pt 2, pp. 104–14.

Horne, G. & Aiston, G. 1924. *Savage life in Central Australia*. Macmillan, London.

Horstman, M. & Wightman, G. 2001. Karparti ecology: recognition of Aboriginal ecological knowledge and its application to management in north-western Australia. *Ecological Management & Restoration*. Vol. 2, part 2, pp. 99–109.

Howitt, A.W. 1904. *Native tribes of south-east Australia*. Macmillan, London.

Howitt, A.W. & Siebert, O. 1904. Legends of the Dieri and kindred tribes of Central Australia. *Journal of the Anthropological Institute of Great Britain & Ireland*. Vol. 34, pp. 100–29.

Hudson, K. 1975. *A social history of museums*. Humanities Press, Atlantic Highlands, New Jersey.

Hughes, R.E. 1975. James Lind and the cure of scurvy: an experimental approach. *Medical History*. Vol. 19, no. 4, pp. 342–51.

Hunter, J. 1793. *An historical journal of the transactions at Port Jackson and Norfolk Island with the discoveries which have been made in New South Wales ....* John Stockdale, London.

Hutchinson, R.C. 1958. *Food for the people of Australia*. Angus & Robertson, Sydney.

Hyam, G.N. 1943. Living off the land. *Victorian Naturalist*. Vol. 59, pp. 171–3.

Iredale, T. 1955. Bellingshausen in Australia. *Proceedings of the Royal Zoological Society of New South Wales*. 1953–54 volume, pp. 34–6.

Isaacs, J. 1987. *Bush food. Aboriginal food and herbal medicine*. Weldons, Sydney.

Jackes, B.R. 1990. Retracing the botanical steps of Leichhardt and Gilbert in June 1845. Pp. 165–9 in P.S. Short (ed.) *History of systematic botany in Australasia. Proceedings of a symposium held at the University of Melbourne*. Australian Systematic Botany Society, Melbourne.

Jain, S.K. (ed.) 1987. *A manual of ethnobotany*. Scientific Publishers, Jodhpur.

Jenkin, G. 1979. *Conquest of the Ngarrindjeri. The story of the Lower Lakes tribes*. Rigby, Adelaide.

Jessop, J.P. (ed.) 1981. *Flora of Central Australia*. The Australian Systematic Botany Society. Reed Books, Sydney.

Jessop, J.P. & Toelken, H.R. 1986. *Flora of South Australia. Parts 1–4*. South Australian Government Printing Division, Adelaide.

Jewett, F.L. & McCausland, C.L. 1958. *Plant hunters*. Houghton Mifflin, Boston.

Johnston, T. Harvey 1916. Frederick Manson Bailey. *Proceedings of the Royal Society of Queensland*. Vol. 28, pp. 3–10.

Johnston, T. Harvey 1939. Pituri. *Mankind*. Vol. 2, no. 7, pp. 224–5.

Johnston, T. Harvey & Cleland, J.B. 1933–34. The history of the Aboriginal narcotic, pituri. *Oceania*. Vol. 4, no. 2, pp. 201–23; no. 3, pp. 268–89.

Johnston, T. Harvey & Cleland, J.B. 1942. Aboriginal names and uses of plants in the Ooldea region, South Australia. *Transactions of the Royal Society of South Australia*. Vol. 66, pp. 93–103.

Johnston, T. Harvey & Cleland, J.B. 1943. Native names and uses of plants in the north-eastern corner of South Australia. *Transactions of the Royal Society of South Australia*. Vol. 67, part 1, pp. 149–73.

Jones, D.L. 1993. *Cycads of the world*. Reed Books Australia, Sydney.

Jones, D.L. 1995. *Palms throughout the world*. Reed Books Australia, Sydney.

Jones, D.S., Mackay, S., & Pisani, A.M. 1997. Patterns in the valley of the Christmas Bush. *Victorian Naturalist*. Vol. 114, pt. 5, pp. 246–9.

Jones, P.G. 1990. Ngapamanha: a case study in the population history of north-eastern South Australia. Pp. 157–73 in P. Austin, R.M.W. Dixon, T. Dutton & I. White (eds) *Language and history: Essays in honour of Luise Hercus*. Pacific Linguistics, C–116. Australian National University, Canberra.

Jones, P.G. 1996. *'A box of native things': Ethnographic collectors and the South Australian Museum, 1830s – 1930s*. Postgraduate thesis. University of Adelaide, Adelaide.

Jones, P.G. 2007. *Ochre and rust. Artefacts and encounters on Australian frontiers*. Wakefield Press, Adelaide.

Jones, P.G. & Sutton, P. 1986. *Art and land. Aboriginal sculptures of the Lake Eyre basin*. South Australian Museum & Wakefield Press, Adelaide.

Jones, R.M. 1969. Firestick farming. *Australian Natural History*. Vol. 16, pp. 224–8.

Jones, R.M. 1974. Appendix on Tasmanian Tribes. Pp. 317–54 in N.B. Tindale (1974) *Aboriginal tribes of Australia. Their terrain, environmental controls, distribution, limits, and proper names*. Australian National University Press, Canberra.

Jordan, C. 2005. *Picturesque pursuits: colonial women artists and their amateur tradition*. Melbourne University Press, Melbourne.

Jorgenson, J. 1837. A narrative of the habits, manners and customs of the Aborigines of Van Diemen's Land. Pp. 47–131 in N.J.B. Plomley (ed.) 1991 *Jorgen Jorgenson and the Van Diemen's Land*. Blubber Head Press, Hobart.

Kaberry, P. M. 1939. *Aboriginal woman. Sacred and profane*. Routledge, London.

Kalma, J.D. & McAlpine, J.R. 1983. Climate and man in the Centre. Pp. 46–69 in G. Crook (ed.) *Man in the Centre*. CSIRO Division of Groundwater Research, Perth.

Kamminga, J. 1988. Wood artefacts: a checklist of plant species utilised by Australian Aborigines. *Australian Aboriginal Studies*. No. 2, pp. 26–55.

Kean, J. 1991. Aboriginal–Acacia relationships in Central Australia. *Records of the South Australian Museum*. Vol. 24, part 2, pp. 111–24.

Keen, I. 1994. *Knowledge and secrecy in an Aboriginal religion. Yolngu of north-east Arnhem Land*. Oxford University Press, Melbourne.

Kempe, H. 1880–82. Plants indigenous to the neighbourhood of Hermannsburg on the River Finke, Central Australia. *Transactions of the Royal Society of South Australia*. Vol. 3, pp. 120–89; Vol. 5, pp. 19–23.

Kempe, H. 1891. A grammar and vocabulary of the language spoken by the Aborigines of the Macdonnell Ranges, South Australia. *Transactions of the Royal Society of South Australia*. Vol. 14, pp. 1–54.

Kemsley, D. 1951. Life in the Dead Heart. *Wild Life*. May 1951 issue, pp. 434–40.

Kenny, J. (ed.) 1973. *Bennelong, first notable Aboriginal. A report from original sources*. Royal Australian Historical Society in association with the Bank of New South Wales, Sydney.

Kimber, R.G. 1984. Resource use and management in central Australia. *Australian Aboriginal Studies*. No. 2, pp. 12–23.

King, P. P. 1827. *Narrative of a survey of the inter-tropical and western coasts of Australia. Performed between the years 1818 and 1822 with an appendix containing various subjects relating to hydrography and natural history*. Two volumes. Murray, London.

Koch, M. 1898. A list of plants collected on Mt. Lyndhurst Run, S. Australia. *Transactions of the Royal Society of South Australia*. Vol. 22, pp. 101–18.

Koch. M. 1900. Supplementary list of plants from Mount Lyndhurst Run. *Transactions of the Royal Society of South Australia*. Vol. 24, pp. 81–5.

Kodicek, E.H. & Young, F.G. 1969. Captain Cook and scurvy. *Notes & Records of the Royal Society of London*. Vol. 24, pp. 43–63.

Kohlstedt, S.G. 1983. Australian museums of natural history. Public priorities and scientific initiatives in the 19th century. *Historical Records of Australian Science*. Vol. 5, part 4, pp. 1–29.

Kolig, E. 1989. *Dreamtime politics: religion, world view and Utopian thought in Australian Aboriginal society*. Reimer, Berlin.

Kuper, A. 1983. *Anthropology and anthropologists. The modern British school*. Revised edn. Routledge & Kegan Paul, London.

Kynaston, E. 1981. *A man on edge. A life of Baron Sir Ferdinand von Mueller*. Allen Lane, Melbourne.

Kyriazis, S. 1995. *Bush medicine of the northern peninsula area of Cape York*. Nai Beguta Agama Aboriginal Corporation, Bamaga, Queensland.

Lakoff, G. 1987. *Women, fire, and dangerous things. What categories reveal about the mind*. University of Chicago Press, Chicago.

Lampert, R.J. & Sanders, F. 1973. Plants and men on the Beecroft Peninsula, New South Wales. *Mankind*. Vol. 9, no. 2, pp. 96–108

Lands, M. (ed.) 1987. *Mayi. Some bush fruits of Dampierland*. Magabala Books, Kimberley Aboriginal Law & Culture Centre, Broome, WA.

Lang, A. 1905. *The secret of the totem*. Longmans, Green & Co., London.

Laramba Community Women 2003. *Anmatyerr ayey arnang-akert. Anmatyerr plant stories*. Compiled J. Green. Institute for Aboriginal Development Press, Alice Springs, Northern Territory.

Lassak, E.V. & McCarthy, T. 1983. *Australian medicinal plants*. Methuen, Melbourne.

Latz, P. 1974. Central Australian species of Nicotiana. *Australian Plants*. March issue.

Vol. 7, pp. 280–3.

Latz, P. 1995. *Bushfires and bushtucker. Aboriginal plant use in Central Australia.* Institute of Aboriginal Development, Alice Springs, Northern Territory.

Laurie, A. 1966. Davis, James (1808–1889). Pp 294–5 in D. Pike (ed.) *Australian dictionary of biography. Volume 1. 1788–1850, A–H.* Melbourne University Press, Melbourne.

Lawrence, R. 1968. *Aboriginal habitat and economy.* Geography Occasional Paper No. 6. Australian National University, Canberra.

Lee, I. 1925. *Early explorers in Australia. From the log-books and journals, including the diary of Allan Cunningham, botanist, from March 1, 1817, to November 19, 1818.* Methuen & Co., London.

Lees, E.H. 1915. What is nardoo? *Victorian Naturalist.* Vol. 31, pt.9, pp. 133–5.

Leichhardt, F.W.L. 1813–48. *Explorer at rest. Ludwig Leichhardt at Port Essington and on the homeward voyage 1845–1846.* Introduction & annotations by E.M. Webster, 1986. Melbourne University Press, Melbourne.

Leichhardt, F.W.L. 1842–48. *Dr. Ludwig Leichhardt's letters from Australia during the years March 23, 1842, to April 3, 1848.* Collected & translated from the German, French & Italian by M. Aurousseau. 1968. Published for the Hakluyt Society by Cambridge University Press, London.

Leichhardt, F.W.L. 1847a. *Journal of an overland expedition in Australia, from Moreton Bay to Port Essington, a distance of upward of 3000 Miles, during the years 1844–1845.* T. & W. Boone, London. Reprinted in 2000 by Corkwood Press, Adelaide.

Leichhardt, F.W.L. 1847b. Lectures on the geology, botany, natural history, and capabilities of the country between Moreton Bay and Port Essington. *The Tasmanian Journal of Natural Science.* Vol. 3, no. 2, pp. 147–149.

Leigh, W.H. 1839. *Reconnoitering voyages and travels with adventures in the new colonies of South Australia.* Smith, Elder & Co., London.

Lemmon, K. 1968. *The golden age of plant hunters.* Phoenix House, London.

Levitt, D. 1981. *Plants and people. Aboriginal uses in plants on Groote Eylandt.* Aboriginal Studies Press, Canberra.

Lewis, D. 2006. The fate of Leichhardt. *Historical Records of Australian Science.* Vol. 17, no. 1, pp. 1–30.

Lindley, J. 1840. A sketch of the vegetation of the Swan River colony ... together with an alphabetical and systematical index to the first twenty-three volumes of Edwards's botanical register. James Ridgway. London.

Lines, W.J. 1994. *An all consuming passion. Origins, modernity, and the Australian life of Georgiana Molloy.* Allen & Unwin, Sydney.

Linnaeus, C. 1753. *Species plantarum.* Facsimile printed 1957, with an introduction by W.T. Stearn. 2 volumes. Ray Society, London.

Liversidge, A. 1880. The alkaloid from pituri. *Journal & Proceedings of the Royal Society of New South Wales.* Vol. 14, pp. 123–32.

Lounsberry, A. 1899. *A guide to the wild flowers.* Frederick A. Stokes, New York.

Lounsberry, A. 1900. *Guide to trees.* Frederick A. Stokes, New York.

Lounsberry, A. 1901. *Southern wild flowers and trees.* Frederick A. Stokes, New York.

Lourandos, H. 1997. *Continent of hunter-gatherers. New perspectives in Australian prehistory.* Cambridge University Press, Cambridge.

Low, T. 1988. *Wild food plants of Australia.* Angus & Robertson, Sydney.

Low, T. 1989. *Bush tucker. Australia's wild food harvest.* Angus & Robertson, Sydney.

Low, T. 1991a. *Wild food plants of Australia.* Revised edn. Angus & Robertson, Sydney.

Low, T. 1991b. *Wild herbs of Australia and New Zealand.* Angus & Robertson, Sydney.

Low, T. 1999. *Feral future. The untold story of Australia's exotic invaders.* Viking, Melbourne.

Lowe, D. 1995. *Forgotten rebels: black Australians who fought back.* Permanent Press, Melbourne.

Lowe, P. 2002. *Hunters and trackers of the Australian desert.* Rosenberg Publishing, Sydney.

Lubbock, J. 1870. *The origin of civilisation and the primitive condition of man.* Longmans, Green & Co., London.

Lumholtz, C.S. 1889. *Among cannibals. An account of four years' travels in Australia and of camp life with the Aborigines of Queensland.* Translated by R.B. Anderson. Charles Scrubner's Sons, New York.

Lycett, J. 1990. *The Lycett album. Drawings of Aborigines and Australian scenery.* The execution of the original drawings dated at between 1820 & 1822. National Library of Australia, Canberra.

Lyte, C. 1980. *Sir Joseph Banks. 18th century explorer, botanist and entrepreneur.* A.H. & A.W. Reed, Sydney.

Lyte, C. 1983. *The plant hunters.* Orbis, London.

McBryde, I. 1986. Exchange in south eastern Australia: an ethnohistorical perspective. *Aboriginal History.* Vol. 8, pt 2, pp. 132–53.

McBryde, I. 1987. Goods from another country: exchange networks and the people of the Lake Eyre Basin. Chapter 13. Pp. 253–459 in D.J. Mulvaney & J.P. White (eds) *Australians to 1788.* Fairfax, Syme & Weldon Associates, Sydney.

McCann, J. 2005. *Maize and grace. Africa's encounter with a new world Crop, 1500–2000.* Harvard University Press, Cambridge, MA.

McCarthy, F.D. 1938–40. 'Trade' in Aboriginal Australia and 'trade' relationships with Torres Strait, New Guinea and Malaya. *Oceania.* Vol. 9, pp. 405–38; vol. 10, pp. 80–104, 171–95.

McCarthy, F.D. 1940. Aboriginal Australian material culture: causative factors in its composition. *Mankind.* Vol. 2, pp. 241–269, 294–320.

McCarthy, F.D. 1966. Bungaree (– 1830). P. 177 in D. Pike (ed.) *Australian dictionary of biography. Volume 1. 1788–1850, A–H.* Melbourne University Press, Melbourne.

McCleary, B.V. & Chick, B.F. 1977. The purification and properties of a thiaminase 1 enzyme from nardoo (*Marsilea drummondii*). *Phytochemistry.* Vol. 16, pp. 207–13.

McConnel, U.H. 1930. The Wik-Munkan tribe of Cape York Peninsula. *Oceania* reprint (originally *Oceania.* Vol. 1, part 1, pp. 97–104, 181–205).

McConnel, U.H. 1953. Native arts and industries on the Archer, Kendall and Holroyd Rivers, Cape York Peninsula, north Queensland. *Records of the South Australian Museum.* Vol. 11, part 1, pp. 1–42.

McCourt, T. 1975. *Aboriginal artefacts.* Rigby, Adelaide.

MacDonald, H. 2005. *Human remains. Episodes in human dissection.* Melbourne University Press, Melbourne.

McEntee, J.C., McKenzie, P & McKenzie, J. 1986. *Witi-ita-nanalpila. Plants and birds of the northern Flinders Ranges and adjacent plains with Aboriginal names.* The authors, South Australia.

McGillivray, D.J. 1969. The first botanical studies in Australia. *Australian Natural History.* Vol. 16, no. 8, pp. 251–4.

McIntosh, T.P. 1927. *Potato. Its history, varieties, culture and diseases.* Oliver & Boyd, Edinburgh.

Macintyre, S. 1999. *A concise history of Australia.* Cambridge University Press, Melbourne.

MacKenzie, J.M. 1991. Plant collecting and imperialism. Pp. 8–14 in J. Illingworth & J. Routh (eds) *Reginald Farrer. Dalesman, planthunter, gardener.* Centre for North-west Regional Studies, University of Lancaster, Lancaster.

Mackerras, M.J. 1969. Bancroft, Joseph (1836–1894). Pp. 84–5 in D. Pike (ed.) *Australian dictionary of biography. Volume 3. 1851–1890, A–C.* Melbourne University Press, Melbourne.

Macknight, C.C. 1976. *The voyage to Marege. Macassan trepangers in northern Australia.* Melbourne University Press, Melbourne.

McKnight, D. 1999. *People, countries, and the rainbow serpent. Systems of classification among the Lardil of Mornington Island.* Oxford University Press, New York.

McLaren, G. 1996. *Beyond Leichhardt. Bushcraft and the exploration of Australia.* Fremantle Arts Centre Press, Fremantle.

McMinn, W.G. 1970. *Allan Cunningham. Botanist and explorer.* Melbourne University Press, Melbourne.

McMinn, W.G. 1971. Botany and geography in early Australia: a case study. *Records of the Australian Academy of Science.* Vol. 2, no. 1, pp. 1–9.

McNicol, S. & Hosking, D. 1994. Wiradjuri. Pp. 79–99 in N. Thieberger & W. McGregor (eds) *Macquarie Aboriginal words.* Macquarie Library, Macquarie University, Sydney.

McPhee, C. 1996. The botanist at Como: Mueller and the Armytage family. *Victorian Naturalist.* Vol. 113, pt.4, pp. 227–8.

MacPherson, J. 1925. The gum-tree and wattle in Australian Aboriginal medical practice. *Australian Nurses Journal.* December 15. Vol. 23, part 12, pp. 588–96.

MacPherson, J. 1939. The Eucalyptus in the daily life and medical practice of the Australian Aborigines. *Mankind.* Vol. 2, part 6, pp. 175–80.

Maddock, K. 1982. *The Australian Aborigines. A portrait of their society.* Second edn. Penguin Books, Melbourne.

Magarey, A.T. 1894–95. Aboriginal water quest. *Proceedings of the Royal Geographical Society of Australasia. South Australian Branch.* Session 1894–5, pp. 3–15.

Magarey, A.T. 1899. Tracking by the Australian Aborigines. *Proceedings of the Royal Geographical Society of Australasia. South Australian Branch.* Vol. 3, pp. 119–26.

Maggiore, P. 1993. Analysis of Australian Aboriginal bush foods. *Australian Aboriginal Studies.* No. 1, pp. 55–8.

Maher, P. 1999. A review of traditional Aboriginal health beliefs. *Australian Journal of Rural Health.* Vol. 7, pp. 229–36.

Maiden, J.H. 1889. *The useful native plants of Australia.* Trubner, London.

Maiden, J.H. 1890. Spinifex resin. *Proceedings of the Linnean Society of New South Wales.* Vol. 14, pt. 3, pp. 639–40.

Maiden, J.H. 1891. *Wattles and wattle-barks, being hints on the conservation and cultivation of wattles together with particulars of their value.* Technical Education Series No. 6. Second edn. George Stephen Chapman, Acting Government Printer, Sydney.

Maiden, J.H. 1892. *A bibliography of Australian economic botany. Part 1.* Technical Education Series No. 10. Government Printer, Sydney.

Maiden, J.H. 1896. Appendix. Notes on some vegetable exudations. Pp. 195–7 in W.B. Spencer (ed.) *Report on the work of the Horn Scientific Expedition to Central Australia. Part III.–geology & botany.* Melville, Mullen & Slade, Melbourne.

Maiden, J.H. 1900. Native food plants. *Agricultural Gazette of New South Wales.* Vol. 10, part 2, pp. 117–30.

Maiden, J.H. 1903. George Caley. Botanical collector in New South Wales, 1800–1810. *Agricultural Gazette of New South Wales.* 16 October 1903, pp. 988–96.

Maiden, J.H. 1909. *Sir Joseph Banks. The 'Father of Australia'.* William Applegate Gullick, Government Printer, Sydney.

Maiden, J.H. 1910. Records of the earlier French botanists as regards Australian plants. *Journal & Proceedings of the Royal Society of New South Wales.* Vol. 44, part 1, pp. 123–55.

Maiden, J.H. 1912. Presidential address. *Journal & Proceedings of the Royal Society of New South Wales.* Vol. 46, part 1, pp. 1–73.

Maiden, J.H. 1918. Notes on *Eucalyptus,* No. VI. *Journal & Proceedings of the Royal Society of New South Wales.* Vol. 52, pp. 486–519.

Maiden, J.H. 1921a. Records of Australian botanists. *Journal & Proceedings of the Royal Society of New South Wales.* Vol. 55, pp. 150–69.

Maiden, J.H. 1921b. Australian botany a century ago. *Journal & Proceedings of the Royal Society of New South Wales.* Vol. 55, pp. xxix–xxxii.

Maiden, J.H. 1928. History of the Sydney Botanic Gardens. Part 1. *Journal & Proceedings of the Royal Australian Historical Society*. Vol. 14, pt. 1, pp. 1–42.

Maiden, J.H. 1931. History of the Sydney Botanic Gardens. Part 2. *Journal & Proceedings of the Royal Australian Historical Society*. Vol. 17, pts 2–3, pp. 126–62.

Maiden, J.H. & Cambage, R.H. 1909. Botanical, topographical and geological notes on some routes of Allan Cunningham. *Journal & Proceedings of the Royal Society of New South Wales*. Vol. 43, pp. 123–38.

Malinowski, B. 1913. *The family among the Australian Aborigines. A sociological study*. Reprinted in 1963. Schocken Books, New York.

Marks, E.N. 1969. Bailey, Frederick Manson (1827–1915). Pp 73–4 in D. Pike (ed.) *Australian dictionary of biography. Volume 3. 1851–1890, A-C*. Melbourne University Press, Melbourne.

Maroske, S. 1996. Introduction to the Mueller issue. *Victorian Naturalist*. Vol. 113, pt. 4, pp. 128–30.

Maroske, S. 2006. Ferdinand Mueller and the shape of nature. *Historical Records of Australian Science*. Vol. 17, no. 2, pp. 147–68.

Martin, G.J. 1995. *Ethnobotany. A methods manual*. Chapman & Hall, London.

Maslin, B.R. 2001. *Wattle: Acacias of Australia*. Electronic resource. Australian Biological Resources Study, Department of Conservation and Land Management, Canberra.

Mathews, R.H. 1903. Languages of the New England Aborigines, New South Wales. *Proceedings of the American Philosophical Society*. Vol. 42, no. 173, pp. 249–63.

Mathews, R.H. 1904. Ethnological notes on the Aboriginal tribes of New South Wales and Victoria. *Journal of the Royal Society of New South Wales*. Vol. 38, pp. 203–381.

Matthews, D.J. 1997. The quandong (*Santalum acuminatum*). *Australian Bushfoods Magazine*. Vol. 1, pp. 14–5.

May, D. 1994. *Aboriginal labour and the cattle industry. Queensland from white settlement to the present*. Cambridge University Press, Cambridge.

Meagher, S.J. 1974. The food sources of the Aborigines of the south-west of Western Australia. *Records of the Western Australian Museum*. Vol. 3, pt. 1, pp. 14–65.

Meagher, S.J. & Ride, W.D.L. 1980. Use of natural resources by the Aborigines of south-western Australia. Pp. 66–80 in R.M. Berndt & C.H. Berndt (eds) *Aborigines of the west. Their past and their present*. University of Western Australia Press, Perth.

Meston, A. 1889. *Report of the Government Scientific Expedition to Bellenden-Ker Range upon the flora and fauna of that part of the colony*. Government Printer, Brisbane:

Meyer, H.A.E. 1843. *Vocabulary of the language spoken by the Aborigines of South Australia*. Allen, Adelaide.

Meyer, H.A.E. 1846. Manners and customs of the Aborigines of the Encounter Bay tribe, South Australia. Reprinted as pp. 183–206 in J.D. Woods (ed.), 1879, *The native tribes of South Australia*. E.S. Wigg, Adelaide.

Mills, E.G. 1999–2000. A rose by any other name would smell as sweet. *Australian Bushfoods Magazine*. No. 14 (December 1999–January 2000), pp. 23–4.

Mitchell, T.L. 1838. *Three expeditions into the interior of eastern Australia*. Two volumes. T. & W. Boone, London.

Mitchell, T.L. 1848. *Journal of an expedition into the interior of tropical Australia, in search of a route from Sydney to the Gulf of Carpentaria*. Longman, Brown, Green & Longmans, London.

Moore, C. 1884. Notes on the genus *Macrozamia*. *Journal and Proceedings of the Royal Society of New South Wales*. Vol. 17, pp. 115–22.

Moore, D.R. 1979. *Islanders and Aborigines at Cape York*. Australian Institute of Aboriginal Studies, Canberra & Humanities Press, New Jersey.

Moore, D.T. 2001. Introduction. Pp. 1–24 in T.G. Vallance, D.T. Moore & E.W. Groves (compilers) *Nature's investigator. The diary of Robert Brown in Australia, 1801–1805*. Australian Biological Resources Study, Canberra.

Moore, H.P. 1924. Notes on the early settlers in South Australia prior to 1836. *Proceedings of Royal Geographical Society of Australasia. South Australian Branch*. Vol. 25, pp. 81–135.

Moorehead, A. 1963. *Cooper's Creek*. Hamish Hamilton, London.

Moorehead, A. 1968. *The fatal impact*. Penguin Books, Melbourne.

Moorehead, A. 1974. King, John (1841–1872). Pp. 28–9 in D. Pike (ed.) *Australian dictionary of biography. Volume 5. 1851–1890, K–Q*. Melbourne University Press, Melbourne.

Moran, R.C. 2004. Dispatches from the fern frontier. Plants with an ancient pedigree are yielding their family secrets to molecular approaches. *Natural History*, October issue, pp. 52–7.

Morgan, J. 1852. *The life and adventures of William Buckley: thirty-two years a wanderer amongst the Aborigines of the unexplored country round Port Phillip*. Republished in 1980 by Australian National University Press, Canberra.

Morley, B.D. & Toelken, H.R. (eds) 1983. *Flowering plants in Australia*. Rigby Publishers, Sydney.

Morphy, H. 1984. *Journey to the crocodile's nest. An accompanying monograph to the film Madarrpa Funeral at Gurka'wuy*. Australian Institute of Aboriginal Studies, Canberra.

Morphy, H. 1988. The original Australians and the evolution of anthropology. Pp. 48–61 in H. Morphy & E. Edwards (eds) *Australians in Oxford*. Monograph 4. Pitt Rivers Museum, University of Oxford, Oxford.

Morrill, J. 1864. *Sketch of a residence among the Aboriginals of northern Queensland for seventeen years*. Republished in 2006 as *Seventeen years wandering among the Aboriginals* by David M. Welch, Virginia, Northern Territory.

Morris, D. 1974. Mueller, Sir Ferdinand Jakob Heinrich von [Baron von Mueller] (1825–1896). Pp. 306–8 in D. Pike (ed.) *Australian dictionary of biography. Volume 5. 1851–1890, K–Q*. Melbourne University Press, Melbourne.

Morton, S.R. & Mulvaney, D.J. (eds) 1996. *Exploring Central Australia. Society, the environment and the 1894 Horn Expedition*. Surrey Beatty & Sons, Sydney.

Moyal, A.M. 1976. *Scientists in nineteenth century Australia. A documentary history*. Cassell Australia, Melbourne.

Moyal, A.M. 1986. *'a bright & savage land.' Scientists in colonial Australia*. Collins, Sydney.

Moyal, A.M. 2001. *Platypus*. Allen & Unwin, Sydney.

Mühlhäusler, P. 1996a. Post-contact languages in mainland Australia. Pp. 11–6 in S.A. Wurm, P. Mühlhäusler & D.T. Tryon (eds). *Atlas of languages of intercultural communication in the Pacific, Asia, and the Americas*. 3 volumes. Mouton de Gruyter, Berlin.

Mühlhäusler, P. 1996b. The diffusion of pidgin English in Australia. Map 15 & pp. 144–6 in S.A. Wurm, P. Mühlhäusler & D.T. Tryon (eds). *Atlas of languages of intercultural communication in the Pacific, Asia, and the Americas*. 3 volumes. Mouton de Gruyter, Berlin.

Mühlhäusler, P. 2003. *Language of environment, environment of language. A Course in Ecolinguistics*. Battlebridge, London.

Mühlhäusler, P. & Fill, A. (eds) 2001. *The ecolinguistics reader. Language, ecology, and environment*. Continuum, London.

Mulvaney, D.J. 1985. The Darwinian perspective. Pp. 68–75 in I. & T. Donaldson (eds) *Seeing the first Australians*. Allen & Unwin, Sydney.

Mulvaney, D.J. 1990. Spencer, Sir Walter Baldwin (1860–1929). Pp. 33–6 in J. Ritchie (ed.) *Australian dictionary of biography. Volume 12. 1891–1939, Smy–Z*. Melbourne University Press, Melbourne.

Mulvaney, D.J. 1994. The Namoi bunyip. *Australian Aboriginal Studies*. No. 1, pp. 36–8.

Mulvaney, D.J. 1996. 'A splendid lot of fellow': achievements and consequences of the Horn Expedition. Pp. 3–12 in S.R. Morton & D.J. Mulvaney *Exploring Central Australia. Society, the environment and the 1894 Horn Expedition*. Surrey Beatty & Sons, Sydney.

Mulvaney, D.J. 2002. 'Difficult to found an opinion': 1788 Aboriginal population estimates. Pp. 1–8 in G. Briscoe & L. Smith (eds) *The Aboriginal population revisited. 70 000 years to the present*. Aboriginal History Monograph No. 10. Aboriginal History, Canberra.

Mulvaney, D.J. & Kamminga, J. 1999. *Prehistory of Australia*. Allen & Unwin, Sydney.

Mulvaney, D.J., Morphy, H. & Petch, A. (eds) 1997. *My dear Spencer. The letters of F.J. Gillen to Baldwin Spencer*. Hyland House, Melbourne.

Mulvaney, D.J, with Petch, A. & Morphy, H. (eds) 2000. *From the frontier. Outback letters to Baldwin Spencer*. Allen & Unwin, Sydney.

Murgatroyd, S. 2002. *The dig tree. The story of Burke and Wills*. Text Publishing, Melbourne.

Museums Australia. 2005. *Continuous cultures, ongoing responsibilities. A comprehensive policy document and guidelines for Australian museums working with Aboriginal and Torres Strait Islander cultural heritage*. Museums Australia, Canberra.

Mutitjulu Community & Baker, L. 1996. *Mingkiri. A natural history of Uluru by the Mutitjulu community*. Institute for Aboriginal Development Press, Alice Springs, Northern Territory.

Nash, D. 1997. Comparative flora terminology of the central Northern Territory. Pp. 187–206 in P. McConvell & N. Evans (eds) *Archaeology and linguistics. Aboriginal Australia in global perspective*. Oxford University Press, Melbourne.

Nelson, E.C. 1990a. '... and flowers, for our amusement': the early collecting and cultivation of Australian plants in European and the problems encountered by today's taxonomists. Pp. 285–96 in P.S. Short (ed.) *History of systematic botany in Australasia. Proceedings of a symposium held at the University of Melbourne*. Australian Systematic Botany Society, Melbourne.

Nelson, E.C. 1990b. James and Thomas Drummond: their Scottish origins and curatorships in Irish botanic gardens (ca 1808–ca 1831). *Archives of Natural History*. Vol. 17, part 1, pp. 49–65.

Newland, S. 1889. *Parkengees or Aboriginal tribes on the Darling River*. Royal Geographical Society of Australasia, South Australian Branch. H.F. Leader, South Australian Government Printer, Adelaide.

Newland, S. 1922. The annual address of the President. *Proceedings of the Royal Geographical Society of Australasia. South Australian Branch*. Vol. 22, pp. 1–64.

Ngaanyatjarra Pitjantjatjara Yankunytjatjara Women's Council Aboriginal Corporation. 2003. *Ngangkari work – Anangu way. Traditional healers of Central Australia*. Ngaanyatjarra Pitjantjatjara Yankunytjatjara Women's Council Aboriginal Corporation, Alice Springs, Northern Territory.

Nicholson, P.H. 1981. Fire and the Australian Aborigine. Pp. 55–76 in A.M. Gill, R.H. Groves & I.R. Noble (eds) *Fire and the Australian biota*. Australian Academy of Science, Canberra.

Noye, R.J. 1972. Foelsche, Paul Heinrich Matthias (1831–1914). Pp. 192–3 in D. Pike (ed.) *Australian dictionary of biography. Volume 4. 1851–1890, D–J*. Melbourne University Press, Melbourne.

Nunn, J.M. 1989. *This southern land. A social history of Kangaroo Island 1800–1890*. Investigator Press, Adelaide.

O'Connell, J.F., Latz, P. K. & Barnett, P. 1983. Traditional and modern plant use among the Alyawara of Central Australia. *Economic Botany*. Vol. 37, part 1, pp. 80–109.

O'Grady, G.N. 1956. A secret language of Western Australia - a note. *Oceania*.Vol. 27, no. 2, pp. 158–9.

Ohlendorf, W. 1996. Domestication and crop development of *Duboisia* spp. (Solanaceae). In *Domestication and commercialization of non-timber forest products in agroforestry systems*. Non-Wood Forest Products 9. Available at http://www.fao.org/docrep/w3735e/w3735e23.htm#P3_0 (accessed 21 September 2007). Food and Agriculture Organization of the United Nations, Rome, Italy.

Orchard, A.E. 1999. A history of systematic botany in Australia. Pp. 11–103 in A.E. Orchard (ed.) *Australian biological resources study. Flora of Australia.* Volume 1. Introduction. Second edn. Bureau of Fauna & Flora, Canberra.

Orchard, K. 1997. Regional botany in mid-nineteenth-century Australia: Mueller's Murray River collecting network. *Historical Records of Australian Science.* Vol. 11, part 3, pp. 389–405.

Oxley, J. 1820. *Journals of two expeditions into the interior of New South Wales. Undertaken by order of the British government in the years 1817–18.* John Murray, London.

Paddle, R.N. 1996. Mueller's magpies and marsupial wolves: a window into 'What might have been'. *Victorian Naturalist.* Vol. 113, pt.4, pp. 215–8.

Palmer, E. 1884. On plants used by the natives of North Queensland, Flinders and Mitchell Rivers, for food, medicine, &c. &c. *Journal and Proceedings of the Royal Society of New South Wales.* Vol. 17, pp. 93–113.

Parkin, A.K. 1996a. Mueller and the North Australia Expedition. *Victorian Naturalist.* Vol. 113, pt.4, pp. 169–70.

Parkin, A.K. 1996b. Mueller, acclimatiser and seed merchant. *Victorian Naturalist.* Vol. 113, pt.4, pp. 213–4.

Parkin, A.K. 1996c. Review of 'Emigrant Eucalypts. Gum Trees as Exotics' by Robert Fyfe Zacharin. *Victorian Naturalist.* Vol. 113, pt.4, p. 229.

Parsons, V. 1966. Cunningham, Richard (1793–1835?). Pp. 268–9 in D. Pike (ed.) *Australian dictionary of biography. Volume 1. 1788–1850, A–H.* Melbourne University Press, Melbourne.

Pate, J.S. & Dixon, K.W. 1982. *Tuberous, cormous and bulbous plants. Biology of an adaptive strategy in Western Australia.* University of Western Australia Press, Nedlands.

Pearn, J.H. 1983. French doctors at Sydney Cove. Gallic contact in the second decade after Phillip. Pp. 45–61 in J.H. Pearn & C. O'Carrigan (eds) *Australia's quest for colonial health. Some influences on early health and medicine in Australia.* Department of Child Health, Royal Children's Hospital, Brisbane.

Pearn, J.H. 1990. *Medicine and botany. An Australian cadaster: Australian flora named after those whose lives have served medicine and health.* Amphion Press, Brisbane.

Pearn, J.H. 2001. *A doctor in the garden. Nomen medici in botanicis: Australian flora and the world of medicine.* Amphion Press, Brisbane.

Pearn, J.H. & O'Carrigan, C. (eds) 1983. *Australia's quest for colonial health. Some influences on early health and medicine in Australia.* Department of Child Health, Royal Children's Hospital, Brisbane.

Peron, F. 1802. King Island and the sealing trade. A translation by H.M. Micco of chapters XXII & XXIII of *The voyage of discovery to the southern lands undertaken in the corvettes Le Geographe, La Naturaliste and the schooner Casuarina, during the years 1800 to 1804, under the command of Captain Nicolas Baudin.* Published in 1971 by the Roebuck Society, Canberra.

Perry, T.M. 1966. Cunningham, Allan (1791–1839). Pp. 265–7 in D. Pike (ed.) *Australian dictionary of biography. Volume 1. 1788–1850, A–H.* Melbourne University Press, Melbourne.

Peterson, N. 1977. Aboriginal uses of Australian Solanaceae. Pp. 171–88 in J.G. Hawkes, R.N. Lester & A.D. Skelding (eds) *The biology and taxonomy of Solanaceae.* Academic Press, London.

Peterson, N. & Rigsby, B. (eds) 1998. *Customary marine tenure in Australia.* Oceania Monograph No. 48. University of Sydney, Sydney.

Phillip, A. 1789. *The voyage of Governor Phillip to Botany Bay, with an account of the establishment of the colonies of Port Jackson and Norfolk Island...* J. Stockdale, London.

Pickering, W.G. 1929. The letters of Georgina Molloy. *Journal & Proceedings of the Western Australian Historical Society.* Vol. 1, pp. 30–84.

Plomley, N.J.B. 1966. *Friendly mission. The Tasmanian journals and papers of George Augustus Robinson. 1829–1834.* Tasmanian Historical Research Association, Hobart.

Plomley, N.J.B. 1976. *A word-list of the Tasmanian Aboriginal languages.* Author & Government of Tasmania, Launceston.

Plomley, N.J.B. 1987. *Weep in silence. A history of the Flinders Island Aboriginal settlement with the Flinders Island journal of George Augustus Robinson 1835–1839.* Blubber Head Press, Hobart.

Plomley, N.J.B. & Cameron, M. 1993. Plant foods of the Tasmanian Aborigines. *Records of the Queen Victoria Museum.* Vol. 101, pp. 1–27.

Plomley, N.J.B. & Henley, K.A. 1990. *The sealers of Bass Strait and the Cape Barren Island community.* Blubber Head Press, Hobart. An offprint of the *Tasmanian Historical Research Association Papers & Proceedings*. 1990, vol. 37, nos 2–3.

Pollan, M. 2002. *The botany of desire. A plant's-eye view of the world.* Bloomsbury Publishing, London.

Ponsonby, L. 1998. Sir William's legacy. *Kew* (Spring), pp. 16–9.

Portman, P. 1989. Register of Australian winter cereal cultivars. *Australian Journal of Experimental Agriculture.* Vol. 29, no. 1, p. 143.

Powell, J.M. 1990. Early impressions of the vegetation of the Sydney region: exploration and plant use by the First Fleet officers. Pp. 87–96 in P.S. Short (ed.) *History of systematic botany in Australasia. Proceedings of a symposium held at the University of Melbourne.* Australian Systematic Botany Society, Melbourne.

Puruntatameri, J. 2001. *Tiwi plants and animals. Aboriginal flora and fauna knowledge from Bathurst and Melville Islands, northern Australia.* Parks & Wildlife Commission of the Northern Territory & Tiwi Land Council, Darwin.

Pyke, M. 1968. *Food and society.* John Murray, London.

Pyne, S.J. 1991. *Burning bush. A fire history of Australia.* Holt, New York.

Rae, C.J., Lamprell, V.J., Lion R.J. & Rae, A.M. 1982. The role of bush foods in contemporary Aboriginal diets. *Proceedings of the Nutrition Society of Australia.* Vol. 7, pp. 45–9.

Ramson, W.S. 1966. *Australian English. An historical study of the vocabulary 1788–1898.* Australian National University Press, Canberra.

Ramson, W.S. (ed.). 1988. *The Australian national dictionary. A dictionary of Australianisms on historical principles.* Oxford University Press, Oxford.

Reid, A. 1995a. *Banksias and bilbies. Seasons of Australia.* Gould League, Melbourne.

Reid, A. 1995b. A plan for all seasons. *Habitat Australia.* Vol. 23, no. 2, pp. 14–15.

Reid, E. 1977. *The records of Western Australian plants used by Aboriginals as medicinal agents.* Reprinted 1986. School of Pharmacy, Western Australian Institute of Technology, Perth.

Reid, J.C. 1979. Health as harmony, sickness as conflict. *Hemisphere.* Vol. 23, pt 4, pp. 194–9.

Reid, J.C. (ed.) 1982. *Body, land and spirit. Health and healing in Aboriginal society.* University of Queensland Press, St Lucia.

Reid, J.C. 1983. *Sorcerers and healing spirits. Continuity and change in an Aboriginal medical system.* Australian National University Press, Canberra.

Rennie, E.H. 1880. On the acids of the native currant (*Leptomeria acida*). *Journal & Proceedings of the Royal Society of New South Wales.* Vol. 14, pp. 119–21.

Rennie, E.H. 1886. Notes on the sweet principle of *Smilax glycyphylla. Journal & Proceedings of the Royal Society of New South Wales.* Vol. 20, pp. 211–12.

Reuther, J.G. 1981. *The Diari.* Volumes 1 to 13. Translated by P.A. Scherer, T. Schwarzchild & L. Hercus. Original in the S.A. Museum Archives. Australian Institute of Aboriginal Studies, Canberra.

Reynolds, B. 1988. Roth, Walter Edmund (1861–1933). Pp. 463–4 in G. Searle (ed.) *Australian dictionary of biography. Volume 11. 1891–1939, Nes–Smi.* Melbourne University Press, Melbourne.

Reynolds, H. 1989. *Dispossession. Black Australians and white invaders.* Allen & Unwin, Sydney.

Reynolds, H. 1990. *With the white people.* Penguin, Melbourne.

Reynolds, H. 1995. *Fate of a free people.* Penguin, Melbourne.

Riches, P. 1964. Nardoo – the clover fern. *South Australian Naturalist.* Vol. 38, part 4, pp. 63–4, 66.

Ritchie, R. 1989. *Seeing the rainforests in nineteenth century Australia.* Rainforest Publishing, Sydney.

Rix, C.E. 1978. *Royal Zoological Society of South Australia 1878–1978.* Royal Zoological Society of South Australia, Adelaide.

Roberts, J., Fisher, C.J., Gibson, R. & Popp, T. 1995. *A guide to traditional Aboriginal rainforest plant use by the Kuku Yalanji of the Mossman Gorge.* Bamanga Bubu Ngadimunku Inc., Mossman, Queensland.

Roberts, K. 1972. Hunt, Charles Cooke (1833–1868). P. 446 in D. Pike (ed.) *Australian dictionary of biography. Volume 4. 1851–1890, D–J.* Melbourne University Press, Melbourne.

Robertson, E. 1979. Black, John McConnell (1855–1951). P. 304 in B. Nairn & G. Searle (eds) *Australian dictionary of biography. Volume 7. 1891–1939, A–Ch.* Melbourne University Press, Melbourne.

Roderick, C. 1988. *Leichhardt, the dauntless explorer.* Angus & Robertson, Sydney.

Rolls, E. 1997. The nature of Australia. Pp. 35–45 in T. Griffiths & L. Robin (eds) *Ecology and empire. Environmental history of settler societies.* Melbourne University Press, Melbourne.

Rolls, E. 2002. *Visions of Australia. Impressions of the landscape 1642–1910.* Lothian Books, Melbourne.

Rose, D. Bird 1987. *Bush medicines. A Ngarinman and Bilinara pharmacopoeia.* Australian Institute of Aboriginal Studies, Canberra.

Rose, D. Bird (ed.) 1995. *Country in flames. Proceedings of the 1994 symposium on biodiversity and fire in north Australia.* Biodiversity Unit, Department of the Environment, Sport & Territories & the North Australia Research Unit, Canberra & Darwin.

Rose, F.G.G. 1994. Ludwig Leichhardt and the Australian Aborigines. *EAZ Ethnogr.–Achaeol.* Vol. 35, pp. 165–72.

Ross, J.H. 1996. The legacy of Mueller's collections. *Victorian Naturalist.* Vol. 113, pt. 4, pp. 146–50.

Roth, H. Ling 1899. *The Aborigines of Tasmania.* Second edn. 1968 facsimile. Fullers Bookshop, Hobart.

Roth, W.E. 1897. *Ethnological studies among the north-west-central Queensland Aborigines.* Queensland Government Printer, Brisbane.

Roth, W.E. 1901a. *North Queensland ethnography. String, and other forms of strand. Basketry, woven bag and net-work. Bulletin 1.* Government Printer, Brisbane.

Roth, W.E. 1901b. *North Queensland ethnography. Food. Its search, capture, and preparation. Bulletin 3.* Government Printer, Brisbane.

Roth, W.E. 1902. *North Queensland ethnography. Games, sports and amusements. Bulletin 4.* Government Printer, Brisbane.

Roth, W.E. 1903a. *North Queensland ethnography. Superstition, magic and medicine. Bulletin 5.* Government Printer, Brisbane.

Roth, W.E. 1903b. Notes of savage life in the early days of West Australian settlement. *Proceedings of the Royal Society of Queensland.* Vol. 17, pp. 45–69.

Roth, W.E. 1904. *North Queensland ethnography. Domestic implements, arts, and manufactures. Bulletin 7.* Government Printer, Brisbane.

Roth, W.E. 1909. *North Queensland ethnography. Fighting weapons. Bulletin 13.* Government Printer, Brisbane.

Roth, W.E. 1910. *North Queensland ethnography. Decoration, deformation, and clothing. Bulletin 15.* Government Printer, Brisbane.

Rowan, E. 1898. *The flower hunter. The adventures, in northern Australia and New Zealand, of flower painter Ellis Rowan.* Angus & Robertson, Sydney.

Rowland, M.J. 2002. Geophagy: an assessment of implications for the development of Australian Indigenous plant processing technologies. *Australian Aboriginal Studies.* No. 1, pp. 51–66.

Rowley, C.D. 1972a. *The destruction of Aboriginal society.* Pelican, Melbourne.

Rowley, C.D. 1972b. *The remote Aborigines.* Pelican, Melbourne.

Rowley, C.D. 1972c. *Outcasts in white Australia.* Pelican, Melbourne.

Rudder, J.C. 1978–79. Classification of the natural world among the Yolngu, Northern Territory, Australia. *Ethnomedicine.* 1978/79, pp. 349–60.

Ruhe, E.L. 1986. Poetry in the older Australian landscape. Pp. 20–51 in P.R. Eaden & F.H. Mares (eds) *Mapped but not known. The Australian landscape of the imagination. Essays and poems presented to Brian Elliott LXXV 11 April 1985.* Wakefield Press, Adelaide.

Ryan, L. 1996. *The Aboriginal Tasmanians.* Second edn. Allen & Unwin, Sydney.

Salier, C.W. 1931. Thomas Livingston Mitchell, explorer, surveyor-general, and savant. *Journal & Proceedings of the Royal Australian Historical Society.* Vol. 17, pt 1, pp. 1–43.

Salter, E. 1971. *Daisy Bates. The great white queen of the never never.* Angus & Robertson, Sydney.

Sampson, H.C. 1935. The Royal Botanic Gardens, Kew and empire agriculture. *Journal of the Royal Society of Arts.* March 15, 1935. Pp. 404–19.

Samuel, H.J. 1961. *Wild flower hunter. The story of Ellis Rowan.* Constable, London.

Sayers, A. 1996. *Aboriginal artists of the nineteenth century.* Oxford University Press, Melbourne.

Schmidt, A. 1993. *The loss of Australia's Aboriginal language heritage.* Aboriginal Studies Press, Canberra.

Schulze, L. 1891. The Aborigines of the upper and middle Finke River: their habits and customs, with introductory notes on the physical and natural-history features of the country. *Transactions, Proceedings & Report of the Royal Society of South Australia.* Vol. 14, pp. 210–46.

Schürmann, C.W. 1844. *Vocabulary of the Parnkalla language.* Dehane, Adelaide.

Schürmann, C.W. 1846. *The Aboriginal tribes of Port Lincoln in South Australia, their mode of life, manners, customs ...* Reprinted in J.D. Woods (ed.) 1879. *The Native Tribes of South Australia.* Dehane, Adelaide.

Scott, M.P. 1972. Some Aboriginal food plants of the Ashburton district, Western Australia. *Western Australian Naturalist.* Vol. 12, part 4, pp. 94–6.

Semple, S.J. 1998. The antiviral properties of traditional Australian Aboriginal medicines. *Proceedings of the Nutritional Society of Australia.* Vol. 22, pp. 1–6.

Semple, S.J., Reynolds, G.D., O'Leary, M.C. & Flower, R.L.P. 1998. Screening of Australian medicinal plants for antiviral activity. *Journal of Ethnopharmacology.* Vol. 60, pp. 163–72.

Serle, P. 1949. *Dictionary of Australian biography.* 2 volumes. Angus & Robertson, Sydney.

Sharp, N. 1993. *Stars of Tagai. The Torres Strait Islanders.* Aboriginal Studies Press, Canberra.

Sharp, N. 2002. *Saltwater people. The waves of memory.* Allen & Unwin, Sydney.

Sharpe, M. 1994. Bundjalung. Pp. 1–22 in N. Thieberger & W. McGregor (eds) *Macquarie Aboriginal words.* Macquarie Library, Macquarie University, Sydney

Sharr, F.A. 1996. *Western Australia plant names and their meanings. A glossary.* Enlarged edn. University of Western Australia Press, Nedlands.

Short, P.S. (ed.) 1990. *History of systematic botany in Australasia. Proceedings of a symposium held at the University of Melbourne.* Australian Systematic Botany Society, Melbourne.

Short, P.S. 2003. *In pursuit of plants. Experiences of nineteenth and early twentieth century plant collectors.* University of Western Australia Press, Perth.

Shteir, A.B. 1996. *Cultivating women, cultivating science. Flora's daughters and botany in England, 1760 to 1860.* John Hopkins University Press, Baltimore.

Sivarajan, V.V. 1991. *Introduction to the principles of plant taxonomy.* Second edn, edited by N.K.P. Robson. Cambridge University Press, New York.

Slater, P. 1978. *Rare and vanishing birds.* Rigby, Adelaide.

Smith, B.W. 1992. *Imagining the Pacific in the wake of the Cook voyages.* Melbourne University Press at the Miegunyah Press, Melbourne.

Smith, J.E. 1793–95. *A specimen of the botany of New Holland.* J. Sowerby, London.

Smith, K. & Smith, I. 1999. *Grow your own bushfoods.* New Holland Publishers, Sydney.

Smith, K.V. 1992. *King Bungaree. A Sydney Aborigine meets the great south Pacific explorers, 1799–1830.* Kangaroo Press, Sydney.

Smith, K.V. 2001. *Bennelong. The coming in of the Eora. Sydney Cove 1788–1792.* Kangaroo Press, Sydney.

Smith, K.V. 2005a. Gooseberry, Cora (c. 1777–1852). Pp. 148 in C. Cunneen (ed.) *Australian dictionary of biography, supplementary volume, 1580–1980.* Melbourne University Press, Melbourne.

Smith, K.V. 2005b. Moowattin, Daniel (c.1791–1816). Pp. 286–7 in C. Cunneen (ed.) *Australian dictionary of biography, supplementary volume, 1580–1980.* Melbourne University Press, Melbourne.

Smith, L.R. 1980. *The Aboriginal population of Australia.* Aborigines in Australia Society No. 14. Academy of the Social Sciences. Australian National University Press, Canberra.

Smith, M. 1982. Late Pleistocene zamia exploitation in southern Western Australia. *Archaeology in Oceania.* Vol. 17, pp. 117–21.

Smith, M. 1996. Revisiting Pleistocene Macrozamia. *Australian Archaeology.* Vol. 42, pp. 52–3.

Smith, M. & Kalotas, A.C. 1985. Bardi plants: an annotated list of plants and their use by the Bardi Aborigines of Dampierland, in north-western Australia. *Records of the Western Australian Museum.* Vol. 12, part 3, pp. 317–59.

Smith, N.M. 1991. Ethnographic field notes from the Northern Territory, Australia. *Journal of the Adelaide Botanical Gardens.* Vol. 14, part 1, pp. 1–65.

Smith, N.M. & Wightman, G.M. 1990. *Ethnobotanical notes from Belyuen, Northern Territory, Australia.* Northern Territory Botanical Bulletin No. 10. Conservation Commission of the Northern Territory, Darwin.

Smyth, R. Brough 1878. *The Aborigines of Victoria.* 2 volumes. Victorian Government Printer, Melbourne.

Southgate, G.W. 1967. *The British empire and commonwealth.* J.M. Dent & Sons, London.

Sparrman, A. 1772–76. *A voyage to the Cape of Good Hope towards the Antarctic Polar Circle round the world and to the country of the Hottentots and the Caffres.* Republished in 1975–76. Introduced & edited by V.S. Forbes. Translated from the Swedish by J. & I. Rudner. Van Riebeeck Society, Cape Town.

Specht, R.L. 1958. An introduction to the ethnobotany of Arnhem Land. Pp. 479–503 in R.L. Specht & C.P. Mountford (eds) *Records of the American-Australian Scientific Expedition to Arnhem Land.* Volume 3. Melbourne University Press, Melbourne.

Specht, J. & White, J.P. 1978. Trade and exchange in Oceania and Australia. *Mankind.* Vol. 11, pt 3, pp. 161–435.

Spencer, W.B. (ed.) 1896. *Report on the work of the Horn Scientific Expedition to Central Australia.* 4 volumes. Melville, Mullen & Slade, Melbourne.

Spencer, W.B. 1918. What is nardoo? *Victorian Naturalist.* Vol. 14, pp. 170–2; Vol. 15, pp. 8–15.

Spencer, W.B. & Gillen, F.J. 1899. *The native tribes of Central Australia.* Macmillan, London.

Spencer, W.B. & Gillen, F.J. 1904. *The northern tribes of Central Australia.* Macmillan, London.

Spencer, W.B. & Gillen, F.J. 1912. *Across Australia.* 2 volumes. Macmillan, London.

Spencer, W.B. & Gillen, F.J. 1927. *The Arunta. A study of a stone age people.* 2 volumes. Macmillan, London.

Spruce, R. 1908. *Notes of a botanist on the Amazon and Andes.* 2 volumes. Macmillan & Co., London.

Stack, E.M. 1989. Aboriginal pharmacopoeia. *Occasional Papers (Northern Territory Library Service).* Vol. 10, pp. 1–7.

Stanner, W.E.H. 1953 (1979). The Dreaming. Pp. 23–40 in *White man got no Dreaming.* Australian National University Press, Canberra.

Stanner, W.E.H. 1972. Howitt, Alfred William (1830–1908). Pp. 432–5 in D. Pike (ed.) *Australian dictionary of biography. Volume 4. 1851–1890, D–J.* Melbourne University Press, Melbourne.

Starbuck, A. 1878. *History of the American whale fishery.* 2 volumes. Reprinted in 1964 by Argosy-Antiquarian, New York.

Stearn, W.T. 1957. An introduction to the *Species Plantarum* and cognate botanical works of Carl Linnaeus. Pp. 1–176, vol. 1, in C. Linnaeus, *Species plantarum,* 1753, facsimile edn, Ray Society, London.

Stearn, W.T. 1969. A Royal Society appointment with Venus in 1769: the voyage of Cook and Banks in the *Endeavour* in 1768–1771 and its botanical results. *Notes & Records of the Royal Society of London.* Vol. 24, no. 1, pp. 64–90.

Stearn, W.T. 1974. Sir Joseph Banks (1743–1820) and Australian botany. *Records of the Australian Academy of Science.* Vol. 2, no. 4, pp. 7–24.

Stearn, W.T. 1983. *Botanical Latin. History, grammar syntax, terminology and vocabulary.* Third edn. David & Charles, London.

Steele, J.G. 1984. *Aboriginal pathways in southeast Queensland and the Richmond River.* University of Queensland Press, St Lucia.

Stephens, E. 1890. The Aborigines of Australia. *Journal & Proceedings of the Royal Society of New South Wales.* Vol. 23, pp. 476–503.

Stephens, S.E. 1974. Meston, Archibald (1851–1924). Pp. 243–4 in D. Pike (ed.) *Australian dictionary of biography. Volume 5. 1851–1890, K–Q.* Melbourne University Press, Melbourne.

Stewart, K. & Percival, B. 1997. *Bush foods of New South Wales. A botanical record and an Aboriginal oral history.* Royal Botanic Gardens, Sydney.

Stirling, E.C. 1896. Anthropology. Pp. 1–157 in W.B. Spencer (ed.) *Report on the work of the Horn Scientific Expedition to Central Australia. Part IV.–Anthropology.* Melville, Mullen & Slade, Melbourne.

Stone, A.C. 1911. Aborigines of Lake Boga. *Proceedings of the Royal Society of Victoria.* Vol. 23, pp. 433–68.

Strehlow, C. 1907–1920. *Die Aranda und Loritja-stamme in Zentral-Australien.* 5 volumes. Joseph Baer, Frankfurt.

Sturt, C. 1833. *Two expeditions into the interior of southern Australia.* 2 volumes. Smith, Elder & Co., London.

Sturt, C. 1849. *An account of the sea coast and interior of South Australia, with observations on various subjects connected with its interests.* Republished as a facsimile in *Journal of the Central Australian Expedition,* 1984, pp. 115–264. Edited & introduced by J. Waterhouse. Caliban Books, London.

Sutton, P. 1994. Material culture traditions of the Wik people, Cape York Peninsula. *Records of the South Australian Museum.* Vol. 27, part 1, pp. 31–52.

Sutton, P. 1995. *Wik-Ngathan dictionary.* Caitlin Press, Adelaide.

Swayne, G.C. 1868. *Lake Victoria: A narrative of explorations in search of the source of the Nile. Compiled from the memoirs of Captains Speke and Grant.* W. Blackwood, Edinburgh.

Sweeney, G. 1947. Food supplies of a desert tribe. *Oceania*. Vol. 7, no. 4, pp. 289–99.

Symon, D.E. 2005. Native tobaccos (Solanaceae: *Nicotiana* spp.) in Australia and their use by Aboriginal peoples. *The Beagle, Records of the Museums & Art Galleries of the Northern Territory*. Vol. 21, pp. 1–10.

Symon, D.E. & Jusaitis, M. 2007. *Sturt pea. A most splendid plant*. Board of the Botanic Gardens & State Herbarium, and Department for Environment & Heritage, Adelaide.

Symons, P. & Symons, S. 1994. *Bush heritage. An introduction to the history of plant and animal use by Aboriginal people and colonists in the Brisbane and Sunshine Coast areas*. The authors, Nambour, Queensland.

Symons, P. & Symons, S. 1997. Ludwig Leichhardt: on the bushfood trail. *Australian Bushfoods Magazine*. Issue 1, pp. 22–3.

Taplin, G. 1859–79. Journals. Typescript. Mortlock Library, Adelaide.

Taplin, G. 1874. *The Narrinyeri*. Reprinted as pp. 1–156 in J.D. Woods (ed.) 1879. *The native tribes of South Australia*. E.S. Wiggs, Adelaide.

Taplin, G. (ed.) 1879. *Folklore, manners, customs and languages of the South Australian Aborigines*. South Australian Government Printer, Adelaide.

Tate, R. (ed.) 1882. Miscellaneous contributions to the natural history of South Australia. *Transactions, Proceedings & Report of the Royal Society of South Australia*. Vol. 4, pp. 135–8.

Tate, R. 1896. Botany. Pp. 117–94 in W.B. Spencer (ed.) *Report on the work of the Horn Scientific Expedition to Central Australia. Part III.–geology & botany*. Melville, Mullen & Slade, Melbourne.

Teichelmann, C.G. & Schürmann, C.W. 1840. *Outlines of a grammar... of the Aboriginal language of South Australia*. 2 parts. Thomas & Co., Adelaide.

Telford, I.R.H. 1990. Moving mountains – Allan Cunningham and the mountains of southern Queensland. P. 157 in P.S. Short (ed.) *History of systematic botany in Australasia. Proceedings of a symposium held at the University of Melbourne*. Australian Systematic Botany Society, Melbourne.

Tench, W. 1788–92. *A narrative of the expedition to Botany Bay & a complete account of the settlement at Port Jackson*. Edited by T. Flannery. 1996. Text Publishing, Melbourne.

Thieberger, N. & McGregor, W. (eds) 1994. *Macquarie Aboriginal words*. Macquarie Library, Macquarie University, Sydney.

Thieret, J.W. 1958. Economic botany of the cycads. *Economic Botany*. Vol. 12, pp. 3–41.

Thomas, E.K. (ed.) 1925. *The diary and letters of Mary Thomas*. Thomas & Co., Adelaide.

Thomas, N.W. 1906. *The natives of Australia*. Archibald Constable & Co., London.

Thomson, D.F. 1939. The seasonal factor in human culture: illustrated from the life of a contemporary nomadic group. *Prehistorical Society Proceedings*. Vol. 5, no. 2, pp. 209–21.

Thomson, D.F. 1949. *Economic structure and the ceremonial exchange cycle in Arnhem Land*. Macmillan, Melbourne.

Thomson, D.F. 1962. The Bindibu expedition. Exploration among the desert Aborigines of Western Australia. *The Geographical Journal*. Vol. 78, part 1, pp. 1–14; part 2, pp. 143–57; part 3, pp. 262–78.

Thorne, A. & Raymond, R. 1989. *Man on the rim. The peopling of the Pacific*. Angus & Robertson, Sydney.

Thozet, A. 1866. *Notes on some of the roots, tubers, bulbs, and fruits used as vegetable food by the Aboriginals of northern Queensland*. W.H. Buzacott, Bulletin, Rockhampton.

Threlkeld, L.E. 1824–1859. *Australian reminiscences and papers of L.E. Threlkeld. Missionary to the Aborigines, 1824–1859*. Edited by Neil Gunson. Published in 1974. Australian Aboriginal Studies No. 40. Ethnohistory Series No. 2. Two volumes. Australian Institute of Aboriginal Studies, Canberra.

Tietkens, W.H. 1891. *Journal of Mr. W.H. Tietkens' Central Australian exploring expedition*. South Australian Government Printer, Adelaide.

Tilbrook, L. 1983. *Nyungar tradition. Glimpses of Aborigines of south-western Australia 1829–1914*. University of Western Australia Press, Perth.

Tindale, N.B. 1925. Natives of Groote Eylandt and of the west coast of the Gulf of Carpentaria. Part 2. *Records of the South Australian Museum*. Vol. 3, no. 1, pp. 103–34.

Tindale, N.B. 1941a. A list of plants collected in the Musgrave Range and Mann Ranges, South Australia, 1933. *The South Australian Naturalist*. Vol. 21, no. 1, pp. 8–12.

Tindale, N.B. 1941b. Survey of the half-caste problem in South Australia. *Proceedings of the Royal Geographical Society of Australasia. South Australian Branch*. pp. 66–161.

Tindale, N.B. 1952. Some Australian Cossidae including witjiti (witchetty) grub. *Transactions of the Royal Society of South Australia*. Vol. 76, pp. 56–65.

Tindale, N.B. 1966. Insects as food for the Australian Aborigines. *Australian Natural History*. Vol. 15, pp. 179–83.

Tindale, N.B. 1974. *Aboriginal tribes of Australia. Their terrain, environmental controls, distribution, limits, and proper names*. 4 maps enclosed. Australian National University Press, Canberra.

Tindale, N.B. 1977. Adaptive significance of the Panara or grass seed culture of Australia. Pp. 345–9 in R.V.S. Wright (ed.) *Stone tools as cultural markers*. Australian Institute of Aboriginal Studies, Canberra.

Tindale, N.B. 1978. Notes on a few Australian Aboriginal concepts. Pp. 156–63 in L.R. Hiatt (ed.) *Australian Aboriginal concepts*. Australian Institute of Aboriginal Studies, Canberra.

Tindale, N.B. 1981. Desert Aborigines and the southern coastal peoples: some comparisons. Pp. 1855–84 in A. Keast (ed.) *Ecological biogeography of Australia*. Junk, The Hague.

Tolcher, H.M. 2003. *Seed of the coolibah. A history of the Yandruwandha and Yawarrawarrka people*. The author, Adelaide.

Tolmer, A. 1882. *Reminiscences of an adventurous and chequered career at home and at the Antipodes*. 2 volumes. Sampson Low, Marston, Searle & Rivington, London.

Troy, J. 1990. *Australian Aboriginal contact with the English language in New South Wales. 1788 to 1845*. Pacific Linguistics, Series B, No. 103. Department of Linguistics, Research School of Pacific Studies, Australian National University, Canberra.

Troy, S. 1993a. Language contact in early colonial New South Wales 1788 to 1791. Pp. 33–50 in M. Walsh & C. Yallop (eds) *Language and culture in Aboriginal Australia*. Aboriginal Studies Press, Canberra.

Troy, J. 1993b. *King plates. A history of Aboriginal gorgets*. Aboriginal Studies Press for the Australian Institute of Aboriginal and Torres Strait Islander Studies, Canberra.

Troy, J. 1994. The Sydney language. Pp. 61–78 in N. Thieberger & W. McGregor (eds) *Macquarie Aboriginal words*. Macquarie Library, Macquarie University, Sydney.

Tunbridge, D. 1985a. Language as heritage: vityurna (dried meat) and other stored food among the Adnyamathanha. *Journal of the Anthropological Society of South Australia*. Vol. 23, part 7, pp. 10–15.

Tunbridge, D. 1985b. Language as heritage: flora in place names. A record of survival in the Gammon Ranges. *Journal of the Anthropological Society of South Australia*. Vol. 23, part 8, pp. 3–15.

Tunbridge, D. 1987. Aboriginal place names. *Australian Aboriginal Studies*. No. 2, pp. 2–13.

Turner, D.H. 1974. *Tradition and transformation: a study of Aborigines in the Groote Eylandt area, northern Australia*. Australian Aboriginal Studies no. 53. Social Anthropology Series no. 8. Australian Institute of Aboriginal Studies, Canberra.

Turner, G.W. 1972. *The English language in Australia and New Zealand*. Second edn. Longmans, London.

Turner, M. 1994. *Arrernte foods. Food from Central Australia*. Institute for Aboriginal Development Press, Alice Springs, Northern Territory.

Turrill, W.B. 1963. *Joseph Dalton Hooker. Botanist, explorer and administrator*. Nelson, London.

Urban, A. 2001. *Wildflowers and plants of inland Australia*. Paul Fitzsimons, Alice Springs, Northern Territory.

Urry, J. & Walsh, M. 1981. The lost 'Macassar language' of Northern Australia. *Aboriginal History*. Vol. 5, pt 2, pp. 90–108.

Vallance, T.G., Moore, D.T. & Groves, E.W. 2001. *Nature's investigator. The diary of Robert Brown in Australia, 1801–1805*. Australian Biological Resources Study, Canberra.

Van Lohuizen, J. 1967. Vlamingh, Willem de (fl. 1697). P. 556 in D. Pike (ed.) *Australian dictionary of biography. Volume 2. 1788–1850, I–Z*. Melbourne University Press, Melbourne.

Von Brandenstein, C.G. 1988. *Nyungar anew. Phonology, text samples and etymological and historical 1500-word vocabulary of an artificially re-created Aboriginal language in the south-west of Australia*. Pacific Linguistics Series C, No. 99. Department of Linguistics, Research School of Pacific Studies, Australian National University, Canberra.

Von Mueller, F. 1825–96. *Regardfully yours. Selected correspondence of Ferdinand von Mueller*. Edited by R.W. Home, A.M. Lucas, S. Maroske, D.M. Sinkora, J.H. Voigt & M. Wells. 1998–2006. 3 volumes. Peter Lang, New York.

Von Mueller, F. 1856. Account of the *gunyang*: a new indigenous fruit of Victoria. *Hooker's Journal of Botany & Kew Garden Miscellany*. Vol. 8, pp. 336–8.

Von Mueller, F. 1858a. On a general introduction of useful plants into Victoria. *Transactions of the Philosophical Institute of Victoria*. Vol. 2, part 2, pp. 93–109.

Von Mueller, F. 1858b. Botanical report on the North-Australia Expedition, under command of A.C. Gregory, Esq. *Journal of the Linnean Society (Botany)*. Vol. 2, pp. 137–63.

Von Mueller, F. 1876. *Select plants readily eligible for industrial culture or naturalisation in Victoria*. McCarron, Bird & Co., Melbourne.

Von Mueller, F. 1888. *Select extra-tropical plants, readily eligible for industrial culture or naturalisation, with indications of their native countries and some of their uses*. Revised edn. R.S. Brain, Government Printer, Sydney.

Waddy, J.A. 1979. Ethnobiology of Groote Eylandt. A progress report. *Australian Institute of Aboriginal Studies Newsletter*. Vol. 11, pp. 46–50.

Waddy, J.A. 1982. Biological classification from a Groote Eylandt Aborigine's point of view. *Journal of Ethnobiology*. Vol. 2, pt 1, pp. 63–77.

Waddy, J.A. 1988. *Classification of plants and animals from a Groote Eylandt Aboriginal point of view*. 2 volumes. Australian National University, North Australia Research Unit, Darwin.

Wakefield, N. 1959. Baron von Mueller gave us blackberries. *Victorian Naturalist*. Vol. 76, pt.2, p. 33.

Wakefield, N. 1961. Baron von Mueller and the blackberries. *Victorian Naturalist*. Vol. 77, pt.9, p. 258.

Walker, J.B. 1888–99. Notes on the Aborigines of Tasmania, extracted from the manuscript journals of George Washington Walker, with an introduction by James B. Walker, F.R.G.S. Pp. 238–87 in *Early Tasmania. Papers read before the Royal Society of Tasmania during the years 1888 to 1899 by James Backhouse Walker*. M.C. Reed, Government Printer, Tasmania.

Walker, J.B. 1889. The founding of Hobart, by Lieut.-Governor Collins. Pp. 60–85 in *Early Tasmania. Papers read before the Royal Society of Tasmania during the years 1888 to 1899 by James Backhouse Walker*. M.C. Reed, Government Printer, Tasmania.

Walker, J.B. 1900. *The Tasmanian Aborigines*. John Vail, Government Printer, Hobart.

Walsh, M. 1993. Classifying the world in an Aboriginal language. Pp. 107–22 in M. Walsh & C. Yallop (eds) *Language and culture in Aboriginal Australia*. Aboriginal Studies Press, Canberra.

Warner, W.L. 1958. *A black civilization. A study of an Australian tribe*. Revised edn. Harper & Row, New York.

Watson, P. 1983. *This precious foliage. A study of the Aboriginal psycho-active drug pituri*. Oceania Monograph No. 26. University of Sydney, Sydney.

Watson, P. 1994. Bush medicine. Pp. 170–1 in D. Horton (ed.) *The encyclopaedia of Aboriginal Australia*. 2 volumes. Aboriginal Studies Press for the Australian Institute of Aboriginal & Torres Strait Islander Studies, Canberra.

Webb, J.B. 1995. *George Caley. Nineteenth century naturalist*. Surrey Beatty & Sons, Sydney.

Webb, J.B. 2003. *The botanical endeavour. Journey towards a flora of Australia*. Surrey Beatty & Sons, Sydney.

Webb, L.J. 1948. *Guide to the medicinal and poisonous plants of Queensland*. Bulletin No. 232. Council for Scientific & Industrial Research, Melbourne.

Webb, L.J. 1960. Some new records of medicinal plants used by the Aborigines of tropical Queensland and New Guinea. *Proceedings of the Royal Society of Queensland*. Vol. 71, pt 6, pp. 103–10.

Webb, L.J. 1969. Australian plants and chemical research. Pp. 82–90 in L.J. Webb, D. Whitelock & J. Le Gay Brereton (eds) *The last of lands*. Jacaranda Press, Brisbane.

Webb, L.J. 1973. 'Eat, die, and learn'. The botany of the Australian Aborigines. *Australian Natural History*. March 17, vol. 9, pp. 290–5.

Webb, L.J. 1977. Ethnobotany: the co-operative approach to research. *Australian Institute of Aboriginal Studies Newsletter. New Series*. January 1977. Pp. 43–5.

Welch, D.M. 2006. Introduction. Pp. 1–6 in the republication of J. Morrill, 1864, *Seventeen years wandering among the Aboriginals*. David M. Welch, Virginia, Northern Territory.

Wells, L.A. 1902. *Journal of the Calvert Scientific Exploring Expedition, 1896–7*. Western Australian Parliamentary Paper No. 46. William Alfred Watson, Government Printer, Perth.

Wharton, W.J.L. 1893. Sketch of Cook's Life. Pp. xvii–liv, as an introduction to J. Cook, *Captain Cook's journal during his first voyage round the world made in M.M. Bark 'Endeavour' 1768–71*. Edited by W.J.L. Wharton. Elliot Stock, London.

White, C.T. 1950. F.M. Bailey: his life and work. *Proceedings of the Royal Society of Queensland*. Vol. 61, no. 8, pp. 105–14.

White, J. 1790. *Journal of a voyage to New South Wales. With sixty-five plates of non descript animals, birds, lizards, serpents, curious cones of trees and other natural productions*. Printed for J. Debrett, London.

White, M.E. 1994. *After the greening. The browning of Australia*. Kangaroo Press, Sydney.

Whitehead, P. J. 1969. Captain Cook's role in natural history. *Australian Natural History*. Vol. 16, no. 8, pp. 242–6.

Whiting, M.G. 1963. Toxicity of cycads. *Economic Botany*. Vol. 17, pp. 271–301.

Whitley, G.P. 1933. Some early naturalists and collectors in Australia. *Journal & Proceedings of the Royal Australian Historical Society*. Vol. 19, pt 5, pp. 291–323.

Whitley, G.P. 1969. The 'Endeavour's' naturalists in Australia. *Australian Natural History*. Vol. 16, no. 8, pp. 247–50.

Whittle, T. 1970. *The plant hunters*. Heinemann, London.

Wickens, G.E. 1990. What is economic botany? *Economic Botany*. Vol. 44, no. 1, pp. 12–28.

Wickens, G.E. 1993. Two centuries of economic botanists at Kew. *Kew Magazine*. Vol. 10, pp. 84–94, 132–8.

Wightman, G.M., Dixon, D., Williams, L.L.V & Injimadi Dalywaters. 1992. *Mudburra ethnobotany. Aboriginal plant use from Kulumindini (Elliot), northern Australia*. Northern Territory Botanical Bulletin No. 14. Conservation Commission of the Northern Territory, Darwin.

Wightman, G.M., Jackson, D.M. & Williams, L.L.V. 1991. *Alawa ethnobotany. Aboriginal plant use from Minyerri, northern Australia*. Northern Territory Botanical Bulletin No. 11. Conservation Commission of the Northern Territory, Darwin.

Wightman, G.M., Roberts, J.G. & Williams, L.L.V. 1992. *Mangarrayi ethnobotany. Aboriginal plant use from the Elsey area, northern Australia*. Northern Territory Botanical Bulletin No. 15. Conservation Commission of the Northern Territory, Darwin.

Wightman, G.M. & Smith, N.M. 1989. *Ethnobotany, vegetation and floristics of Milingimbi, northern Australia*. Northern Territory Botanical Bulletin No. 6. Conservation Commission of the Northern Territory, Darwin.

Wilhelmi, J.F.C. 1861. Manners and customs of the Australian natives in particular of the Port Lincoln district. *Transactions of the Royal Society of Victoria*. Volume 5, pp. 164–203.

Wilkinson, G.B. 1848. *South Australia – its advantages and resources – being a description of that colony and a manual of information for emigrants*. John Murray, London.

Williams, G. 1985. Reactions to Cook's voyage. Pp. 35–50 in I. & T. Donaldson (eds) *Seeing the first Australians*. George Allen & Unwin, Sydney.

Williams, M. 1974. *The making of the South Australian landscape. A study in the historical geography of Australia*. Academic Press, London.

Williams, M. 1997. Ecology, imperialism and deforestation. Pp. 169–84 in T. Griffiths & L. Robin (eds) *Ecology and empire. Environmental history of settler societies*. Melbourne University Press, Melbourne.

Willis, J.H. 1962. The botany of the Victoria Exploring Expedition (September 1860–June 1861) and of relief contingents from Victoria (July 1861–November 1862). *Proceedings of the Royal Society of Victoria*. Vol. 75, pp. 247–68.

Willis, J.H. 1990. Melbourne: a focal point for early botanical activity. Pp. 1–3 in P.S. Short (ed.) *History of systematic botany in Australasia. Proceedings of a symposium held at the University of Melbourne*. Australian Systematic Botany Society, Melbourne.

Willis, M. 1949. *By their fruits. A life of Ferdinand von Mueller, botanist and explorer*. Angus & Robertson, Sydney.

Wills, W.J. 1863. *Successful exploration through the interior of Australia, from Melbourne to the Gulf of Carpentaria. From the journals and letters of William John Wills*. Edited by W. Wills. Richard Bentley, London.

Willsteed, T., Smith, K. & Bourke, A. 2006. *Eora: mapping Aboriginal Sydney, 1770–1850*. State Library of New South Wales, Sydney.

Wiminydji & Peile, A.R. 1978. A desert Aborigine's view of health and nutrition. *Journal of Anthropological Research*. Vol. 34, pt 4, pp. 497–523.

Wood, H.W. 1981. *Bushman born*. Artlook Books Trust, Perth.

Woods, J.D. (ed.) 1879. Introduction. Pp. vii–xxxviii in J.D. Woods (ed.) *The native tribes of South Australia*. E.S. Wigg & Son, Adelaide.

Woolmington, E. 1972. The Australian environment as a problem area. Pp. 22–36 in A. Rapoport (ed.) *Australia as human setting*. Angus & Robertson, Sydney.

Worsnop, T. 1897. *The prehistoric arts, manufactures, works, weapons, etc., of the Aborigines of Australia*. South Australian Government Printer, Adelaide.

Wreck Bay Community & Renwick, C. 2000. *Geebungs and snake whistles. Koori people and plants of Wreck Bay*. Aboriginal Studies Press, Canberra.

Wyndham, W.T. 1890 Australian Aborigines: varieties of food and methods of obtaining it. *Journal & Proceedings of the Royal Society of New South Wales*. Vol. 24, pp. 112–20.

Yallop, C. 1982. *Australian Aboriginal languages*. Andre Deutsch, London.

Yarwood, A.T. 1967. Marsden, Samuel (1765–1838). Pp. 207–12 in D. Pike (ed.) *Australian dictionary of biography. Volume 2. 1788–1850, I–Z*. Melbourne University Press, Melbourne.

Young, E. 1992. Hunter-gatherer concepts of land and its ownership in remote Australia and North America. Pp. 255–72 in K.J. Anderson & F. Gale (eds) *Inventing places. Studies in cultural geography*. Longman Cheshire, Melbourne.

Yunupingu, B. 1995. *Rirratjinu ethnobotany. Aboriginal plant use from Yirrkala, Arnhem Land, Australia*. Conservation Commission of the Northern Territory, Darwin.

Zarkawi, M., Al-Masri, M. R. & Khalifa, K. 2005. Nutritive value of *Sesbania aculeata* grown on salty soil and its effect on reproductive parameters of Syrian Awassi ewes. *Australian Journal of Agricultural Research*. Vol. 56, part 8, pp. 819–25.

Zola, N. & Gott, B. 1992. *Koorie plants. Koorie people. Traditional Aboriginal food, fibre and healing plants of Victoria*. Koorie Heritage Trust, Melbourne.

Common Plant Names Index

Scientific Plant Names Index

Muehlenbeckia gunnii (coastal lignum or coastal sarsaparilla or Macquarie Harbour grape) 30, 35, *35*
Musa species (bush banana & plantation banana) 148
Myoporum insulare (blueberry tree or boobialla or native juniper) 51, *51*
Myristica insipida (Australian nutmeg) 23

Nauclea orientalis (Leichhardt tree) 105, 116
Nelumbo nucifera (lotus lily or sacred lotus) 99, 100, 102, *102*
Nicotiana species (wild tobacco) 55
Nicotiana gossei (rock pituri tobacco) 55
Nicotiana rosulata subspecies *ingulba* (sandhill pituri tobacco) 55
Nicotiana tabacum (commercial tobacco varieties such as broad-leaved Virginian tobacco) 18, 20, 72, 79, 96, 112
Nitraria billardierei (dillon or giant saltbush or nitre bush) 51, *52*
Nothofagus cunninghamii (myrtle beech) 80
Nuytsia floribunda (Western Australian Christmas tree) 87, *87*
Nymphaea species (water lily) 46-7, 99-100
Nymphaea gigantea (giant water lily) 47, *100*, 116

Ocimum tenuiflorum (sacred basil) 41
Owenia acidula (emu apple) 30
Owenia reticulata (native walnut) 43
Owenia venosa (sour plum or crows apple) 35, 44
Oxalis species (Australian wood sorrel) 40

Pandanus spiralis (pandanus or screw palm) 100, *101*, 116
Pandorea species (particular species are spearbush or spearwood) 45, *45*
Parinari nonda (nonda plum) 44, 116
Pastinaca sativa (parsnip) 79, 91
Pelargonium species (pelargonium) 60
Pemphis acidula (digging-stick tree) 45
Peripentadenia mearsii (grey quandong) 54
Persoonia species (geebung) 29, 49
Persoonia falcata (wild pear) 43
Petroselinum crispum (parsley) 26, 40, 79
Phaseolus species (long-padded bean) 79
Phaseolus coccinea (scarlet runner bean) 79
Phoenix dactylifera (date palm) 119
Physalis minima (wild gooseberry) 43
Pimelea argentea (silvery leaved pimelea) 85
Pisum sativum (marrowfat pea) 79
Pittosporum angustifolium (native apricot or native willow) 44
Planchonella australis (Australian black apple) 11
Platysace effusa (little-kidney) 84
Pleiogynium timorense (Burdekin plum) 11, 30
Podaxis pistillaris (stalked puffball fungus) 47, 156 endnote 40
Podocarpus species (plum pine) 44
Polyporus eucalyptorum (bracket fungus or punk) 27, *27*, 90
Portulaca oleracea (common pigweed or munyeroo or portulacca or purslane) 32, 38, 55, 104, 116, 123, *124*, 127, 148
Protea cynaroides (king protea) 60
Prunus armeniaca (apricot) 44, 79
Prunus domestica (domestic plum) 44
Prunus insititia (bullace or damson) 11
Prunus persica (peach & nectarine cultivars) 30
Psydrax latifolia (wild currant) 43, 114
Pteridium esculentum (bracken) 27, *86*
Pterocaulon sphacelatum (fruit salad bush or horehound) 40, *40*
Pterostylis species (greenhood orchid) 90, *90*
Ptychosperma elegans (solitaire palm or cabbage palm) 74, 99, 157 endnote 80

Quercus robur (English oak) 79, 160 endnote 79

Rhagodia parabolica (fragrant saltbush) 32
Ribes species (cultivated currant) 26
Rorippa species (watercress) 26
Rorippa palustris (besser marsh watercress or native cabbage) 31
Rubus species (various fruit cultivars such as dewberry, loganberry& raspberry) 54, 103, 117
Rubus fruticosus (blackberry) 117
Rubus gunnianus (mountain raspberry) 118
Rubus moluccanus (native bramble or native raspberry) 29, 43, 117-8, 164 endnote 70
Rubus moorei (native bramble or native raspberry) 29, 43, 117-8, 164 endnote 70
Rubus parvifolius (native bramble or native raspberry) 29, 43, 117-8, 164 endnote 70
Rubus rosifolius (native raspberry or rose-leaf bramble) 29, 43, 117-8
Rubus ursinus hybrid (boysenberry) 117

Salvia officinalis (sage) 31
Santalum acuminatum (quandong or native peach) 34, *35*, 54, *76*, 91
Santalum lanceolatum (northern sandalwood) 23
Santalum murrayanum (bitter quandong or ming) 54-5, 91
Santalum spicatum (Australian sandalwood) 142
Sarcostemma australe (caustic-vine or milk-bush) 24, 40, 110, *111*
Sarcozona praecox (pigface or sarcozona or wild fig) 29, 127, *127*
Scaevola brookeana (heart-leaved fanflower) 108
Scleroglossum wooroonooran (scleroglossum) 144
Sesbania aculeata (sesbania) 128-9
Smilax australis (Australian sarsaparilla or sweet sarsaparilla) 35, 40, 43, 162 endnote 26
Smilax glyciphylla (sweet tea) 35, 39-40, 140
Smilax officinalis (sarsaparilla) 35, 40, *95*
Solanum aviculare (gunyang or kangaroo apple) 53, 107, 157 endnote 105
Solanum centrale (desert raisin) 43
Solanum chippendalei (bush tomato) 43, *44*
Solanum laciniatum (gunyang or kangaroo apple) 53, 107, 157 endnote 105
Solanum simile (oondoroo) 30
Solanum tuberosum (potato) 41-2, 87
Solanum vescum (gunyang or kangaroo apple) 53, 107, 157 endnote 105
Sonchus oleraceus (sow thistle or milk thistle) 99
Spinacia oleracea (spinach) 26, 31, 74, 79
Spinifex species (spinifex grass) 9
Stellaria media (chickweed) 31
Sterculia quadrifida (peanut tree) 44
Sterculia species (kurrajong) 50, 101, 157 endnote 72
Strelitzia reginae (bird of paradise flower) 60
Swainsona formosa (Sturt desert pea) 9, *11*, 47
Synaphea spinulosa (synaphea) 9, 152 endnote 4
Syzygium species (eugenia or lilly pilly or scrub cherry) 29, *99*, 154 endnote 42
Syzygium armstrongii & some other *Syzygium* species (bush apple) *29*
Syzygium cormiflorum (white apple) *29*
Syzygium forte (flaky-barked satinash or white apple) *113*
Syzygium tierneyanum (Bamaga satinash) 30

Tacca leontopetaloides (Polynesian arrowroot) 74
Tasmannia lanceolata (Tasmanian pepper tree) 30
Telopea speciosissima (tulip tree or waratah) 49, *60*, 64
Telopea truncata (Tasmanian waratah or tulip tree) 49, *60*, 64
Terminalia species (wild plum) 44, 99
Terminalia catappa (tropical almond) 43
Terminalia ferdinandiana (billy goat plum or Kakadu plum) 44, 119
Terminalia hadleyana subspecies *carpentariae* (gulf plum) 44, 163 endnote 73
Terminalia muelleri (Mueller's damson) 119
Terminalia sericocarpa (damson plum) 44
Tetragonia species (wild spinach) 127, 166 endnote 58
Tetragonia moorei (native spinach) 127
Tetragonia tetragonioides (New Zealand spinach or warrigal cabbage, also sometimes along with various species of saltbush as Botany Bay greens or Sydney greens or warrigal greens) 31, *31*
Trachymene ceratocarpa (creeping carrot) 142
Triodia species (spinifex grass) 9
Triodia basedowii (hard spinifex grass or porcupine grass) 114, *115*
Triodia pungens (soft spinifex grass) 146, 167 endnote 74
Tropaeolum majus (nasturtium) *43*, 44
Tulipa species (tulip) 49, 58
Typha species (bulrush or cats-tail or cumbungi or flag or reedmace) 28, 48, 52, 82, 116

Verticordia cunninghamii (feather-plant) 47
Vigna lanceolata (pencil yam & small yam) 43, 123
Vitex acuminata (black plum or kerosene wood) 44
Vitex glabrata (black plum or kerosene wood) 44

Xanthorrhoea species (blackboy or grasstree) 45, 53, 82, *83*, 84
Xylomelum pyriforme (woody pear) 57

Zea mays (maize) 41
Zieria arborescens (stinkwood) 40

General Index

The references to figures are *in italics*)